Y. Zhao

BEAM DYNAMICS AND TECHNOLOGY ISSUES FOR $\mu^+\mu^-$ COLLIDERS

9th Advanced ICFA Beam Dynamics Workshop

BEAM DYNAMICS AND TECHNOLOGY ISSUES FOR μ⁺μ⁻ COLLIDERS

9th Advanced ICFA Beam Dynamics Workshop

Montauk, NY October 15–20, 1995

EDITOR
Juan C. Gallardo
Center for Accelerator Physics
Brookhaven National Laboratory

**AIP CONFERENCE
PROCEEDINGS 372**

American Institute of Physics   **Woodbury, New York**

L.C. Catalog Card No. 96-84189
ISBN 1-56396-554-2
DOE CONF- 9510295

Printed in the United States of America

CONTENTS

II. APPENDICES

PREFACE

The *9th Advanced ICFA Beam Dynamics Workshop: Beam Dynamics and Technology Issues for* $\mu^+\mu^-$ *Colliders* was held at Montauk, New York on October 15–20, 1995. It was sponsored by the Center for Accelerator Physics (CAP), Brookhaven National Laboratory, and supported by the U.S. Department of Energy, Advanced Technology R&D Branch.

About fifty-five scientists from the U.S., Russia, Japan, and various countries from Europe discussed, in a lively and vigorous exchange of ideas, the status and progress of the $\mu^+\mu^-$ Collider studies toward a high-energy (2+2 TeV), high-luminosity machine ($\mathscr{L}=2\times10^{35}\ cm^{-2}s^{-1}$). Among the participants were: A. Skrinsky (BINP), the first to propose a muon muon collider using ionization cooling; D. Neuffer (CEBAF/FERMILAB), who contributed to the further understanding of the concept; R. Palmer, author of a compelling and exhaustive scenario for the collider and whose technical ideas and intellectual force have enlivened the field; A. Tollestrope (FERMILAB), and A. Sessler (LBNL) both of whom have instilled on the participants the need for discipline and concrete goals.

In the first three days, invited speakers presented several possible alternatives to the various components of a $\mu^+\mu^-$ Collider in consonance with the scenario put forward by R. Palmer (BNL).

This was followed by four working groups:

- Muon Production, led by F. Mills (Fermilab) and T. Roser (BNL)

- Muon Cooling, led by R. Fernow (BNL) and A. Skrinsky (BINP)

- Machine design, led by D. Neuffer (CEBAF/Fermilab) and K. Hirata (KEK)

- Detectors, led by D. Lissauer (BNL)

The workshop closed with summaries, conclusions, and recommendation of tasks for the future.

This book contains some of the invited lectures and contributions by the participants. The articles cover different aspects of the collider: the proton driver, pion capture, decay channel, accelerator, lattice, and final focus of the storage ring; detector and background issues; the unique physics opportunities opened by a high-luminosity machine.

I would like to acknowledge the contribution of Kathleen Tuohy and Patricia Tuttle; their long hours of preparation, organization, and attention to the most minimal details contributed greatly to the success of the Workshop. I also want to thank all the participants for a very stimulating and productive workshop. We hope this workshop represents a positive contribution toward the R&D of a practical high energy, high luminosity $\mu^+\mu^-$ Collider.

Juan C. Gallardo

Scientific Advisory Committees

B. Autin, CERN
V. Barger, University of Wisconsin
W. Barletta, LBNL
P. Chen, SLAC
D. Cline, UCLA
H. Hirata, KEK
Y. Mori, KEK
D. Neuffer, CEBAF/FERMILAB
R. Noble, FERMILAB
C. Pellegrini, UCLA
A. Ruggiero, BNL
A. Sessler, LBNL
A. Skrinsky, BINP
A. Tollestrup, FERMILAB
W. Willis, Columbia University

International Advisory Committees
ICFA Beam Dynamics Panel

A. Ando, SPRING8
V. Balbekov, IHEP (Protvino)
H. Hirata, KEK
A. Hofmann, CERN
C. Hsue, SRRC
J. Laclare, ESFR
A. Lebedev, LPI
S. Lee, Indiana University
L. Palumbo, INFN-LNF
C. Pellegrini, UCLA
E. Perelstein, JINR
D. Pestrikov, BINP
R. Siemann, SLAC
F. Willeke, DESY
C. Zhang, IHEP (Beijing)

I. INVITED AND CONTRIBUTED ARTICLES

MUON COLLIDERS

R. B. Palmer[1,2], A. Sessler[3], A. Skrinsky[4], A. Tollestrup[5],
A. J. Baltz[1], P. Chen[2], W-H. Cheng[3], Y. Cho[6], E. Courant[1],
R. C. Fernow[1], J. C. Gallardo[1], A. Garren[3,7], M. Green[3],
S. Kahn[1], H. Kirk[1], Y. Y. Lee[1], F. Mills[5], N. Mokhov[5],
G. Morgan[1], D. Neuffer[5,8], R. Noble[5], J. Norem[6], M. Popovic[5],
L. Schachinger[3], G. Silvestrov[4], D. Summers[9], I. Stumer[1],
M. Syphers[1], Y. Torun[1,10], D. Trbojevic[1], W. Turner[3],
A. Van Ginneken[5], T. Vsevolozhskaya[4], R. Weggel[11], E. Willen[1],
D. Winn[12], J. Wurtele[13]

Abstract. Muon Colliders have unique technical and physics advantages and disadvantages when compared with both hadron and electron machines. They should thus be regarded as complementary. Parameters are given of 4 TeV and 0.5 TeV high luminosity $\mu^+\mu^-$ colliders, and of a 0.5 TeV lower luminosity demonstration machine. We discuss the various systems in such muon colliders, starting from the proton accelerator needed to generate the muons and proceeding through muon cooling, acceleration and storage in a collider ring. Problems of detector background are also discussed.

[1] Brookhaven National Laboratory, Upton, NY 11973-5000, USA
[2] Stanford Linear Accelerator Center, Stanford, CA 94309, USA
[3] Lawrence Berkeley National Laboratory, Berkeley, CA 94720, USA
[4] BINP, RU-630090 Novosibirsk, Russia
[5] Fermi National Accelerator Laboratory, Batavia, IL 60510, USA
[6] Argonne National Laboratory, Argonne, IL 60439-4815, USA
[7] UCLA, Los Angeles, CA 90024-1547, USA
[8] CEBAF, Newport News, VA 23606, USA
[9] Univ. of Mississippi, Oxford, MS 38677, USA
[10] SUNY, Stony Brook, NY 11974, USA
[11] Francis Bitter National Magnet Laboratory, MIT, Cambridge, MA 02139, USA
[12] Fairfield University, Fairfield, CT 06430-5195, USA
[13] UC Berkeley, Berkeley, CA 94720-7300, USA

1 INTRODUCTION

1.1 Technical Considerations

The possibility of muon colliders was introduced by Skrinsky et al. [1], Neuffer [2], and others. More recently, several workshops and collaboration meetings have greatly increased the level of discussion [3], [4]. In this paper we discuss scenarios for 4 TeV and 0.5 TeV colliders based on an optimally designed proton source, and for a lower luminosity 0.5 TeV demonstration based on an upgraded version of the AGS. It is assumed that a demonstration version based on upgrades of the FERMILAB machines would also be possible (see second Ref. [4]).

Hadron collider energies are limited by their size and technical constraints on bending magnetic fields. At very high energies it would also be difficult to obtain the required luminosities, which must rise as the energy squared. e^+e^- colliders, because they undergo simple, single-particle interactions, can reach higher energy final states than an equivalent hadron machine. However, extension of e^+e^- colliders to multi-TeV energies is severely performance-constrained by beamstrahlung, and cost-constrained because two full energy linacs are required [5] to avoid the excessive synchrotron radiation that would occur in rings. Muons ($\frac{m_\mu}{m_e} = 207$) have the same advantage in energy reach as electrons, but have negligible beamstrahlung, and can be accelerated and stored in rings, making the possibility of high energy $\mu^+\mu^-$ colliders attractive. There are several major technical problems with μ's:

- they decay with a lifetime of 2.2×10^{-6} s. This problem is partially overcome by rapidly increasing the energy of the muons, and thus benefitting from their relativistic γ factor. At 2 TeV, for example, their lifetime is 0.044 s: sufficient for approximately 1000 storage-ring collisions;

- another consequence of the muon decays is that the decay products heat the magnets of the collider ring and create backgrounds in the detector;

- they are created through pion decay into a diffuse phase space and this phase space cannot be reduced by conventional stochastic or synchrotron cooling. It can, to some extent, be dealt with by ionization cooling, but the final emittance of the muon beams will remain larger than that possible in an e^+e^- collider.

Despite these problems it appears possible that high energy muon colliders might have luminosities comparable to or higher than those in e^+e^- colliders at the same energy [6]. And because the $\mu^+\mu^-$ machines would be much smaller [7], and require much lower precision (the final geometric emittances are about 5 orders of magnitude larger, and the spots are about three orders of magnitude larger), they may be significantly less expensive. It must be remembered,

4

FIGURE 1. Schematic of a Muon Collider.

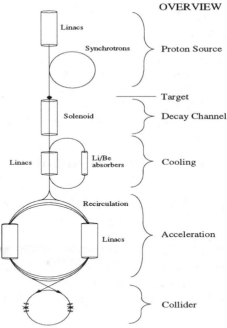

however, that a $\mu^+\mu^-$ collider remains a new and untried concept, and its study has just began; it cannot yet be compared with the more mature designs for an e^+e^- collider.

1.2   Overview of Components

The basic components of the $\mu^+\mu^-$ collider are shown schematically in Fig.1 and Tb.1 shows parameters for the candidate designs.

A high intensity proton source is bunch compressed and focussed on a heavy metal target. The pions generated are captured by a high field solenoid and transferred to a solenoidal decay channel within a low frequency linac. The linac serves to reduce, by phase rotation the momentum spread of the pions, and of the muons into which they decay. Subsequently, the muons are cooled by a sequence of ionization cooling stages. Each stage consists of energy loss, acceleration, and emittance exchange in energy absorbing wedges in the presence of dispersion. Once they are cooled the muons must be rapidly accelerated to avoid decay. This can be done in recirculating accelerators (à la CEBAF) or in fast pulsed synchrotrons. Collisions occur in a separate high field collider storage ring with very low beta insertion.

		4 TeV	0.5 TeV	Demo
Beam energy	TeV	2	0.25	0.25
Repetition rate	Hz	15	15	2.5
Muons per bunch	10^{12}	2	4	4
Bunches of each sign		2	1	1
Norm. rms emittance ϵ^N	mm mrad	50	90	90
β^* at intersection	mm	3	8	8
Luminosity [units 10^{35}]	cm^{-2}s^{-1}	1	0.05	0.006

TABLE 1. Summary of Parameters of $\mu^+\mu^-$ Colliders

1.3   Physics Considerations

There are at least two physics advantages of a $\mu^+\mu^-$ collider, when compared with a e^+e^-:

- Because of the lack of beamstrahlung, a $\mu^+\mu^-$ collider can be operated with an energy spread of as little as 0.01 %. It is thus possible to use the $\mu^+\mu^-$ collider for precision measurements of masses and widths, that would be hard, if not impossible, with an e^+e^- collider.

- The direct coupling of a lepton-lepton system to a Higgs boson has a cross section that is proportional to the square of the mass of the lepton. As a result, the cross section for direct Higgs production from the $\mu^+\mu^-$ system is 40,000 times that from an e^+e^- system.

However, there are liabilities:

- It is relatively hard to obtain both polarization and good luminosity in a $\mu^+\mu^-$ collider, whereas good polarization can be obtained in an e^+e^- collider without any loss in luminosity.

- because of the decays of the muons, there will be a considerable background of photons, muons and neutrons in the detector. This background may be acceptable for some experiments, but it is certainly not as clean as an in e^+e^- collider.

1.4   Conclusion

It is thus reasonable, from both technical and physics considerations, to consider $\mu^+\mu^-$ colliders as complementary to e^+e^- colliders, just as e^+e^- colliders are complementary to hadron machines.

		4 TeV	.5 TeV	Demo
Proton energy	GeV	30	30	24
Repetition rate	Hz	15	15	2.5
Protons per bunch	10^{13}	2.5	2.5	2.5
Bunches		4	2	2
Long. phase space/bunch	eV s	4.5	4.5	4.5
Final *rms* bunch length	ns	1	1	1

TABLE 2. Proton Driver Specifications

2 SYSTEM COMPONENTS

2.1 Proton Driver

The specifications of the proton drivers are given in Tb.2. In the examples, the μ-source driver is a high-intensity (2.5×10^{13} protons per pulse) 30 GeV proton synchrotron. The preferred cycling rate would be 15 Hz, but for the demonstration using the AGS [8], the repetition rate is limited to 2.5 Hz and to 24 GeV. For the lower energy machines, 2 final bunches are employed (one to make μ^-'s and the other to make μ^+'s). For high energy collider, four are used (two μ bunches of each sign).

Earlier studies had suggested that the driver could be a 10 GeV machine with the same charge per bunch, but a repetition rate of 30 Hz. This specification was almost identical to that studied [9] at ANL for a spallation neutron source. Studies at FNAL [10] have further established that such a specification is not unreasonable. But in order to reduce the cost of the muon phase rotation section and for minimizing the final muon longitudinal phase space, it appears now that the final proton bunch length should be 1 ns (or even less). This appears difficult to achieve at 10 GeV, but possible at 30 GeV. If it is possible to obtain such short bunches with a 10 GeV source, or if future optimizations allow the use of the longer bunches, then the use of a lower energy source could be preferred.

In order to achieve the required 1 ns (rms) bunch length, an RF sequence must be designed to phase rotate the bunch prior to extraction. The total final momentum spread, based on the ANL parameters (95% phase space of 4.5 eVs per bunch), is 6 %, (2.5 % rms), and the space charge tune shift just before extraction would be approx 0.5. It might be necessary to perform this bunch compression in a separate high field ring to avoid space charge problems.

2.2 Target and Pion Capture

Predictions of the nuclear Monte-Carlo program ARC [11] suggest that π production is maximized by the use of heavy target materials, and that the

FIGURE 2. ARC forward π^+ production vs proton energy and target material.

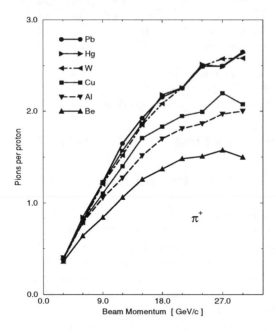

production is peaked at a relatively low pion energy ($\approx 100\,\mathrm{MeV}$), substantially independent of the initial proton energy. Fig. 2 shows the forward π^+ production as a function of proton energy and target material; the π^- distributions are similar. Fig.3 shows the energy distribution for π^+ and π^- for 24 GeV protons on Hg.

The target could be either Cu (approximately 24 cm long by 2 cm diameter), or Hg (approximately 14 cm long by 2 cm diameter). A Hg target is being studied for the European Spallation Source and would be cooled by circulating the liquid. The Cu target would require water cooling.

Pions are captured from the target by a high-field (20 T) water cooled Bitter solenoid that surrounds it. Such a magnet is estimated [12] to require about 14 MW: a significant but not unreasonable power. The π's are then matched, using a suitable tapered field [13], into a periodic superconducting solenoidal decay channel (5 T and radius = 15 cm).

Monte Carlo studies indicate a yield of 0.6 muons, of each sign, per initial proton, captured in the decay channel. But these pions have an extremely broad energy spectrum so that only about half of them (0.3 μ/p) can be used.

FIGURE 3. ARC energy distribution for 24 GeV protons on Hg.

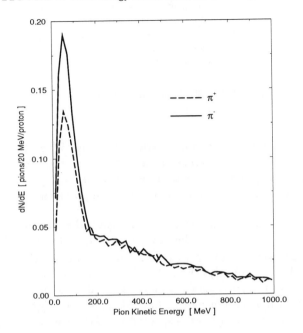

2.3 Capture and Use of Both Signs

Protons on the target produce pions of both signs, and a solenoid will capture both, but the required subsequent RF systems will have opposite effects on each. One solution is to break the proton bunch into two, target them on the same target one after the other, and adjust the RF phases such as to act correctly on one sign of the first bunch and on the other sign of the second. This is the solution assumed in the parameters of this paper.

A second solution is to separate the pions of each charge prior to the use of RF, and feed the beams of each charge into different channels. A third possibility would be to separate the charges, delay the particles of one charge, recombine the charges, and feed them into a single channel with appropriate phases of RF.

After the target, and prior to the use of any RF or cooling, the beams have very large emittances and energy spread. Conventional charge separation using a dipole is not practical. But if a solenoidal channel is bent, then the particles trapped within that channel will drift [14] in a direction perpendicular to the bend. With our parameters this drift is dominated by a term (curvature drift) that is linear with the forward momentum of the particles, and has a direction that depends on the sign of the charges. It has been suggested [15] that if sufficient bend is employed, the two charges could then be separated

9

Linac	Length m	Frequency MHz	Gradient MeV/m
1	3	60	5
2	29	30	4
3	5	60	4
4	5	37	4

TABLE 3. Parameters of Phase Rotation Linacs

by a septum and captured into two separate channels. When these separate channels are bent back to the same forward direction, then the momentum dispersion is separately removed in each new channel.

Although this idea is very attractive, it has some problems:

- If the initial beam has a radius r=0.15 m, and if the momentum range to be accepted is $F = \frac{p_{max}}{p_{min}} = 3$, then the required height of the solenoid just prior to separation is $2(1+F)r=1.2$ m. Use of a lesser height will result in particle loss. Typically, the reduction in yield for a curved solenoid compared to a straight solenoid is about 25 % (due to the loss of very low and very high momentum pions), but this must be weighed against the fact that both charge signs are captured for each proton on target.

- The system of bend, separate, and return bend will require significant length and must occur prior to the start of phase rotation (see below). Unfortunately, it appears that the cost of the phase rotation RF appears to be strongly dependent on keeping this distance as short as possible. On the other hand it may be advisable to separate the remnant proton beam and other charged debris exiting the target before the RF cavities. A curved solenoid would accomplish this as well as charge-separate pions.

Clearly, compromises will be involved, and more study of this concept is required.

2.4   Phase Rotation Linac

The pions, and the muons into which they decay, have an energy spread from about 0 - 3 GeV, with an rms/mean of $\approx$ 100%, and with a peak at about 100 MeV. It would be difficult to handle such a wide spread in any subsequent system. A linac is thus introduced along the decay channel, with frequencies and phases chosen to deaccelerate the fast particles and accelerate the slow ones; i.e. to phase rotate the muon bunch. Tb.3 gives an example of parameters of such a linac. It is seen that the lowest frequency is 30 MHz: a low but not impossible frequency for a conventional structure.

10

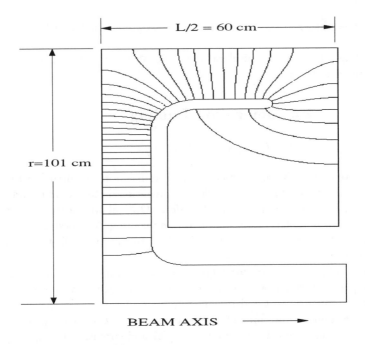

FIGURE 4. 30 MHz cavity for use in phase rotation and early stages of cooling.

A design of a reentrant 30 MHz cavity is shown in Fig. 4. Its parameters are given in Tb. 4. It has a diameter of approximately 2 m: only about one third that of a conventional pill-box cavity. To keep its cost down, it would be made of Al. Multipactoring would probably be suppressed by stray fields from the 5 T focusing coils, but could also be controlled by an internal coating of titanium nitride.

After this phase rotation, a bunch can be selected with mean energy 150 MeV, rms bunch length 1.7 m, and rms momentum spread 20 %. The number of muons per initial proton in this selected bunch is 0.3. Unfortunately, the linacs cannot phase rotate both signs in the same bunch: hence the need for

Cavity Radius	cm	101
Cavity Length	cm	120
Beam Pipe Radius	cm	15
Accelerating Gap	cm	24
Q		18200
Average Acceleration Gradient	MV/m	3
Peak RF Power	MW	6.3
Average Power (15 Hz)	KW	18.2
Stored Energy	J	609

TABLE 4. Parameters of 30 MHz RF Cavity

two bunches. The phases are set to rotate the μ^+'s of one bunch and the μ^-'s of the other.

Prior to cooling, the bunch is accelerated to 300 MeV, at which energy the momentum spread is 10 %.

2.5 Ionization Cooling

2.5.1 Cooling Theory

For collider intensities, the phase-space volume must be reduced within the μ lifetime. Cooling by synchrotron radiation, conventional stochastic cooling and conventional electron cooling are all too slow. Optical stochastic cooling [16], electron cooling in a plasma discharge [17] and cooling in a crystal lattice [18] are being studied, but appear very difficult. Ionization cooling [19] of muons seems relatively straightforward.

In ionization cooling, the beam loses both transverse and longitudinal momentum as it passes through a material medium. Subsequently, the longitudinal momentum can be restored by coherent reacceleration, leaving a net loss of transverse momentum. Ionization cooling is not practical for protons and electrons because of nuclear interactions (p's) and bremsstrahlung (e's) effects in the material, but is practical for μ's because of their low nuclear cross section and relatively low bremsstrahlung.

The equation for transverse cooling (with energies in GeV) is:

$$\frac{d\epsilon_n}{ds} = -\frac{dE_\mu}{ds}\frac{\epsilon_n}{E_\mu} + \frac{\beta_\perp (0.014)^2}{2\,E_\mu m_\mu\,L_R}, \tag{1}$$

where ϵ_n is the normalized emittance, $\beta_\perp$ is the betatron function at the absorber, dE_μ/ds is the energy loss, and L_R is the material radiation length. The first term in this equation is the coherent cooling term, and the second term is the heating due to multiple scattering. This heating term is minimized if $\beta_\perp$ is small (strong-focusing) and L_R is large (a low-Z absorber).

From Eq.1 we find a limit to transverse cooling, which occurs when heating due to multiple scattering balances cooling due to energy loss. The limits are $\epsilon_n \approx 0.6\ 10^{-2}\ \beta_\perp$ for Li, and $\epsilon_n \approx 0.8\ 10^{-2}\ \beta_\perp$ for Be.

The equation for energy spread (longitudinal emittance) is:

$$\frac{d(\Delta E)^2}{ds} = -2\,\frac{d\left(\frac{dE_\mu}{ds}\right)}{dE_\mu} < (\Delta E_\mu)^2 > + \frac{d(\Delta E_\mu)^2_{straggling}}{ds} \tag{2}$$

where the first term is the cooling (or heating) due to energy loss, and the second term is the heating due to straggling.

Cooling requires that $\frac{d(dE_\mu/ds)}{dE_\mu} > 0$. But at energies below about 200 MeV, the energy loss function for muons, dE_μ/ds, is decreasing with energy and there is thus heating of the beam. Above 400 MeV the energy loss function increases gently, thus giving some cooling, though not sufficient for our application.

In the long-path-length Gaussian-distribution limit, the heating term (energy straggling) is given by [20]

$$\frac{d(\Delta E_\mu)^2_{straggling}}{ds} = 4\pi \left(r_e m_e c^2\right)^2 N_o \frac{Z}{A} \rho \gamma^2 \left(1 - \frac{\beta^2}{2}\right), \qquad (3)$$

where N_o is Avogadro's number and ρ is the density. Since the energy straggling increases as γ^2, and the cooling system size scales as γ, cooling at low energies is desired.

Energy spread can also be reduced by artificially increasing $\frac{d(dE_\mu/ds)}{dE_\mu}$ by placing a transverse variation in absorber density or thickness at a location where position is energy dependent, i.e. where there is dispersion. The use of such wedges can reduce energy spread, but it simultaneously increases transverse emittance in the direction of the dispersion. Six dimensional phase space is not reduced, but it does allow the exchange of emittance between the longitudinal and transverse directions.

2.5.2   Cooling System

We require a reduction of the normalized transverse emittance by almost three orders of magnitude (from 1×10^{-2} to 5×10^{-5} m-rad), and a reduction of the longitudinal emittance by more than one order of magnitude. This cooling is obtained in a series of cooling stages. In general, each stage consists of three components with matching sections between them:

1. a lattice consisting of spaced axial solenoids with alternating field directions and lithium hydride absorbers placed at the centers of the spaces between them, where the $\beta_\perp$'s are minimum.

2. a lattice consisting of more widely separated alternating solenoids, and bending magnets to generate dispersion. At the location of maximum dispersion, wedges of lithium hydride are introduced to interchange longitudinal and transverse emittance;

3. a short linac to restore the energy lost in the absorbers.

At the end of this sequence of cooling stages, the transverse emittance can be reduced to about 10^{-3} m, still a factor of ≈ 20 above the emittance goals of Tb.1. The longitudinal emittance, however, can be cooled to a value nearly three orders of magnitude less than is required. The additional reduction of transverse emittance can then be obtained by a reverse exchange of transverse and longitudinal phase-spaces. Again this is done by the use of wedged absorbers in dispersive regions between solenoid elements.

13

Throughout this process appropriate momentum compaction and RF fields must be used to control the bunch, in the presence of space charge, wake field and resistive wall effects.

In a few of the later stages, current carrying lithium rods might replace item (1) above. In this case the rod serves simultaneously to maintain the low $\beta_\perp$, and attenuate the beam momenta. Similar lithium rods, with surface fields of 10 T , were developed at Novosibirsk and have been used as focusing elements at FNAL and CERN [21]). It is hoped [22] that liquid lithium columns, can be used to raise the surface field to 20 T and improve the resultant cooling.

It would be desirable, though not necessarily practicable, to economize on linac sections by forming groups of stages into recirculating loops.

2.5.3   Example

A *model* example has been generated that uses no lithium rods and no recirculating loops. Individual components of the lattices used have been defined, but a complete lattice has not yet been specified and no Monte Carlo study of its performance has yet been performed. Spherical aberration due to solenoid end effects, wake fields, and second order RF effects have not yet been included.

The phase advance in each cell of the lattice is made as close to $\pi/2$ as possible in order to minimize the β's at the location of the absorber, but it is kept somewhat less than this value so that the phase advance per cell should never exceed $\pi/2$. The following effects are included:

1. the maximum space charge transverse defocusing

2. a 3 σ fluctuation of momentum

3. a 3 σ fluctuation in amplitude

In all but the final stages of cooling it is assumed that both charges will use the same channel. Bending magnets are introduced to generate dispersion, but the dispersion is kept equal to zero at the center of all solenoids. The maximum allowed beam angle with respect to the axis, due to dispersion, is 45 degrees.

In the early stages, the solenoids are relatively large and their fields are limited to 7 T. In later stages the emittance has decreased, the apertures are smaller and the fields are increased to 10 T. The maximum bending fields used are 7 T, but most are at 3 T.

The emittances, transverse and longitudinal, as a function of stage number, are shown in Fig.5, together with the beam energy. In the first 15 stages, relatively strong wedges are used to rapidly reduce the longitudinal emittance, while the transverse emittance is reduced relatively slowly. The object here is to reduce the bunch length, thus allowing the use of higher frequency and

14

FIGURE 5. $\epsilon_\perp$, ϵ_L and E_μ [GeV], vs stage number in the cooling sequence.

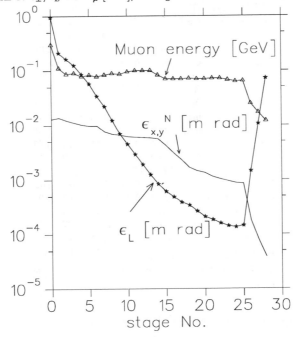

higher gradient RF in the reacceleration linacs. In the next 10 stages, the emittances are reduced close to their asymptotic limits. The charges are now separated for the last two stages. In these stages, the transverse and longitudinal emittances are again exchanged, but in the opposite direction: lowering the transverse and raising the longitudinal. During this exchange the energy is allowed to fall to 10 MeV in order to minimize the β, and thus limit the emittance dilution.

The total length of the system is 880 m, and the total acceleration used is 2.7 GeV. The fraction of muons remaining at the end of the cooling system is calculated to be 43 %.

2.6 Acceleration

Following cooling and initial bunch compression the beams must be accelerated to full energy (2 TeV, or 250 GeV). A sequence of linacs would work, but would be expensive. Conventional synchrotrons cannot be used because the muons would decay before reaching the required energy. The conservative solution is to use a sequence of recirculating accelerators (similar to that used at CEBAF). A more economical solution would be to use fast rise time pulsed magnets in synchrotrons, or synchrotrons with rapidly rotating permanent

		Linac	#1	#2	#3	#4
initial energy	GeV	0.20	1	8	75	250
final energy	GeV	1	8	75	250	2000
nloop			1	12	18	18	18
freq.	MHz	100	100	400	1300	2000
linac V	GV	0.80	0.58	3.72	9.72	97.20
grad			5	5	10	15	20
dp/p initial	%	12	2.70	1.50	1	1
dp/p final	%	2.70	1.50	1	1	0.20
σ_z initial	mm	341	333	82.52	14.52	4.79
σ_z final	mm	303	75.02	13.20	4.36	3.00
η	%	1.04	0.95	1.74	3.64	4.01
N_μ	10^{12}	2.59	2.35	2.17	2.09	2
τ_{fill}	μs	87.17	87.17	10.90	s.c.	s.c.
beam t	μs	0.58	6.55	49.25	103	805
decay survival		0.94	0.91	0.92	0.97	0.95
linac len	km	0.16	0.12	0.37	0.65	4.86
arc len	km	0.01	0.05	0.45	1.07	8.55
tot circ	km	0.17	0.16	0.82	1.72	13.41
phase slip	deg	0	38.37	7.69	0.50	0.51

TABLE 5. Parameters of Recirculating Accelerators

magnets interspersed with high field fixed magnets.

2.6.1 Recirculating Acceleration

Tb.5 gives an example of a possible sequence of recirculating accelerators. After an initial linac from $0.2 \to 1$ GeV, there are two conventional RF recirculating accelerators taking the muons up to 75 GeV, then two superconducting recirculators going up to 2000 GeV. Criteria that must be considered in picking the parameters of such accelerators are:

- The wavelengths of rf should be chosen to limit the loading, η, (it is restricted to below 4 % in this example) to avoid excessive longitudinal wakefields and the resultant emittance growth.

- The wavelength should also be sufficiently large compared to the bunch length to avoid second order effects (in this example: 10 times).

- For power efficiency, the cavity fill time should be long compared to the acceleration time. When conventional cavities cannot satisfy this condition, superconducting cavities are required.

- In order to minimize muon decay during acceleration (in this example 73% of the muons are accelerated without decay), the number of recirculations at each stage should be kept low, and the RF acceleration voltage

FIGURE 6. A cross section of a 9 aperture sc magnet.

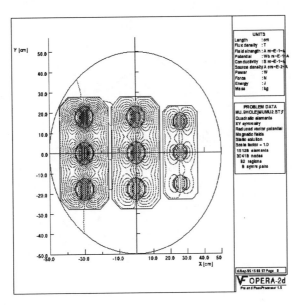

correspondingly high. But for the minimum cost the number of recirculations appears to be of the order of 20 - a relatively high number. In order to avoid a large number of separate magnets, multiple aperture magnets can be designed (see Fig.6).

Note that the linacs see two bunches of opposite signs, passing through in opposite directions. In the final accelerator in the 2 TeV case, each bunch passes through the linac 18 times. The total loading is then $4 \times 18 \times \eta = 288\%$. With this loading, assuming 60% klystron efficiencies and reasonable cryogenic loads, one could probably achieve 35% wall to beam power efficiency, giving a wall power consumption for the RF in this ring of 108 MW.

A recent study [23] tracked particles through a similar sequence of recirculating accelerators and found a dilution of longitudinal phase space of the order of 15%.

2.6.2   Pulsed Magnet Synchrotron Alternatives

An alternative to recirculating accelerators for stages #2 and #3 would be to use pulsed magnet synchrotrons. The cross section of a pulsed magnet for this purpose is shown in Fig. 7. If desired, the number of recirculations could be higher in this case, and the needed RF voltage correspondingly lower,

FIGURE 7. Cross section of pulsed magnet for use in the acceleration to 250 GeV.

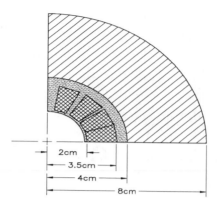

but the loss of particles from decay would be somewhat more. The cost for a pulsed magnet system appears to be significantly less than that of a multi-hole recirculating magnet system, but its power consumption seems impractical at energies above 250 GeV.

2.6.3 Pulsed and Rotating Magnet Alternatives for the Final Accelerator

For the final acceleration to 2 TeV in the high energy machine, the power consumed by pulsed magnets would be excessive. A recirculating accelerator is still usable, but there are two other, possibly cheaper, alternatives being considered:

a) A sequence of two hybrid accelerators (0.25-1, and 1-2 TeV), each employing superconducting fixed magnets (e.g. 10 T) alternating with pairs of counter-rotating permanent magnets [24]. The power consumption would be negligible, but its practicality is not yet clear.

b) A similar sequence of two hybrid accelerators (0.25-1, and 1-2 TeV), each again employing alternate superconducting fixed magnets (e.g. 10 T), but instead of pairs of rotating magnets, pulsed warm magnets (whose fields might swing from -1.5 T to +1.5 T) would be used. The power consumption

		4 TeV	.5 TeV	Demo.
Beam energy	TeV	2	.25	.25
Beam γ		19,000	2,400	2,400
Repetition rate	Hz	15	15	2.5
Muons per bunch	10^{12}	2	4	4
Bunches of each sign		2	1	1
Normalized rms emittance ϵ^N	mm mrad	50	90	90
Bending Field	T	9	9	8
Circumference	Km	7	1.2	1.5
Average ring mag. field B	T	6	5	4
Effective turns before decay		900	800	750
β^* at intersection	mm	3	8	8
rms beam size at I.P.	μm	2.8	16	16
Luminosity	$cm^{-2}s^{-1}$	10^{35}	$5\ 10^{33}$	$6\ 10^{32}$

TABLE 6. Parameters of Collider Rings

would be considerable, but the initial cost might be significantly less than that for a recirculating accelerator, and it might be more practical than the rotating magnet scheme.

2.7 μ Storage Ring

After acceleration, the μ^+ and μ^- bunches are injected into a separate storage ring. The highest possible average bending field is desirable, to maximize the number of revolutions before decay, and thus maximize the luminosity. Collisions would occur in one, or perhaps two, very low-β^* interaction areas. Parameters of the rings are given in Tb.6.

The bunch populations decay exponentially, yielding an integrated luminosity equal to its initial value multiplied by an *effective* number of turns $n_{effective} = 150\ B$, where B is the mean bending field in T.

2.7.1 Bending Magnet Design

The magnet design is complicated by the fact that the μ's decay within the rings ($\mu^- \rightarrow e^- \overline{\nu_e} \nu_\mu$), producing electrons whose mean energy is approximately 1/3 that of the muons. These electrons travel to the inside of the ring dipoles, radiating a substantial fraction of their energy as synchrotron radiation towards the outside of the ring. Fig.8 shows the attenuation of the heating produced as a function of the thickness of a warm tungsten liner. If conventional superconductor is used, then the thicknesses required in the three cases would be as given in Tb.7. If high Tc superconductors could be used then these thicknesses could probably be halved.

		2TeV	0.5 TeV	Demo
Unshielded Power	MW	17	4	.5
Liner thickness	cm	4.5	3	2
Power leakage	KW	170	150	50

TABLE 7. Required Thickness of Shielding in Collider Magnets.

2.7.2 Lattice Design

Studies [25] of the resistive wall impedance instabilities indicate that the required muon bunches (eg for 2 TeV: $\sigma_z = 3 \, mm$, $N_\mu = 2 \times 10^{12}$) would be unstable in a conventional ring. In any case, the RF requirements to maintain such bunches would be excessive. It is thus proposed to use an isochronous lattice of the type discussed by S.Y. Lee *et al* [26]. The elements of such a lattice have been designed [28], and are being incorporated into a full ring.

It had been feared that amplitude dependent anisochronicity generated in the insertion would cause bunch growth in an otherwise purely isochronous design. It has, however, been pointed out [27] that if chromaticity is corrected in the ring, then amplitude dependent anisochronicity is automatically removed.

FIGURE 8. Energy attenuation vs the thickness of a tungsten liner.

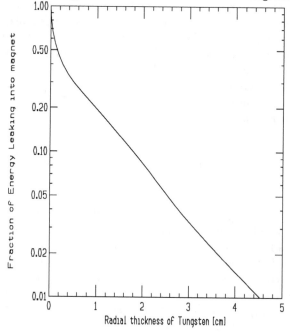

20

	field (T)	4 MeV		0.5 MeV	
		L(m)	R(cm)	L(m)	R(cm)
drift		6.5		1.99	
focus	6	6.43	6	1.969	5.625
drift		4.0		1.2247	
defocus	6.4	13.144	12	4.025	11.25
drift		4.0		1.2247	
focus	6.4	11.458	12	3.508	11.25
drift		4.0		1.2247	
defocus	6.348	4.575	10	1.400	9.375
drift		80		24.48	

TABLE 8. Final Focus Quadrupoles; L and R are the length and the radius respectively.

2.7.3 Low β Insertion

In order to obtain the desired luminosity we require a very low beta at the intersection point: $\beta^* = 3\,$mm for 4 TeV, $\beta^* = 8\,$mm for .5 TeV. A possible final focusing quadruplet design is shown in Fig.9. The parameters of the quadrupoles for this quadruplet are given in Tb.8. With these elements, the maximum beta's in both x and y are of the order of 400 km in the 4 TeV case, and 14 km in the 0.5 TeV machine. The chromaticities $(1/4\pi \int \beta dk)$ are approximately 6000 for the 4 TeV case, and 600 for the .5 TeV machine. Such chromaticities are too large to correct within the rest of a conventional ring and therefore require local correction of [29].

A preliminary *model* design [31] of local chromatic correction (for the 4 TeV case) has been presented. Fig.10 shows the horizontal dispersion and beta functions for this design. Fig.11 shows the tune shift as a function of momentum. It is seen that this design has a momentum acceptance of $\pm 0.3\,\%$. The second order amplitude dependent tune shifts appear acceptable in this design, but the bending fields are unrealistic. It is expected that these limitations will soon be overcome, and that more sophisticated designs [32] should do even better. It is hoped to achieve a momentum acceptance of $\pm 0.6\,\%$ for use with a clipped rms momentum spread of 0.2 %.

3   COLLIDER PERFORMANCE

3.1   Luminosity, Energy and Momentum Spread

The luminosity is given by:

$$\mathcal{L} = \frac{N^2\, f\, n_c \gamma}{4\pi\,\beta^*\,\epsilon_n} H(A, D) \qquad (4)$$

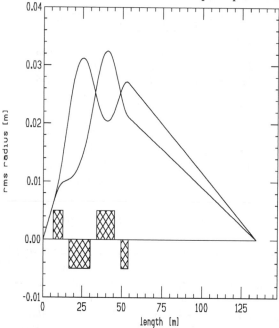

FIGURE 9. *rms* radius of the beam at the last four quadrupoles of the final focus.

where $A = \sigma_z/\beta^*$, $D = \frac{2\sigma_z N}{\gamma \sigma_{x,y}(\sigma_x + \sigma_y)} r_e(\frac{m_e}{m_\mu})$ and the enhancement factor is $H(A,D) \approx 1 + D^{1/4} \left[\frac{D^3}{1+D^3}\right] \left\{\ln\left(\sqrt{D}+1\right) + 2\ln\left(\frac{0.8}{A}\right)\right\}$.

The luminosities given in Tb.1 are those for the design energy and energy spread. At lower energies, or energy spread, the luminosities will be lower.

For a fixed collider lattice, operating at energies lower than the design value, the luminosity will fall as γ^3. One power comes from the γ in the above equation; a second comes from n_e, the effective number of turns, that is proportional to $\frac{\gamma}{2\pi R}$; the third term comes from β^*, which must be increased proportional to γ in order to keep the beam size constant within the focusing magnets. The bunch length σ_z must also be increased so that the required longitudinal phase space is not decreased.

In view of this rapid drop in luminosity with energy, it would be desirable to have separate collider rings at relatively close energy spacings: e.g. not more than factors of two apart.

If it is required to lower the energy spread $\Delta E/E$ at a fixed energy, then, given the same longitudinal phase space, the bunch length σ_z must be increased. If the final focus is retuned to simultaneously increase β^* to maintain the value of A, then the luminosity will be exactly proportional to $\Delta E/E$. But if, instead, the β^* is kept constant, and the parameter A allowed to increase, then the luminosity falls initially at a somewhat lower rate. The luminosity,

There remains some question about the coherent pair production generated by the virtual photons interacting with the coherent electromagnetic fields of the entire oncoming bunch. A simple Weizsäcker-Williams calculation [39] yields a background so disastrous that it would consume the entire beam at a rate comparable with its decay. However, I. Ginzburg [40] and others have argued that the integration must be cut off due to the finite size of the final electrons. If this is true, then the background becomes negligible.

If the coherent pair production problem is confirmed, then there are two possible solutions: 1) one could design a two ring, four beam machine (a μ^+ and a μ^- bunch coming from each side of the collision region, at the same time). In this case the coherent electromagnetic fields at the intersection are canceled and the pair production becomes negligible. 2) Plasma could be introduced at the intersection point to cancel the beam electromagnetic fields [41].

4 CONCLUSION

- The scenario for a 2 + 2 TeV, high luminosity collider is by no means complete. There are many problems still to be examined. Much work remains to be done, but no obvious show stopper has yet been found.

- Many technical components require development: a large high field solenoid for capture, low frequency RF linacs, long lithium lenses, multi-beam pulsed and/or rotating magnets for acceleration, warm bore shielding inside high field dipoles for the collider, muon collimators and background shields, but:

- None of the required components may be described as *exotic*, and their specifications are not far beyond what has been demonstrated.

- If the components can be developed and if the problems can be overcome, then a muon-muon collider may be a viable tool for the study of high energy phenomena, complementary to e^+e^- and hadron colliders.

5 ACKNOWLEDGMENTS

We acknowledge extremely important contributions from our colleagues, especially W. Barletta, A. Chao, D. Cline, D. Douglas, D. Helms, J. Irwin, K. Oide, H. Padamsee, Z. Parsa, C. Pellegrini, A. Ruggiero, W. Willis and Y. Zhao.

This research was supported by the U.S. Department of Energy under Contract No. DE-ACO2-76-CH00016 and DE-AC03-76SF00515.

REFERENCES

1. E. A. Perevedentsev and A. N. Skrinsky, Proc. 12th Int. Conf. on High Energy Accelerators, F. T. Cole and R. Donaldson, Eds., (1983) 485; A. N. Skrinsky and V.V. Parkhomchuk, Sov. J. of Nucl. Physics **12**, (1981) 3; *Early Concepts for $\mu^+\mu^-$ Colliders and High Energy μ Storage Rings, Physics Potential & Development of $\mu^+\mu^-$ Colliders. 2nd Workshop*, Sausalito, CA, Ed. D. Cline, AIP Press, Woodbury, New York, (1995).

2. D. Neuffer, IEEE Trans. **NS-28**, (1981) 2034.

3. *Proceedings of the Mini-Workshop on $\mu^+\mu^-$ Colliders: Particle Physics and Design*, Napa CA, Nucl Inst. and Meth., **A350** (1994) ; Proceedings of the Muon Collider Workshop, February 22, 1993, Los Alamos National Laboratory Report LA- UR-93-866 (1993) and *Physics Potential & Development of $\mu^+\mu^-$ Colliders 2nd Workshop*, Sausalito, CA, Ed. D. Cline, AIP Press, Woodbury, New York, (1995).

4. Transparencies at the *$2 + 2$ TeV $\mu^+\mu^-$ Collider Collaboration Meeting*, Feb 6-8, 1995, BNL, compiled by Juan C. Gallardo; transparencies at the *$2 + 2$ TeV $\mu^+\mu^-$ Collider Collaboration Meeting*, July 11-13, 1995, FERMILAB, compiled by Robert Noble.

5. D. V. Neuffer, R. B. Palmer, Proc. European Particle Acc. Conf., London (1994); M. Tigner, in Advanced Accelerator Concepts, Port Jefferson, NY 1992, AIP Conf. Proc. **279**, 1 (1993).

6. *Overall Parameters and Construction Techniques Working Group report, Proceedings of the Fifth International Workshop on Next-Generation Linear Colliders*, Oct 13-21, 1993, Slac-436, pp.428.

7. R. B. Palmer et al., *Monte Carlo Simulations of Muon Production, Physics Potential & Development of $\mu^+\mu^-$ Colliders 2nd Workshop*, Sausalito, CA, Ed. D. Cline, AIP Press, Woodbury, New York, pp. 108 (1995).

8. T. Roser, *AGS Performance and Upgrades: A Possible Proton Driver for a Muon Collider*, this proceedings.

9. Y. Cho, et al., *A 10-GeV, 5-MeV Proton Source for a Pulsed Spallation Source, Proc. of the 13th Meeting of the Int'l Collaboration on Advanced Neutron Sources*, PSI Villigen, Oct. 11-14 (1995); Y. Cho, et al., *A 10-GeV, 5-MeV Proton Source for a Muon-Muon Collider*, this proceedings.

10. F. Mills, et al., presentation at this Wokshop, unpublished; see also second reference in [4].

11. D. Kahana, et al., *Proceedings of Heavy Ion Physics at the AGS-HIPAGS '93*, Ed. G. S. Stephans, S. G. Steadman and W. E. Kehoe (1993); D. Kahana and Y. Torun, *Analysis of Pion Production Data from E-802 at 14.6 GeV/c using ARC*, BNL Report # 61983 (1995).

12. R. Weggel, private communication; Physics Today, pp. 21-22, Dec. (1994).

13. R. Chehab, J. Math. Phys. **5**, (1978) 19

14. F. Chen, *Introduction to Plasma Physics*, Plenum, New York, pp. 23-26 (9174); T. Tajima, *Computational Plasma Physics: With Applications to Fusion and Astrophysics*, Addison-Wesley Publishing Co., New York, pp. 281-282 (1989).

15. N. Mokhov, R. Noble and A. Van Ginneken, *Target and Collection Optimization for Muon Colliders*, this proceedings.

16. A. A. Mikhailichenko and M. S. Zolotorev, Phys. Rev. Lett. **71**. (1993) 4146; M. S. Zolotorev and A. A. Zholents, SLAC-PUB-6476 (1994)

17. Ady Hershcovitch, Brookhaven National Report AGS/AD/Tech. Note No. 413 (1995)

18. Z. Huang, P. Chen, R. Ruth, SLAC-PUB-6745, *Proc. Workshop on Advanced Accelerator Concepts*, Lake Geneva, WI , June (1994); P. Sandler, A. Bogacz and D. Cline, *Muon Cooling and Acceleration Experiment Using Muon Sources at Triumf, Physics Potential & Development of $\mu^+\mu^-$ Colliders 2nd Workshop*, Sausalito, CA, Ed. D. Cline, AIP Press, Woodbury, New York, pp. 146 (1995).

19. Initial speculations on ionization cooling have been variously attributed to G. O'Neill and/or G. Budker see D. Neuffer, Particle Accelerators, **14**, (1983) 75; D. Neuffer, Proc. 12th Int. Conf. on High Energy Accelerators, F. T. Cole and R. Donaldson, Eds., 481 (1983); D. Neuffer, in Advanced Accelerator Concepts, AIP Conf. Proc. 156, 201 (1987); see also [1].

20. U. Fano, Ann. Rev. Nucl. Sci. 13, 1 (1963).

21. G. Silvestrov, Proceedings of the Muon Collider Workshop, February 22, 1993, Los Alamos National Laboratory Report LA-UR-93-866 (1993); B. Bayanov, J. Petrov, G. Silvestrov, J. MacLachlan, and G. Nicholls, Nucl. Inst. and Meth. **190**, (1981) 9; Colin D. Johnson, Hyperfine Interactions, **44** (1988) 21; M. D. Church and J. P. Marriner, Annu. Rev. Nucl. Sci. **43** (1993) 253.

22. G. Silvestrov, presentation at this Workshop, unpublished.

23. D. Neuffer, presentation at this Workshop, unpublished.

24. D. Summers, presentation at this Workshop, unpublished.

25. M. Syphers, private communication; K.-Y. Ng, *Beam Stability Issues in a Quasi-Isochronous Muon Collider*, this proceedings.

26. S.Y. Lee, K.-Y. Ng, D. Trbojevic, FNAL Report FN595 (1992); Phys. Rev. **E48**, (1993) 3040.

27. K. Oide, private communication.

28. D. Trbojevic, et al., *Design of the Muon Collider Isochronous Storage Ring Lattice, Micro-Bunches Workshop*, BNL Oct. (1995), to be published.

29. K. L. Brown , J Spencer, SLAC-PUB-2678 (1981) presented at the Particle Accelerator Conf., Washington, (1981) and K.L. Brown, SLAC-PUB-4811 (1988), Proc. Capri Workshop, June 1988 and J.J. Murray, K. L. Brown, T.H. Fieguth, Particle Accelerator Conf., Washington, 1987.

30. Bruce Dunham, Olivier Napoly, *FFADA, Final Focus. Automatic Design and Analysis*, CERN Report CLIC Note 222, (1994); Olivier Napoly, it CLIC Final Focus System: Upgraded Version with Increased Bandwidth ans Error Analysis, CERN Report CLIC Note 227, (1994).

31. Juan C. Gallardo, Robert B. Palmer, *Final Focus System for a Muon Collider: A Test Model*, BNL #, CAP 138-MUON-96R (1996)

32. K. Oide, SLAC-PUB-4953 (1989); J. Irwin, SLAC-PUB-6197 and LBL-33276, Particle Accelerator Conf.,Washington, DC, May (1993); R. Brinkmann, *Optimization of a Final Focus System for Large Momentum Bandwidth*, DESY-M-

90/14 (1990).

33. G. W. Foster, and N. V. Mokhov, *Backgrounds and Detector Performance at 2 + 2 TeV $\mu^+\mu^-$ Collider, Physics Potential & Development of $\mu^+\mu^-$ Colliders 2nd Workshop*, Sausalito, CA, Ed. D. Cline, AIP Press, Woodbury, New York, pp. 178 (1995).

34. N. V. Mokhov, *The MARS10 Code System*, Fermilab FN-509 (1989); *The MARS95 Code System (User's Guide)*, to be published as Fermilab Report.

35. I. Stumer, private communication

36. Geant Manual, Cern Program Library V. 3.21, Geneva, Switzerland, 1993.

37. P. Chen presentation at this Workshop, unpublished.

38. L. D. Landau and E. M. Lifshitz, Phys. Zs. Sowjetunion **6**, 244 (1934); V. M. Budnev, I. F. Ginzburg, G. V. Medelin and V. G. Serbo, Phys Rep., **15C**, 181 (1975

39. P. Chen presentation at this Workshop, unpublished.

40. I. Ginzburg, private communication.

41. G. V. Stupakov, P. Chen, *Plasma Suppression of Beam-Beam Interaction in Circular Colliders*, SLAC Report: SLAC-PUB-95-7084 (1995)

A 10-GeV, 5-MW PROTON SOURCE FOR A MUON - MUON COLLIDER*

Y. Cho, Y.-C. Chae, E. Crosbie, H. Friedsam, K. Harkay, D. Horan,
R. Kustom, E. Lessner, W. McDowell, D. McGhee, H. Moe, R. Nielsen,
G. Norek, K. Peterson, Y. Qian, K. Thompson and M. White

Argonne National Laboratory
Argonne, IL 60439

Abstract. The performance parameters of a proton source which produces the required flux of muons for a 2-TeV on 2-TeV muon collider are: a beam energy of 10 GeV, a repetition rate of 30 Hz, two bunches per pulse with 5×10^{13} protons per bunch, and an rms bunch length of 3 nsec (1). Aside from the bunch length requirement, these parameters are identical to those of a 5-MW proton source for a spallation neutron source based on a 10-GeV rapid cycling synchrotron (RCS) (2). The 10-GeV synchrotron uses a 2-GeV accelerator system as its injector, and the 2-GeV RCS is an extension of a feasibility study for a 1-MW spallation source described elsewhere (3-9). A study for the 5-MW spallation source was performed for ANL site-specific geometrical requirements. Details are presented for a site-independent proton source suitable for the muon collider utilizing the results of the 5-MW spallation source study.

1 INTRODUCTION

A feasibility study of a proton driver that generates the required number of muons for a 2-TeV on 2-TeV muon collider is presented. It utilizes the results of a site-specific study of a proton source for a 5-MW pulsed spallation source that was performed at Argonne National Laboratory (ANL) (2). The proton source performance parameters, except for bunch length requirements, are nearly identical to those of the proton driver for the muon-muon collider (1). The site-specific 5-MW proton source design was made site-independent, and was optimized as a proton driver for the muon collider.

The required performance parameters for a proton driver for a 2-TeV on 2-TeV muon collider and for a 5-MW proton source are listed in Table 1.

The 5-MW proton source is based on a 10-GeV, 30-Hz rapid cycling synchrotron (RCS-II) with a 2-GeV injector system. The 2-GeV injector is based on a system previously designed for a 1-MW spallation source that is described elsewhere in detail (3-9). The injector is comprised of a 2-GeV, 30-Hz, 190.4-meter circumference rapid cycling synchrotron (RCS-I), and a 400-MeV

---

* Work supported by the U. S. Department of Energy, Office of Basic Energy Sciences under the Contract W-31-109-ENG-38.

linear accelerator. The linac system consists of an H⁻ ion source, a 2-MeV RFQ, a ramped-gradient drift tube linac matching section (RGDTL), a 70-MeV DTL, a matching section, and a 330-MeV coupled cavity linac (CCL). Linac performance parameters are listed in Table 2.

TABLE 1. Comparison of Accelerator Parameters for Muon Collider Proton Driver and ANL 5-MW Proton Source

Required Performance Parameters	Muon Collider Proton Driver	ANL 5-MW Proton Source
Energy	10 GeV	10 GeV
Bunches/pulse	2	2
Protons/bunch	5×10^{13}	5×10^{13}
Bunch Length	3 nsec (rms)	~ 25 nsec (100 %)
Repetition Rate	30 Hz	30 Hz

TABLE 2. Linac Parameters

Parameters	Value	Units
Ion	H⁻	
Emittance (rms normalized)	1 π	mm mr
Beam Energy	400	MeV
ΔE	±2.5	MeV
Peak Current	50	mA
Average Current	0.5	mA
Beam Power	200	kW
Removed by Chopping	25	%
Macro-Pulse Length	0.5	msec
Linac Beam Duty Factor	1.5	%
Repetition Rate	30	Hz

A key feature of the design of these accelerator systems is that beam losses are minimized from injection to extraction, reducing activation to levels consistent with hands-on maintenance. In order to minimize injection and capture losses in RCS-I, the low energy beam from the ion source is chopped at the synchrotron rf frequency, accelerated, then stacked in the waiting synchrotron buckets. At the end of the multi-turn injection, there are two gaps in the circulating beam, each corresponding to 25 % of an rf wavelength, to facilitate no-loss capture.

RCS-I accelerates 1.04×10^{14} of protons per pulse from 400 MeV to 2 GeV at a 30-Hz repetition rate. The dipole, quadrupole, and sextupole magnets for RCS-I are designed for nominal beam energies of 2 GeV but are capable of handling a maximum value of 2.2 GeV. They are powered by a dual-frequency resonant circuit that excites the magnets at a 20-Hz rate and de-excites them at a 60-Hz rate, resulting in an effective rate of 30 Hz and a reduction of 1/3 in the required peak rf voltage. The rf frequency varies from 2.2 MHz to 3 MHz during the cycle. Detailed descriptions of the RCS-I hardware can be found in Reference (3).

Two beam bunches per pulse are required at the pion production target, thus RCS-I's harmonic number was chosen to be 2. The two RCS-I bunches are separated by 332 nsec and are extracted in a single turn. The 2-GeV injector for the 5-MW proton source differs from the 1-MW spallation source RCS in harmonic number, thus a different rf cavity is required in this case. The 3-MHz RCS-I bunches are transferred into waiting buckets of the 6-MHz RCS-II rf system. The circumference ratio between the 10-GeV ring and the 2-GeV ring is four. As a consequence of this change in the rf frequency and of the ratio of circumferences of the machines, RCS-II has 16 buckets (harmonic number = 16), of which only two are occupied by the beam, separated by an empty bucket. RCS-II magnets are powered by dual-frequency resonant circuits similar to RCS-I resulting in an effective 30-Hz repetition rate. A time-averaged current of 0.5 mA, or 1.04×10^{14} protons per pulse, is achieved by this accelerator system.

The dipole, quadrupole, and sextupole magnets for RCS-II are designed for nominal beam energies of 10 GeV but are capable of handling a maximum value of 11 GeV. The gap geometries were scaled down from the corresponding magnets in the RCS-I ring to fit the smaller vacuum chambers. Calculated field qualities for the RCS-II magnets are, therefore, nearly identical to the RCS-I values. Major parameters for the synchrotron ring magnets can be found in Reference (2), along with other relevant information.

The RCS-II ring magnet vacuum chambers are 99.7% pure alumina ceramic and are similar to those used in RCS-I (3,9). There are rf shields close to the inside surface of each chamber section. A more detailed description of the RCS-II rf system, controls and diagnostics are found in Reference (2). Figure 1 shows the schematic layout of the proposed facility.

2 LATTICES

2.1 Lattice Features

Required features of the lattices for both the 2-GeV and 10-GeV rings are: 1) the transition energy must be larger than the extraction energy so that the lattice has a relatively large slip factor, $\eta = \left| \gamma^{-2} - \gamma_t^{-2} \right|$, 2) there must be enough straight-section length for the radio-frequency cavity system to provide the required energy gain using a gradient of 10 kV/m, and 3) the straight sections should be dispersion-free so that the rf cavities can be placed in dispersion-free regions. In addition, the circumference of the 10-GeV ring should be a multiple of that of the 2-GeV ring in order to have a harmonic relationship between the two synchrotron rf systems during bunch transfer from one machine to the other.

One way to satisfy requirement 1) above is for both lattices to have large horizontal tunes, since the transition energy, γ_t, is proportional to the horizontal tune. A FODO cell lattice with a phase advance of approximately 90° in both

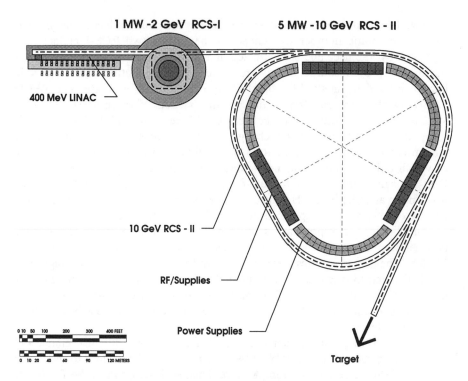

FIGURE 1. Site-Independent 5-MW Proton Source Machine Layout

transverse planes was chosen for the normal cells of both rings in order to obtain a higher tune.

Additional advantages of 90° FODO cells are: 1) Dispersion suppression can be obtained by removing a dipole from a normal cell. 2) Normal cells without both dipoles can be used to construct the straight-section cells. 3) Focusing quadrupoles in the normal cells, the dispersion-suppressor-cells and the straight-section cells have the same strengths, and therefore only one power supply is needed for the quadrupole family. Similarly, defocusing quadrupoles in both rings can also each be powered by a single power supply.

These general design features have been incorporated in the designs of both the 2-GeV and 10-GeV lattices. Figures 2(a) and 2(b) show 1/2 of a super period with reflective symmetry at both ends for the 2-GeV and 10-GeV lattices, respectively. Each cell of the FODO structure has a phase advance of ~ 90° in both transverse planes. The normal cells, dispersion-suppressor cell and the straight-section cells are evident in the figure. Dispersion suppression is achieved by removing a dipole from a normal cell. The missing dipole scheme suppresses the dispersion function when the vertical phase advance is slightly less than 90° while maintaining a horizontal phase advance of 90°. A benefit of this arrangement is that the horizontal tune is about one unit higher than the vertical tune, which is a good

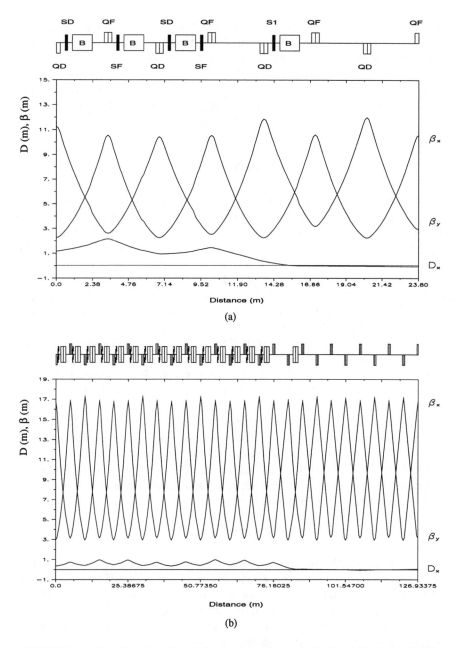

FIGURE 2. Lattice Functions for 1/2 Superperiod for (a) 2-GeV and (b) 10-GeV RCS.

feature. Further study of this lattice showed that the lattice tunes can be adjusted to within ±1/2 unit while maintaining the tune difference of one unit and a vanishing dispersion function through the straight sections.

The superperiodicity of the 2-GeV ring was chosen to be four and the superperiodicity of the 10-GeV ring was chosen to be three. These numbers result from the spallation source design; however, the choice of the periodicity is arbitrary. Table 3 shows the normal cell parameters for both synchrotrons. The overall lattice parameters are shown in Table 4.

2.2 Working Point

In order to find the optimum working point, the tune region where the dynamic apertures are largest was searched while also taking into account the space charge tune shift. As a result of extensive tracking studies, the optimal tune ranges were found that provide enough dynamic aperture in both rings to contain the beam at injection (3,4,10). The lattice tunes were chosen to be $v_x = 6.821$ and $v_y = 5.731$ for RCS-I and $v_x = 19.20$ and $v_y = 18.19$ for RCS-II. The working points, including the maximum space-charge tune spread assuming an elliptic distribution, are shown in Figures 3(a) and 3(b) for RCS-I and RCS-II, respectively. The RCS-I injection mechanism is expected to result in a Kapchinskij-Vladimirskij (K-V) distribution (3). The point Q in Figure 3(a) is the working point for a K-V distribution.

The tune spread around the working point for RCS-II is clear of structure resonances. In RCS-I, the tune spread can cross two nonlinear structure resonances. The $v_x + 3v_y = 24$ resonance can be driven by skew-octupole fields, but such fields arise only from multipole errors and have little effect on the dynamic aperture. The $3v_x = 20$ resonance near the full space charge tune can be driven by chromaticity-correcting sextupoles and can have a deleterious effect on the circulating beam. This effect can be made small by placing harmonic-correcting sextupoles in the dispersion-free sections, a standard technique used for a synchrotron light source Chasman-Green lattice.

2.3 Dynamic Aperture

The lattices of both machines have to provide a large enough dynamic aperture to contain the beam. The dynamic aperture is each case is limited primarily by the chromaticity-correcting sextupoles. In RCS-I, although the tune spread due to $\Delta p/p$ is small, chromaticity correction may be desirable. In RCS-II, the tune spread due to $\Delta p/p$ is larger than the space-charge tune spread, thus chromaticity correction is necessary, especially during injection.

In RCS-I, with the sextupoles energized to the values required to adjust the chromaticities to zero, the dynamic aperture is larger than the beam-stay-clear (BSC), as shown in Figure 4. Detailed calculations of the effects of magnetic field

TABLE 3. Normal Cell Parameters for 2-GeV and 10-GeV RCS

2 GeV, Bρ = 9.288 Tm			10 GeV, Bρ = 36.352 Tm			
Elements	Length (m)	Strength	Elements	Length (m)	Strength	Units
QD	0.25	-6.765	QD	0.4	-13.086	T/m
D1	0.3		DQS	0.2		
SD	0.2	-39.15	SD	0.4	-118.05	T/m^2
D1	0.3		DSB	0.675		
B	1.3	1.403	B	1.727	1.377	T
D2	0.8		DBQ	1.275		
QF	0.5	7.685	QF	0.8	13.536	T/m
D1	0.3		DQS	0.2		
SF	0.2	28.42	SF	0.4	78.52	T/m^2
D1	0.3		DSB	0.675		
B	1.3	1.403	B	1.727	1.377	T
D2	0.8		DBQ	1.275		
QD	0.25	-6.765	QD	0.4	-13.086	T/m

TABLE 4. Lattice Parameters for 2-GeV and 10-GeV RCS

Parameters	2-GeV RCS	10-GeV RCS	Units
Circumference	190.4	761.6	m
Superperiodicity	4	3	-
Total number of cells	28	75	-
Number of normal cells	12	45	-
Number of dispersion-suppressor cells	8	6	-
Number of straight-section cells	8	24	-
Normal cell length	6.8	10.155	m
Bending radius	6.621	26.392	m
Number of dipole magnets	32	96	-
Dipole magnets effective length	1.30	1.727	m
Dipole field strength at extraction	1.40	1.377	m
Dipole field strength at injection	0.48	0.352	m
Number of quadrupole magnets	56	150	-
Quadrupole magnet effective length	0.5	0.8	m
Maximum quadrupole gradient (B')	7.73	13.54	T/m
Number of sextupole magnets (F)	16	42	-
Number of sextupole magnets (D)	16	48	-
Sextupole magnet effective length	0.2	0.4	m
Sextupole field coefficient (B″)	40	118.1	T/m^2
Transition energy, γ_t	5.4	14.74	-
Horizontal tune, ν_x	6.821	19.20	-
Vertical tune, ν_y	5.731	18.19	-
Natural chromaticity, $\xi_x = (\Delta\nu)_x/(\Delta p/p)$	-7.23	-23.90	-
Natural chromaticity, $\xi_y = (\Delta\nu)_y/(\Delta p/p)$	-6.88	-23.06	-
Maximum β_x	12	16.84	m
Minimum β_y	11	17.26	m
Maximum dispersion function	2.2	1.0	m

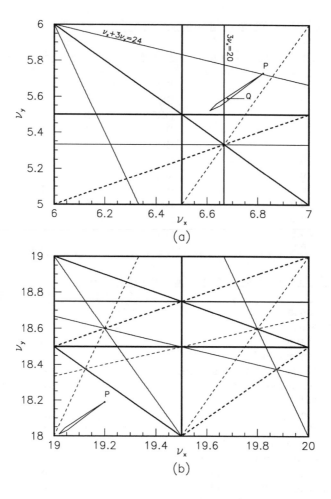

FIGURE 3. Integer and Half-Integer Stop-Bands and Structure Resonances Through Fourth Order Including Working Point (P) and Tune Spread for (a) RCS-I and (b) RCS-II. In (a), the maximum space charge tune shift for a uniform, K-V distribution is labeled Q.

imperfections, misalignments, and orbit corrections are reported in References (3,4).

The lattice dynamic apertures at the design tune and at the depressed tune for the RCS-II are shown in Figure 5. The beam size at injection and the BSC are also shown. In Figure 5(a), we show the dynamic apertures at the chromaticities $\xi_{x,y} = 0$ and -5. The operating range of the chromaticity is expected to be close to $\xi_{x,y} = -5$, where the dynamic aperture is shown to be larger than the BSC. In Figure 5(b), we show the dynamic aperture when the working point is moved to the space-charge depressed tune.

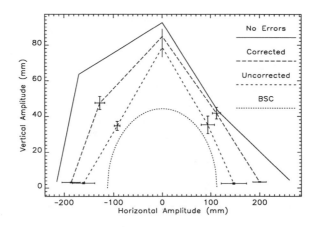

FIGURE 4. Dynamic Aperture of the RCS-I Lattice at the Design Tune. Details of the calculations are found in References (3,4).

3 APERTURES AND ACCEPTANCES

The stacked beam transverse emittance (100%) in RCS-I is 375 π mm mr. The beam is damped to 128 π mm mr at 2 GeV when it is injected into RCS-II. The BSC acceptance for both machines is defined as being twice the emittance and the ± 1 % momentum spread of the incoming beam. Figure 6 shows the vacuum chamber cross-sections of RCS-I (a) and RCS-II (b) at a focusing quadrupole indicating the BSC and the injected beam envelope.

4 INJECTION AND BEAM TRANSFER

For injection into RCS-I, multiple turns are accumulated using charge exchange by stripping the H⁻ linac beam. This, together with programming the closed orbit during injection, allows transverse phase space painting to achieve a K-V-like distribution of the injected particles in RCS-I (3,5,6). The 400-MeV injection energy, a stacked emittance of 375 π mm mr in both transverse planes, and an allowed space charge tune shift of about 0.17 results in a space charge limit of 1.04 x 10^{14} protons.

For injection into RCS-II, bunch-to-bucket transfer is used. Each pulse of the 2-GeV extracted beam contains two bunches. The beam from RCS-I has a single bunch area of 3.7 eV sec; this bunch is transferred into a slightly larger waiting bucket in RCS-II. Matching between the bunch and the bucket is performed using the following algorithm. The rf voltage at the extraction of the 2-GeV beam determines the energy spread, ΔE, and length, Δt, of the bunch. The energy spread

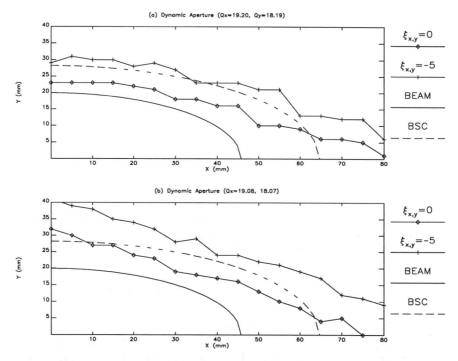

FIGURE 5. Dynamic Aperture of the RCS-II Lattice at Different Chromaticities. (a) Lattice at the Design Tune. (b) Working Point at the Space-Charge Depressed Tune v_x = 19.08 and v_y = 18.07, where a Uniform Charge Distribution is Assumed.

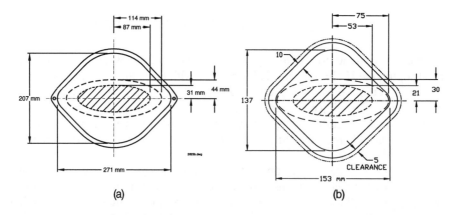

FIGURE 6. Focusing Quadrupole Vacuum Chamber Cross Section in (a) RCS-I and (b) RCS-II Showing Beam-Stay-Clear (dashed line) and Beam Envelopes (area containing diagonal lines). Note that the scale is different in the two drawings.

and bunch length of the incoming beam are matched to a Hamiltonian phase space contour of the waiting bucket. The space charge distortion of the contours is taken into account in these calculations to prevent mismatching and consequent dilution of the beam. Figure 7(a) shows the rf bucket and the bunch population in RCS-I at extraction, when the rf voltage is 200 kV. Figure 7(b) shows the bunch injected into the waiting bucket of RCS-II. The required rf voltage at this time is 700 kV. The dashed line in Figure 7(b) represents the Hamiltonian contour whose height, ΔE, and whose enclosed area are equal to those of the injected beam. This shows that the 3.7 eV sec bunch is matched to a contour of a 3.7 eV sec phase space area within a 5.8 eV sec bucket.

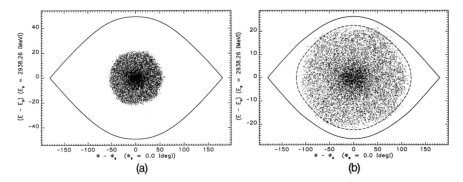

FIGURE 7. (a) Rf Bucket and Phase Space Distribution at Extraction in RCS-I. The Bunch Area is 3.7 eV sec. (b) Rf Bucket and Phase Space Distribution at Injection in RCS-II. The Dotted Line Indicates the Contour Enclosing an Area of 3.7 eV sec.

5 RF VOLTAGE PROGRAM

The key feature of the design studies for both the 1- and 5-MW machines was to determine the rf voltage program that results in no beam loss from injection to extraction. Such a program was obtained for each machine using a Monte Carlo method to track the particles from injection to extraction (11). The rf voltage programs for the accelerating cycles of both rings are depicted in Figures 8(a) and 8(b). The time variation of the bucket and bunch areas are also shown in the figures. The rf voltage program allows the capture and acceleration of 1.04×10^{14} particles per pulse with minimum loss of particles during the beam transfer and acceleration processes.

5.1 Rf Voltage at Injection

RCS-I accepts the 400-MeV beam from the linac and accelerates it to 2-GeV. Capture with minimum losses is obtained by chopping the linac beam such that it occupies 75 % of the ring. A beam with ± 2.5 MeV energy spread is injected in

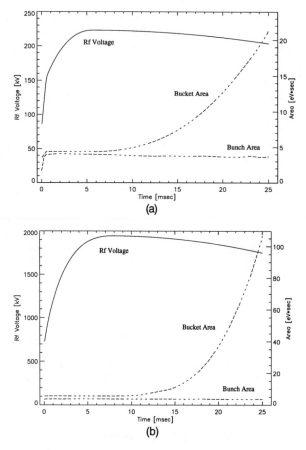

FIGURE 8. Rf Voltage Programs for (a) RCS-I and (b) RCS-II, Showing Time Evolution of the Bucket (solid line) and Bunch (dotted line) Areas Over the Complete Cycle.

561 turns, which corresponds to 0.5 msec. The area of the chopped incoming beam is 1.7 eV sec per bunch. A minimum rf voltage of 81 kV is required to contain the beam during the initial injection. The corresponding per-bucket area is 3.5 eV sec. During injection, the rf voltage must be increased from 81 kV to 134 kV to prevent beam losses. This rapid increase in the voltage is crucial to overcome the space charge forces that increase with particle density as multiple turns are accumulated. At the end of injection, the per-bucket area is 4.5 eV sec, and the single bunch area grows from 1.7 eV sec to 3.7 eV sec, giving a dilution factor of 2.2.

The unloaded bunch-to-bucket matching into RCS-II is straightforward, as discussed in the previous section. However, transient beam loading is expected to be severe, and either a low-impedance cathode-follower or a feed-forward system needs to be investigated.

5.2 Rf Voltage During Acceleration

The rf voltage in RCS-I is raised during the first 6 msec of acceleration to maintain a fixed per-bucket area of 4.5 eV sec and reaches a maximum of 220 kV at this time. The voltage is gradually decreased to 200 kV from 6 msec to the end of acceleration. The per-bucket area grows to 21 eV sec at extraction. The voltage for the latter part of the cycle is maintained high to ensure a synchrotron frequency fast enough to allow the particles in the bunch to follow the rapid change of the synchronous phase near extraction. The Keil-Schnell criterion for longitudinal stability is also met by making the bucket area large. The per-bunch area is 3.7 eV sec throughout the cycle.

In RCS-II, the initial per-bucket area is 5.8 eV sec, as described in the previous section. The per-bucket area is maintained at this initial value for the first 8 msec, and the voltage reaches 1.9 MV at this time. The voltage is decreased to 1.7 MV over the remainder of the cycle, giving a per-bucket area of 106 eV sec at extraction. The voltage for the latter part of the cycle is maintained high for the same reason as in RCS-I. The bunch height is ±102.5 MeV and the full bunch length is 22.6 nsec at extraction from RCS-II. The per-bunch area is 3.7 eV sec. The final bunch length can be shortened by rf manipulations, as explained in detail in the following section.

6 BUNCH LENGTH ADJUSTMENT

The required bunch length at the pion production target is 3 nsec rms, or 15 nsec full length (1). The typical full bunch length at extraction from RCS-II is about 25 nsec. The bunch length can be compressed through rf manipulations just prior to extraction. One way to shorten a bunch is to suddenly increase the rf voltage for a quarter synchrotron period. The initial bunch has $\pm\Delta E_i$ and $\pm\Delta\phi_i$ prior to the sudden voltage jump. The sudden jump of the rf voltage causes the bunch to rotate with a synchrotron frequency derived from the final voltage, V_f. After a quarter synchrotron period, the initial bunch has rotated 90° in phase space so that ΔE_i is projected onto to the ϕ-axis. The following relation is obtained for a stationary bucket when the bunch area is much smaller than the bucket area:

$$\sin^2\left(\Delta\phi_f/2\right) = \frac{\pi h|\eta|}{2\beta^2 eE_s}\frac{\left(\Delta E_i\right)^2}{V_f} \ .$$

In this equation, $\Delta\phi_f$ is the half-length of the final bunch, E_s is the total energy of the synchronous particle, β is the relative velocity, and e in the unit electric charge. The above equation shows that the final bunch length is related to the increased voltage and the initial energy spread. The initial energy spread is determined by the voltage before the sudden increase.

Figure 9 shows the results of tracking studies of this bunch length adjustment process. Figure 9(a) shows the rf voltage program throughout the acceleration cycle. The voltage step from $V_i = 1.2$ MV to $V_f = 3.2$ MV is shown near the end of the cycle. The voltage is increased in 50 μsec, and the higher voltage is maintained for 0.45 msec. Figure 9(b) shows the particle phase space distribution just prior to the voltage increase. The full bunch length is 25 nsec and the bunch has a $\Delta p/p$ of 0.9%. Figure 9(c) shows the phase space distribution after a quarter synchrotron period (0.45 msec) after the voltage increase to V_f. The new full bunch length is 15 nsec, which corresponds to 3 nsec rms, and the new $\Delta p/p$ is 1.5%. This bunch is contained within a BSC momentum acceptance of 4.4% at extraction. The space charge tune shift at extraction for the shortened bunch is about 0.1.

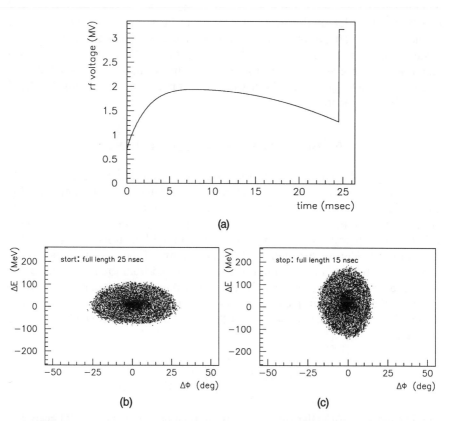

(a)

(b)

(c)

FIGURE 9. Bunch Length Adjustment at Extraction from RCS-II: (a) is the RCS-II Rf Voltage Program Including the Voltage Step, (b) is the Particle Distribution in Phase Space Just Prior to the Voltage Increase and (c) is the Distribution at Extraction.

7 IMPEDANCE AND INSTABILITIES

Longitudinal and transverse instability thresholds were obtained from the estimated coupling impedances. The transverse impedance is dominated by the space charge self-field. The contribution to the longitudinal impedance from space charge dominates in RCS-I. In RCS-II, the space charge impedance is comparable to that from the broadband rf cavity impedance. Beam parameters such as $\Delta p/p$ and peak current were obtained through Monte Carlo studies of beam capture and acceleration using a longitudinal tracking code that includes the effects of space charge (11). The analyses lead to rf voltage profiles and beam parameters that minimized the instabilities as well as beam losses (3,7,8).

Both RCS-I and RCS-II operate below the transition energy, thus the longitudinal microwave instability is not expected to occur unless there is a large resistive component. However, a conservative approach was adopted to ensure that the momentum spread is sufficient to satisfy the Keil-Schnell (KS) stability criterion. In RCS-I, the ceramic vacuum chamber is constructed with a special rf shield, such as that used at the ISIS facility of Rutherford-Appleton Laboratory (12), to minimize the impedance due to space charge. The shield follows the beam envelope at an aperture equal to the BSC, reducing both the longitudinal and transverse space charge impedances by between 30 and 35 % at injection. In RCS-II, the space charge impedance is smaller than in RCS-I by a factor of 5 at injection and 10 at extraction. Therefore, a contour-following rf shield, such as that used in the RCS-I, is not required. The rf shields are close to the inside surface of each chamber section. Studies were performed to choose an rf voltage profile that provides adequate bucket area and momentum spread. Tracking studies show that the RCS-I beam remains in the stable region defined by the KS criterion over the duration of the acceleration cycle (3,7). The KS criterion is more difficult to satisfy in RCS-II because the threshold is dominated by the high peak current. Tracking studies show that the RCS-II beam remains in the stable region through most of the acceleration, where, at extraction energy, the $\Delta p/p$ is about 1% and the bunching factor (peak/circulating current) is about 0.01. The peak current in RCS-II is 160 A at injection and less than 800 A at extraction.

In the transverse plane, the head-tail instability was analyzed in both machines. The growth rate is proportional to the resistive wall and kicker impedances. The head-tail modes are stabilized at a slightly negative chromaticity. The threshold tune spread for transverse microwave instability is 0.06 in RCS-I and 0.11 in RCS-II.

8 SUMMARY

A feasibility study of a proton source that generates the required number of muons in a 2-TeV on 2-TeV muon collider is presented. The results of a 5-MW, 10-GeV proton source study for a pulsed spallation neutron source were used.

The accelerator configuration for the muon collider proton source consists of a 400-MeV H⁻ linac, a 2-GeV RCS-I and a 10-GeV RCS-II, each operating at 30 Hz and an average current of 0.5 mA. RCS-I accepts the 400-MeV H⁻ beam and accelerates it to 2 GeV in an $h = 2$ rf system. The linac beam is chopped at the source at the injection rf frequency so that when multi-turn injection is completed, there are two gaps in the circulating beam, each corresponding to 25% of an rf wavelength. This gap is introduced to facilitate no-loss capture. For both RCS's, low losses are achieved by providing large dynamic apertures in the transverse planes and sufficient bucket areas in the longitudinal space. The extracted pulse for the pion production target consists of two bunches with 5×10^{13} protons each and whose length can be made about 3 nsec rms by the introduction of a sudden jump in the rf voltage. The corresponding $\Delta p/p$ is 1.5%.

9   REFERENCES

1. R. B. Palmer, et. al., "Beam Dynamics Problems in a Muon Collider," BNL Report 61580 (March 16,1995) and references therein.
2. Y. Cho, et. al., "A 10-GeV, 5-MW Proton Source for a Pulsed Spallation Source," *Proc. of 13th Meeting of the Int'l. Collaboration on Advanced Neutron Sources,* PSI, Villigen, to be published (Oct. 11-14, 1995).
3. "IPNS Upgrade - A Feasibility Study", ANL Report ANL-95/13 (April, 1995).
4. E. Lessner, Y.-C. Chae, and S. Kim, "Effects of Imperfections on Dynamic Aperture and Orbit Functions of the IPNS Upgrade Synchrotron," *Proc. of the 1995 Particle Accelerator Conference,* Dallas, Texas, to be published (May 1995).
5. Y.-C. Chae and Y. Cho, "Study of Field Ionization in Charge Exchange Injection for the IPNS Upgrade," *Proc. of the 1995 Particle Accelerator Conference,* Dallas, Texas, to be published (May 1995).
6. E. Crosbie and K. Symon, "Injecting a Kapchinskij-Vladimirskij Distribution into a Proton Synchrotron," *Proc. of the 1995 Particle Accelerator Conference,* Dallas, Texas, May 1995, to be published (May 1995).
7. K. Harkay, Y. Cho, and E. Lessner, "Longitudinal Instability Analysis for the IPNS Upgrade," *Proc. of the 1995 Particle Accelerator Conference,* Dallas, Texas, to be published (May 1995).
8. K. Harkay and Y. Cho, "Transverse Instabilities Analysis for the IPNS Upgrade," *Proc. of the 1995 Particle Accelerator Conference,* Dallas, Texas, to be published (May 1995).
9. Y. Cho, et al., "Feasibility Study of a 1-MW Pulsed Spallation Source," *Proc. of 13th Meeting of the Int'l. Collaboration on Advanced Neutron Sources,* PSI, Villigen, to be published (Oct. 11-14, 1995).
10. Y.-C. Chae and Y. Cho, "Study of the High γ_t Lattice for a 5-MW Proton Synchrotron: FODO-I," ANL Report No. NSA-95-6 (August 11, 1995).
11. Y. Cho, E. Lessner, and K. Symon, "Injection and Capture Simulations for a High Intensity Proton Synchrotron," *Proc. of the European Particle Accelerator Conference,* London, page 1228 (1994).
12. G. H. Rees, "Status Report on ISIS," *Proc. of the IEEE Particle Accelerator Conference* (March 16-19, 1987).

AGS PERFORMANCE AND UPGRADES*
A Possible Proton Driver for a Muon Collider

T. ROSER

Brookhaven National Laboratory, Upton, N.Y. 11973, USA

Abstract

After the successfull completion of the AGS Booster and several upgrades of the AGS, a new intensity record of 6.3×10^{13} protons per pulse accelerated to $24\,GeV$ was achieved. Futher intensity upgrades are being discussed that could increase the average delivered beam intensity by up to a factor of six. The total beam power then reaches almost $1\,MW$ and the AGS can then be considered as a proton driver for a muon collider.

1   Recent AGS High Intensity Performance

The proton beam intensity in the AGS has increased steadily over the 35 year existence of the AGS, but the most dramatic increase occurred over the last few years with the addition of the new AGS Booster[1]. In Fig. 1 the history of the AGS intensity improvements is shown and the major upgrades are indicated. The AGS Booster has one quarter the circumference of the AGS and therefore allows four Booster beam pulses to be stacked in the AGS at an injection energy of $1.5\,GeV$. At this energy space charge forces are much reduced and this in turn allowed for the dramatic increase in the AGS beam intensity.

The beam intensity in the Booster surpassed the design goal of 1.5×10^{13} protons per pulse already to reach a peak value of 2.2×10^{13} protons per pulse. This was achieved by very carefully correcting all the important nonlinear orbit resonances especially at the injection energy of $200\,MeV$, where the space charge tune shift reaches about 0.4, and also by using the extra set of rf cavities, that were installed for heavy ion operation, as a second harmonic rf system. A second harmonic system allows for the creation of a flattened rf bucket which gives longer bunches with lower space charge forces.

The AGS itself also had to be upgraded to be able to cope with the higher beam intensity. During beam injection from the Booster, the AGS needs to store the already transferred beam bunches. During this time the beam is

---

*Work performed under the auspices of the U.S. Department of Energy

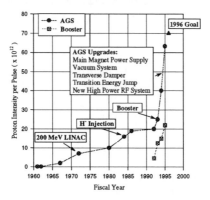

Figure 1: The history of the evolution of the proton beam intensity in the Brookhaven AGS.

exposed to the strong image forces from the vacuum chamber which causes beam loss from coupled bunch beam instabilities within as short a time as a few hundred revolutions. A very powerful feedback system was installed that senses any transverse movement of the beam and compensates with a correcting kick. New more powerful rf power amplifiers were build and installed immediately next to the ten rf cavities. This was needed to deliver to the beam the necessary $400\,kW$ power during acceleration and also to counteract the very large beam loading effects in the rf cavities from the high intensity beam.

During acceleration the AGS beam has to pass through the transition energy after which the revolution time of higher energy protons becomes longer than for the lower energy protons. This potentially unstable point during the acceleration cycle was crossed very quickly with a new powerful transition energy jump system with only minimal losses even at the highest intensities. However at energies above transition, a very rapid, high frequency instability developed which could only be avoided by purposely increasing the bunch length using a $100\,Mhz$ dilution cavity.

The peak beam intensity reached at the AGS extraction energy of $24\,GeV$ was 6.3×10^{13} protons per pulse also exceeding the design goal for this latest round of intensity upgrades. It also represents a world record beam intensity for a proton synchrotron. With a 1.6 second slow-extracted beam spill the average extracted beam current was about $3\,\mu A$. This level of performance was reached quite consistently during the last AGS experimental run of 24 weeks during which more than 10^{20} protons were accelerated in the AGS to $24\,GeV$.

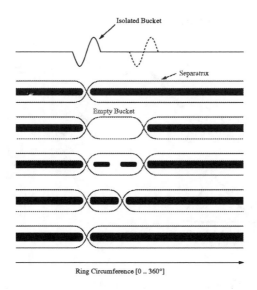

Figure 2: Time domain stacking scheme using a barrier bucket. The evolution of the longitudinal beam structure during the stacking process are shown from top to bottom.

2   Possible Future AGS Intensity Upgrades

Currently the number of Booster beam pulses that can be accumulated in the AGS is limited to four by the fact that the circumference of the AGS is four times the circumference of the Booster. This limits the maximum beam intensity in the AGS to four times the maximum Booster intensity which itself is limited to about 2.5×10^{13} protons per pulse by the space charge forces at Booster injection. To overcome this limitation some sort of stacking would have to be used in the AGS. The most promising scheme is stacking in the time domain. To accomplish this a cavity that produces isolated rf buckets can be used to maintain a partially debunched beam in the AGS and still leave an empty gap for filling in additional Booster beam pulses. The stacking scheme is illustrated in Fig. 2. It makes use of two isolated rf buckets to control the width of this gap. Isolated bucket cavities, also called Barrier Bucket cavities, have been used elsewhere[2]. However, for this stacking scheme, a much higher rf voltage would be needed. An additional important advantage of this scheme is that with the beam partially debunched in the AGS the beam density and therefore space charge forces are reduce by up to a factor of two.

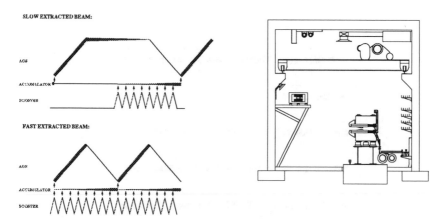

Figure 3: A 1.5 GeV accumulator in the AGS tunnel can be used for both slow-extracted beam (SEB) and fast-extracted beam (FEB) to improve the machine duty cycle. On the left operation scenarios for these two modes are shown. On the right the location of the 1.5 GeV accumulator in the AGS tunnel is shown. The low field combined function magnets are shown on the left side of the tunnel cross section vertically elevated with respect to the AGS.

The increases in beam intensity will then have to be accelerated in the AGS. Presently, the beam intensity that can be accelerated across transition energy is limited by the very large dispersion waves produced by the transition jump quadrupoles. The large dispersion severely limits the momentum aperture. The amplitude of the required dispersion wave could be reduced significantly by using a quadrupole arrangement that generates a transition energy shift that increases linearly with the dispersion wave amplitude[3]. The present system generates a global dispersion wave where all the first order contributions to the change in transition energy cancel[4].

As more Booster beam pulses are accumulated in the AGS the reduction in the overall duty cycle becomes more significant. For fast-extracted beam operation (FEB) already the accumulation of four Booster pulses contributes significantly to the overall cycle time. With the addition of a 1.5 GeV accumulator ring in the AGS tunnel, shown in Fig. 3, this overhead time could be completely avoided. Such a ring could be build rather inexpensively possibly using low field permanent magnets[5]. Fig. 3 also shows the possible running scenarios with an accumulator and a barrier cavity.

3   The AGS Complex as a Proton Driver

For the current scenario of a muon collider the two biggest challenges for the proton driver are delivering the required beam power and producing the required extremely short and intense bunches.

For the 250 on 250 GeV muon collider a proton beam power of about 1 MW is required which could be achieved at the AGS by accumulating four Booster beam pulses of 0.25×10^{14} protons each in the Accumulator at the maximum possible rate of 10 Hz and then accelerate the resulting 1.0×10^{14} protons in the AGS with a repetition rate of 2.5 Hz. This would require an additional upgrade of the AGS main magnet power supply which now limits the AGS repetition rate to about 1 Hz.

The relative high energy of the AGS proton beam is a distinct advantage in producing the required short and intense beam bunches since the space charge forces are significantly reduced. The goal is to produce two bunches with 0.5×10^{14} protons each and a rms length of $1\,ns\,(\,0.3\,m\,)$. The maximum incoherent space charge tune shift of such a bunch for a $100\,\pi\,mm\,mrad$ transverse 95% beam emittance is 0.4 at $24\,GeV$. Although manageable this still very large tune shift illustrates clearly that the relative high proton energy is critical to the production of the required short proton bunches. At lower energy, not only does the space charge tune shift increase rapidly, but also more protons are required to produce the same number of muons per bunch.

The momentum spread of such a short bunch is very large, but again the higher energy can bring it into a manageable range. The current AGS momentum acceptance is at most $\pm 3\,\%$ which requires the longitudinal phase space occupied by one bunch to be less than $4.5\,eV\,s$. This in turn creates more stringent demands on the earlier parts of the accelerator cycle. In particular, Landau damping from the beam momentum spread is used to guard against resistive wall instabilities during AGS injection and also longitudinal microwave instabilities after transition[6]. Beam stability can be restored with a more powerful transverse damping system and possibly a new low impedance vacuum chamber. Also, the transverse microwave instability after transition crossing is predicted to cause beam loss unless damped by Landau damping from incoherent tune spread or possibly high frequency quadrupoles.

In summary, the required upgrades to the AGS complex to achieve the requirements for a 250 on 250 GeV muon collider consist of a 1.5 GeV accumulator and an AGS power supply upgrade to achieve 2.5 Hz operation and also possibly a Barrier Cavity system, a new AGS vacuum chamber, and upgrades to the AGS rf, transition jump, and transverse damping system. Although substantial, the cost of these upgrades is rather modest on the scale of a muon collider project.

References

[1] M. Blaskiewicz et al., High Intensity Proton Operations at Brookhaven, 1995 Particle Accel. Conf. Dallas, Texas, May 1995, to be published.

[2] J.E. Griffin et al., IEEE Trans. on Nucl. Sc. Vol. NS-30, No. 4, (1983) 3502

[3] V. Visnjic, Phys. Rev. Lett. **73** (1994) 2860

[4] W.K. van Asselt et al., The Transition Jump System for the AGS, 1995 Particle Accel. Conf., Dallas, Texas, May 1995, to be published.

[5] K. Bertsche et al., Temperature Considerations in the Design of a Permanent Magnet Storage Ring, 1995 Particle Accel. Conf. Dallas, Texas, May 1995, to be published.

[6] S.Y. Zhang, these proceedings.

The Solenoid Muon Capture System for the MELC Experiment

Rashid M. Djilkibaev and Vladimir M. Lobashev

Institute for Nuclear Research, Russian Academy of Sciences,
60-th Oct. Ann. 7a, Moscow 117312, Russia

Abstract. A solenoid capture system for the MELC experiment in which the efficiency of soft muon generation from the primary proton (600 MeV) is 10^{-4} in comparison with 10^{-8} for ordinary schemes has been proposed. Both signs of muons with an intensity 10^{11} μ^-/sec for negative and 2×10^{11} μ^+/sec for positive component can be generated by a pulse proton beam with an average current up to $\simeq 200$ μA. A detail 3-D calculation of the magnetic field for the MELC setup are presented. Production of muon from pion decay in solenoid capture system is studied. The target life time and radiation condition of the superconducting coil are considered.

INTRODUCTION

The problem of detecting processes violating the Law of lepton flavour conservation is one of the most important in modern elementary particle physics. In the $\mu^- \to e^-$ conversion process the muon and electron family numbers are not conserved, therefore this process is absent in standard theory of electroweak interaction. Discovering a connection between lepton families will prove the existence of new physical phenomena beyond the standard model. At present the $\mu^- \to e^-$ conversion [1] has been studied at a level of B.R. $< 4 \times 10^{-12}$. To advance further, it is necessary to have a much higher intensity of stopped negative muons. The MELC project will make it possible to obtain a sensitivity in exploring the $\mu^- \to e^-$ process as high as B.R. $\simeq 10^{-16}$, using a muon μ^- beam with a stopping rate up to $\simeq 10^{11}$ μ^-/sec.

Schematic of the MELC set-up [2] is shown in Fig.1. A proton beam is injected along the axis of the solenoid, having at the front part a magnetic field $\simeq 2.4$ Tesla, gradually decreasing as low as $\simeq 2$ Tesla. In the vicinity of the solenoid axis there are targets, consisting of thin tungsten (or molybdenum) disks of a small radius $\simeq 1$ cm, the total thickness of the targets along the beam is $\simeq 20$ g/cm^2 .

Pions produced in these discs precess along the magnetic field lines. Then they are reflected partially from the magnetic plug and drift mainly backward along the beam line. After few meters long flight most of soft pions have decayed and the resulting muons, confined inside a cylinder of a small radius $R \simeq 25$ cm go straight or (after reflection from the magnetic plug) back, precessing along magnetic field lines.

The high efficiency of soft muon production backward is determined by the location of targets along the solenoid axis and by spacing of target disks, so that the possibility of secondary crossing by pion or muon of one of the target disks is relatively small. The backward production scheme has advantages due to the low background from high energy neutrons, the simplicity of injection of proton beam along the magnetic field and convenience of the target location.

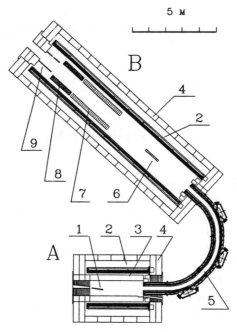

FIGURE 1. Set-up MELC: A – meson-producing part, B – detector part. Tungsten target of the meson-producing part (1), superconducting solenoids (2), solenoid shield (3), steel magnetic circuit (4), solenoid-collimator (5), aluminium target of the detector part (6), coordinate detector (7), total absorption scintillation spectrometer (8), shield against non-interacted muons and pions beam (9).

Owing to the spacing of target disks, and their extended surface it is possible to use radiation cooling for average proton current $\simeq 200$ μA.

It is shown that for the average field $\simeq 2$ Tesla, a collimator diameter $\simeq 25$ cm and 200 μA average proton current the stopping rate in the detector target is 10^{11} particles/sec for the negative and 2×10^{11} particles/sec for the positive muons.

MELC MAGNETIC SYSTEM

The MELC superconducting solenoid system (Fig. 1) consists of a solenoid for the meson producing part and a solenoid for the detector part, linked by transporting solenoid-collimator system. For simplification of the calculation the solenoid-collimator was approximated with a straight system. Solenoids of the meson producing and detecting parts are surrounded by the iron yoke for the magnetic flux return. The solenoid-collimator has no iron yoke.

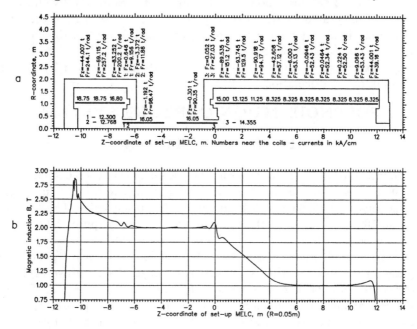

FIGURE 2. a) Axial plane (X,Z) of the solenoid system, b) dependence of B=f(Z) for X=30 cm, Y=0 cm.

To close the magnetic flux in the solenoids of large diameter, a possibility was considered to use an iron yoke of rectangular transverse section, assembled from a set of steel plates. The application of such a construction simplified to manufacture such a yoke.

The program MAGNUS [3] for 3-D magnetic calculation was used. The calculations showed that the use of an iron yoke of the rectangular transverse section does not affect the axial symmetry of the field inside of the solenoid coil and nor its value.

The magnetic calculations have shown that by selecting the form of pole pieces in the areas of solenoid joints and by changing mutual positions of the coil of the large and small solenoids, it is possible to achieve a rather gradual transition of the field lines.

The length of the solenoid coil element is 118.5 cm. Gaps between the elements are 5 cm determined by the thickness of the flanges necessary for joining coil elements.

Fig. 2a gives the cross section of the solenoid system in the (Z,R) plane.

Fig. 2a also shows the current in each element of the solenoid coils and forces acting on a given element are indicated over each element. F_z denotes forces acting along the Z axis (tons), F_r is the radial force with dimensions of ton/radian. $2 \times F_r$ corresponds to the rupture forces for the two halves of the coil (if integrated by the angle from 0 to π). It is seen from the figure that the maximal axis force acting on a particular element of the coil $\simeq$ 90 t, the maximal radial force $\simeq$ 260 t.

Fig. 2b shows the variation of the magnetic induction line B=f(Z) obtained as a result of the calculations.

SOLENOID MUON CAPTURE SYSTEM

The System Efficiency

The meson production target (Fig. 1) is a set of thin (~ 0.015 cm thick) tungsten disks with radius $0.4 \div 1.6$ cm; its total length is 70 cm. which corresponds to a thickness 20 g/cm^2 along the beam axis. The target is tilted at 10° with respect to the solenoid axis, which is required by technical requirements that have to do with the beam injection into the setup.

The pion production cross-section [4] (600 MeV proton) depends only slightly on the pion ejection angle and is 34 μb/sr/MeV on average. The integral cross-section is $\sigma = 4.3$ mb $\times (T_\pi^{max}/10\ MeV)$, where T_π^{max} is the effective maximum kinetic energy of pions that may still produce μ^- that may travel to the detector part of the setup. At a current of 200 μA (1.2 $\times$ 10^{15} protons/sec) 10^{12} π^-/sec will be produced with energies up to 30 MeV. The flux of muons stopped in the detector target N_μ, the efficiency of proton interaction in the target ε_p and the efficiency that a pion produced in the target produces a muon stopped in the detector target $\varepsilon_{\pi\mu}$ were calculated by the GEANT program [5].

TABLE 1. Muon Flux N_μ, Efficiency ε_p and $\varepsilon_{\pi\mu}$

Target radius [cm]	ε_p	$\varepsilon_{\pi\mu}$	Flux N_μ [sec^{-1}]
0.4	0.4	0.15	0.7×10^{11}
0.8	0.7	0.12	0.8×10^{11}
1.2	0.8	0.11	1.0×10^{11}
1.6	0.9	0.08	0.8×10^{11}

The parameters of the detector target are: length 250 cm; disk radius 10 cm; disk thickness 0.02 cm; number of disks 50; density 2.7 g/cm^3, which corresponds to the thickness 2.7 g/cm^2 along the axis. The results of calculations are presented in table 1.

The muon flight time is determined as the muon life time from production to capture in the detector target. The muon kinetic energy and muon time of flight distributions are presented in Fig. 3 (a,b) respectively.

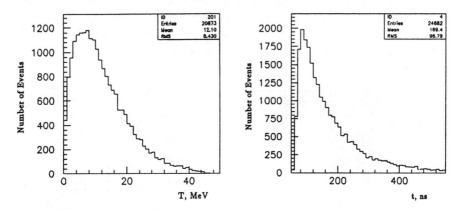

FIGURE 3. a) Muon kunetic energy distribution in detector part (Z = 20cm) b) Muon time of flight distribution until stop in detector target

The mean value of muon kinetic energy is$\simeq$ 12 MeV.

The transverse size of the muon trajectory is determined by three parameters: the helix line radius ($R_\perp$), the distance from Z-axis to the center of helix line (R_{Pole}) and the angle between muon momentum and Z-axis (θ).

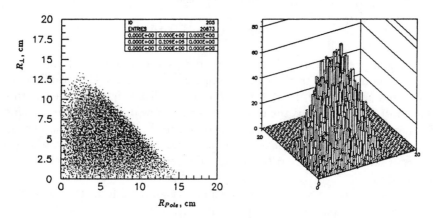

FIGURE 4. Scatteer plot for the muon radius $R_\perp$ versus the distance R_{Pole} in the detector part (Z = 20 cm)

The scatter plot for the radius $R_\perp$ versus R_{Pole} is shown in Fig. 4. The mean value of the radius $R_\perp$ is 5 cm and the mean distance R_{Pole} is 6 cm.

The muons produced by pion decay in flight move along helix lines, embracing the same field lines. The MELC magnetic field is smooth enough to provide adiabatic invariant conservation $sin^2\theta/B = const$, where θ is the angle between particle momentum and the magnetic field line and B is the magnetic field value.

Most of muons are produced in the higher magnetic (field $\simeq$ 2 T) and according to the adiabatic character of particle movement, the angle θ for the muon trajectory should be decreased in the detector part (field $\simeq$ 1 T). This effect is seen from the scatter plot for $cos(\theta)$ versus muon kinetic energy Fig.5.

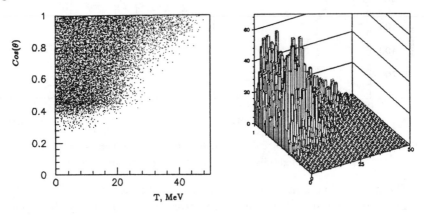

FIGURE 5. Scatter plot for the muon $cos(\theta)$ versus muon kunetic energy in the detector part (Z = 20 cm)

Target Life Time

The aging of the target is caused by tungsten evaporation from the surface of the disks due to their high temperature. The target durability was estimated as the time it takes for 1/10 of the disk thickness to evaporate. The most critical place is the centers of the first few disks of the target, since further downstream, the proton beam broadens due to multiple scattering.

To obtain a target life time of more than 1 year, the evaporation rate should be less than $4.8 \times 10^{-10} \times$ **d** $(g/cm^2/sec)$, where **d** is a distance between the target disks.

The estimates, using data on tungsten evaporation from [6] rates, show that with a proton beam of a few mm size, a distance **d** less than 1 cm, a disk thickness $\simeq$ 0.015 cm and an average proton current of 200 μA, the target life time for the first few disks is more than 1 year.

Solenoid Radiation Conditions

We have estimated the heat release in the coil for different shield materials. The main contribution to the coil heat release is from low-energy neutrons ($\simeq$ 20 MeV) from the target.

In order to determine the heat release in the meson production solenoid coil a Monte-Carlo program [7] that calculates nucleon-meson cascades initiated by 0.02 to 10 GeV hadrons in heterogeneous media has been used.

The protection thickness was assumed to be equal to 55 cm, while the aluminium solenoid coil was 5 cm thick. The results of calculations are presented in the table 2.

TABLE 2. Heat Release in the Solenoid Coil

Shield material	Density $[g/cm^3]$	Heat release [W]
Fe	7.9	6.0
Cu	9.0	5.00
Cu powder	7.1	10.0
Pb	11.3	7.0
U_3O_8	7.5	19.0
UO_2	9.7	14.0

Considering that the energy released in the solenoid should not exceed $\simeq$ 10 W, we see that U_3O_8 and UO_2 cannot be used as a material for protection.

For a 200 μA beam, the total flux of neutrons in the solenoid after the stainless steel shield is $5 \times 10^6/cm^2$. For a running time $\simeq 10^7$ s the activity of the solenoid aluminium coil is $\simeq$ 1 mr/h after 7 days cooling.

CONCLUSION

It is shown that the solenoid muon capture system for the MELC experiment has the advantage over ordinary schemes of a few orders of magnitude for soft muon ($\simeq$ 10 MeV) generation using 600 MeV primary protons.

The beam of both muon signs with an intensity 10^{11} μ^-/sec for negative and 2×10^{11} μ^+/sec for positive muons can be generated assuming the average proton current is 200 μA.

Further increasing the muon intensity is possible by means of increasing of the magnetic field and increasing the target thickness.

The transverse beam size can be decreased by means of absorbers in the form of thin disks placed along solenoid-collimator axis.

ACKNOWLEDGEMENT

We would like to thank D. Bryman, J-M. Poutissou, T. Numao and A. Zelensky for the helpful decussions and remarks.

REFERENCES

1. Ahmad S., et al., Phys. Rev. Lett. 59, 970 (1987).
2. Djilkibaev R.M. and Lobashev V.M., Sov.J.Nucl.Phys. 49(2), 384, (1989).
3. Pissanetzky S., Program MAGNUS, (1986).
4. Crawford J.F., Daum M., et al. SIN Prep. 79 - 010 Oct. (1979).
5. Brun R., Bruyant F. et al., Program GEANT3, CERN, (1987).
6. Zalikman A.N. and Nikitin L.S., Tungsten, Moscow (1978).
7. Ilinov A.S., et al. Program SUPER, Preprint INR Moscow (1985).

Target and Collection Optimization for Muon Colliders

Nikolai V. Mokhov, Robert J. Noble and A. Van Ginneken

Fermi National Accelerator Laboratory*

P.O. Box 500, Batavia, Illinois 60510

Abstract. To achieve adequate luminosity in a muon collider it is necessary to produce and collect large numbers of muons. The basic method used in this paper follows closely a proposed scheme which starts with a proton beam impinging on a thick target (∼ one inter-action length) followed by a long solenoid which collects muons resulting mainly from pion decay. Production and collection of pions and their decay muons must be optimized while keeping in mind limitations of target integrity and of the technology of magnets and cavities. Results of extensive simulations for 8 GeV protons on various targets and with various collection schemes are reported. Besides muon yields results include energy deposition in target and solenoid to address cooling requirements for these systems. Target composition, diameter, and length are varied in this study as well as the configuration and field strengths of the solenoid channnel. A *curved* solenoid field is introduced to separate positive and neg-ative pions within a few meters of the target. This permits each to be placed in separate RF buckets for acceleration which effectively doubles the number of muons per bunch available for collisions and increases the luminosity fourfold.

INTRODUCTION

Interest in a muon collider for future high energy physics experiments has greatly increased recently [1]. Muons suffer far less synchrotron radiation than electrons providing hope that well-known circular machine technology can be extended to much higher energies than presently available—or even contemplated—for lepton colliders. The short muon lifetime and the difficulty and expense of producing large numbers of them makes a useful muon collider luminosity hard to achieve. Because of their short lifetime muons must be generated by a single proton pulse for each new acceleration cycle. Techniques for efficient production and collection of an adequate number of muons are thus needed to make a muon collider viable.

Earlier estimates of muon yield, based on conventional lithium lens and quadru-pole magnet collection methods, indicate that roughly 1000 protons are needed for

*Work supported by the U. S. Department of Energy under contract No. DE-AC02-76CH03000.

every muon delivered to the collider rings [2]. This results from inherent limitations in the momentum acceptance of these systems (typically less than ± 5 percent) which causes most (potential) muons produced to be wasted.

Motivated by neutrino beamline experience, a solenoid collection scheme for pions has been suggested [3]. Cursory simulations indicate significant improvement in muon yields for proton energies below 100 GeV while above this a collection system with two lithium lenses could surpass a solenoid. However the power required for a 15 to 30 Hz rapid-cycling proton synchrotron with 10^{14} protons per pulse becomes expensive above 30 GeV. Along with considerations on space charge limits and pion yields this suggests a kinetic energy of the proton driver between 3 and 30 GeV. Because of interest at Fermilab in upgrading its 15 Hz Booster to higher intensity (5×10^{13} protons per pulse) for the hadron program, a proton beam kinetic energy of 8 GeV is assumed in this study. This choice also might enable experimental verification of results presented here. Actual *numbers* reported here, such as yields and energy densities, may well depend considerably on incident energy. But *intercomparisons* and conclusions derived from them, such as in the optimization of target size or solenoid field with respect to muon yield, are expected to be much less sensitive to incident energy.

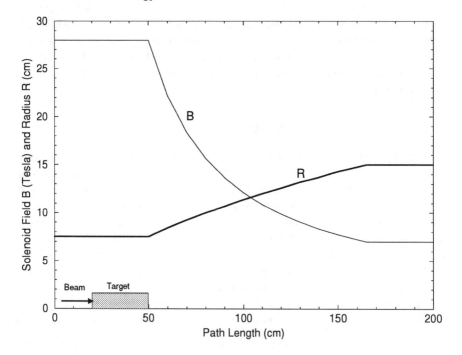

FIGURE 1. Capture solenoid field and inner radius as a function of distance.

The basic collection scheme, as outlined by Palmer et al. [3], is illustrated in Fig. 1 and forms the starting point for the simulations described in this paper. A very high-field hybrid solenoid extends the length of a target upon which a proton beam impinges. Based on near term technology, a field of 28 T and a 7.5 cm radius appears to be achievable for this purpose. This target solenoid collects pions with a large momentum spread and with large angles and guides them downstream into a long 7 T solenoid channel (15 cm radius) where they decay to muons.

The high-field solenoid aperture is chosen to have a large transverse phase space acceptance adequate for a transverse momentum

$$p_\perp^{max} = qBa/2 = 0.314 \ GeV/c \tag{1}$$

where B is the magnetic field, q the particle charge, and a the solenoid radius. The normalized acceptance of this solenoid for pions is

$$A_n = ap_\perp^{max}/ m_\pi c = qBa^2/2m_\pi c. \tag{2}$$

For the given parameters this acceptance is 0.17 m·rad, which is much larger than the intrinsic pion beam emittance at the target, $r_p p_\perp^{max}/m_\pi c = 0.02$ m·rad for a proton beam radius r_p. Hence there is no reason to further reduce proton beam size—which may thus be set by considerations of yield and target heating rather than pion emittance.

The target region is followed immediately by a 115 cm long matching section which reduces the field to 7 T via a $B_0/(1+\alpha z)$ dependence. In this region the pipe radius increases from 7.5 to 15 cm which corresponds to the radius of the lower field solenoid serving as pion decay region. This keeps the product Ba^2 constant and the acceptance unchanged. The parameter $\alpha = (qB_0/2p_\pi)(d\beta_f/dz)$ is chosen such that for a characteristic pion momentum the rate of change of the beta focusing function $(\beta_f = 2p_\pi/qB)$ with distance is less than 0.5, which might still be considered an adiabatic change of the field. In the present case $p_\pi = 0.8$ GeV/c and $\alpha = 2.62$ m^{-1} are chosen.

Below, following a brief description of computational procedures, results are presented on energy deposition in the targets and surrounding solenoid and on pion and muon production from various targets. Next, dependence of yield on the parameters specifying the decay channel solenoids is examined. This includes introduction of a *curved* solenoid to separate particle species by charge downstream. Concluding remarks are in the final section.

TARGET REGION STUDIES

For the collection geometry described in the Introduction, target composition, length and radius are varied and pion yield is studied using particle production and transport simulation codes. The 8 GeV proton beam is assumed to have an emittance of $\epsilon_N^{rms} = 3.8 \times 10^{-5}$ m·rad consistent with a value expected from a high-intensity proton source. The focusing function at the target is conservatively chosen as $\beta^* = 4$ meters to result in a relatively wide beam with $\sigma(x) = \sigma(y) = 0.4$ cm and $\sigma(x') = \sigma(y') = 1$ mrad.

Two computer codes are used in this study. The code MARS [4], developed over many years at IHEP and Fermilab for particle–matter interaction simulations, is used for simulating particle production and transport in thick targets within the solenoid field. MARS is also used to study energy deposition in the target and surrounding solenoid. Calculated pion, kaon, and proton spectra at the target exit for a representative case are shown in Fig. 2. MARS describes all the physics processes, so particle decay, interaction, and transport down the solenoid channel can be simulated within MARS as well. It is found preferable in this case to write a special, fast code for tracking particles after the target, using as input a particle file generated by MARS at the end of the target.

This code keeps track of vectorial positions and momenta of each particle as it traverses the beamline as well as time elapsed since the arrival of the incident proton at the target. In addition the code performs $\pi/K \to \mu$ decay Monte Carlo selection and full kinematics. Muons are progressively downweighted by their decay probability as they traverse the channel. Pions and muons leaving the beampipe are considered lost. In principle there is a small (but presumably negligible) fraction which may scatter back out of the wall or—in the case of a pion—produce a secondary pion which may rejoin the beam. Debuncher cavities for reducing pion/muon momentum spread downstream of the target are at present not included in the simulations but are planned to be added at a later date.

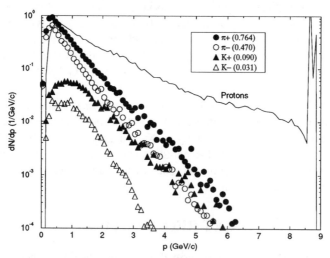

FIGURE 2. Pion, kaon and proton spectra for 8 GeV protons incident on a copper target (1.5 λ length, 1 cm radius). Total meson yields are shown in parentheses.

A large variety of particles is produced by the 8 GeV protons and subsequently by secondary and higher generation particles. For 8 GeV p–p interactions the average charged particle multiplicity is about three [5] with a modest increase expected for p–nucleus collisions. Excluding the incident protons this leaves an average of somewhat in excess of one charged particle produced per interaction—mostly as pions. Of all produced pions one expects roughly one third to be π^0 which decay quickly into γs leading to electromagnetic cascades in the target. For heavier targets the shorter radiation length permits considerable growth of these cascades leading to many low energy electrons and photons. Among the outgoing particles there will also be some nucleons and nuclear fragments which are dislodged from the target nuclei. All these processes are represented in the MARS code along with elastic and quasi elastic scattering of incident and produced particles.

Simulation of π/μ transport in constant solenoidal fields is readily performed using exact helical trajectories. In the matching region, where the field is more complicated, the simulation proceeds by taking small steps (0.1–0.5 cm) and sampling the field along the trajectory. The declining field in the matching region means that according to the $\nabla \cdot \boldsymbol{B} = 0$ condition the field has a radial component: $B_r \simeq -\frac{1}{2}r\partial B_z/\partial z$. For the above z-dependence $\partial B_r/\partial z = -\frac{1}{2}r\partial^2 B_z/\partial z^2 \neq 0$ and it follows [6] from the $\nabla \times \boldsymbol{B} = 0$ condition that B_z must depend on r. This requires that an extra term be present in B_z which—in turn—requires (via $\nabla \cdot \boldsymbol{B} = 0$) an extra term in B_r, etc. For the present simulations the iteration is pursued up to quadratic correction terms:

$$B_z = \frac{B_0}{1+\alpha z}\left[1 - \frac{1}{2}\left(\frac{\alpha r}{1+\alpha z}\right)^2\right]$$

$$B_r = \frac{B_0\alpha r}{2(1+\alpha z)^2}\left[1 - \frac{3}{4}\left(\frac{\alpha r}{1+\alpha z}\right)^2\right]. \tag{3}$$

It should be remarked that the analysis simplifies considerably if B_z is made to decline *linearly* with distance in the matching region: $B_z = B_0(1-az)$. Then $B_r = \frac{1}{2}raB_0$ independent of z and both $\nabla \cdot \boldsymbol{B} = 0$ and $\nabla \times \boldsymbol{B} = 0$ are satisfied. Results of simulations performed with a linear field do not differ significantly from those obtained with the $(1+\alpha z)^{-1}$–dependence.

Target Optimization

A crude target optimization with respect to yield starts by 'tagging' those pions (and kaons) which result in an acceptable muon deep in the decay channel for the case of a copper target followed by the 'standard' geometry as described in the Introduction (see Fig. 1). In excess of 90% of all accepted muons are thus shown to be the progeny of pions in the momentum range 0.2–2.5 GeV/c. The π^+/π^- ratio is about 1.6 at this proton energy. Then for a series of MARS runs, *pion* yield in the above momentum range is determined for various target parameters—without simulation of the collection channel. In addition to contributing little to the muon yield, pions with momenta less than 0.2 GeV/c have velocities below 0.82c and thus will quickly drop far behind the main pulse of faster particles.

Fig. 3(a–c) show momentum versus time scatter plots of pions, kaons and muons for a proton beam with $\sigma_t = 3$ nsec incident on a 22.5 cm copper target. In Fig. 3a the π/K distributions are shown immediately after the target and in Figs. 3b and 3c the π/K and μ distributions are shown 25 meters downstream of the end of the target. In all plots $t = 0$ refers to the center of the proton bunch at the target entrance. Materials investigated as target candidates are carbon, aluminum, copper, tungsten, and iridium. This set spans the Periodic Table and ranges in density from 1.8 to 22.4 g/cm^3. It is found that the optimal target radius needed to maximize the pion yield is about 2.5 times the rms beam size for all target materials and lengths. This corresponds to a 1 cm radius target for the beam used in this study. Almost all studies reported here are carried out with this target radius.

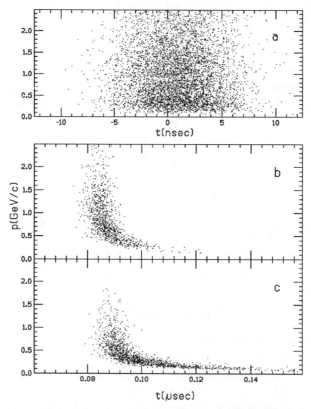

FIGURE 3. (a) π and K momentum *vs* time distribution immediately after the target for 8 GeV proton beam with $\sigma_t = 3$ nsec. (b) π and K distributions 25 meters downstream of target. (c) μ distribution 25 meters downstream of target.

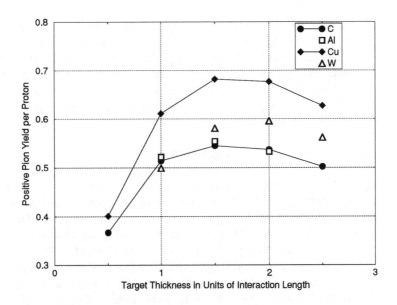

FIGURE 4. π^+ yield per 8 GeV proton for 1 cm radius targets of various materials *vs* target length for π momenta of $0.2 \leq p \leq 2.5$ GeV/c.

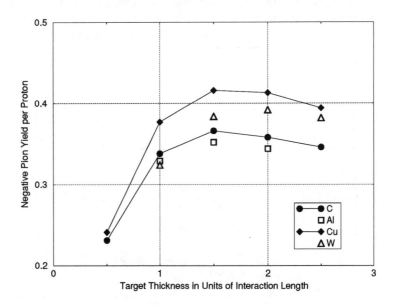

FIGURE 5. π^- yield per 8 GeV proton for 1 cm radius targets of various materials *vs* target length for π momenta of $0.2 \leq p \leq 2.5$ GeV/c.

Target length is varied from 0.5 to 2.5 nuclear interaction lengths (λ_I). Figs. 4 and 5 show positive and negative pion yields at the target exit as a function of target length for four materials (W and Ir are nearly identical). Optimal length is about 1.5 λ_I but yields vary by no more than 10% over a range of 1 to 2.5 λ_I for any target material. Yields are also rather insensitive with respect to target composition. The maximum yield is observed for a (1.5 $\lambda_I = 22.5$ cm) copper target, but the low–Z materials produce only about 20% less.

Target Heating

Beam power deposited in the target varies greatly with composition due mainly to increased electromagnetic shower development in high-Z materials. With 5×10^{13} protons on a 1.5 λ_I, 1 cm radius target, average power dissipation at 30 Hz ranges from 0.39 kW/cm^3 in carbon to 7.6 kW/cm^3 in iridium (Fig. 6). Peak energy deposition (on axis) in the target ranges from 20 J/g (C) to 35 J/g (Ir). This is at least a factor of ten below the shock damage limit. For forced water cooling of solid targets, the maximum surface heat flux (ϕ_{max}) that can be practically removed is about 200 W/cm^3. This implies a maximum target radius r=$2\phi_{max}/P$ where P is the average power density in W/cm^3. Hence—at 30 Hz—a (1 cm radius) carbon or aluminum target appears a viable candidate with proper cooling. Heavier targets probably need to have a larger radius at this beam intensity to lower the power density. Alternatively, at high power densities one may resort to 'microchannel' cooling wherein target wires are interspersed with small diameter cooling channels.

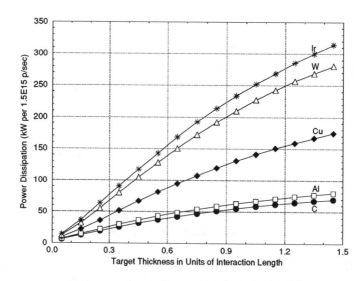

FIGURE 6. Average power dissipation in different 1 cm radius targets due to 8 GeV incident beam of 5×10^{13} protons at 30 Hz.

Another option to alleviate target heating problems is to use elliptically shaped beams on matching targets. This has two advantages: the overall surface area accessible to cooling is increased while the distance from the point of maximum energy density to a cooled surface is decreased. There are some disadvantages connected with preparation of the elliptic beam and with target alignment—particularly for large aspect ratios. Details of this are not further pursued here beyond a brief investigation which compares a round copper target of 1.5 λ_I and 1 cm radius with an elliptical one of equal length measuring 0.25 and 4 cm along the axes. The beam is likewise deformed in the same aspect ratio but retains the same emittance as for the 'round' case. As is confirmed by the simulations this has little effect on either maximum or total energy deposition or on yield. In this example total surface area increases by a factor of about 2.5 while distance between maximum energy deposition and cooling surface decreases by a factor of four.

Fig. 7 shows maximum temperature rise $\Delta T=T-T_0$ relative to room temperature $T_0=27°C$ reached in copper and carbon targets 1.5 λ_I long and 1 cm in radius when irradiated with 8 GeV protons. Results are obtained with the ANSYS code [7] starting with energy deposition distributions generated by MARS. Ideal cooling with $\Delta T=0$ at r=1 cm is assumed. Equilibrium is approached in less than two seconds with a steady-state temperature at the maximum of 347°C in copper and 186°C in carbon. Equilibrium temperature rise versus depth z in the target for a set of $\Delta r=0.2$ cm radial increments is shown in Fig. 8. Below, a 1.5 λ_I, 1 cm radius copper target is used in many instances. In view of the weak dependence of yield on target-Z almost all results pertaining to particle yield—*not* energy deposition—obtained with this target are expected to apply to carbon or aluminum with minor adjustments.

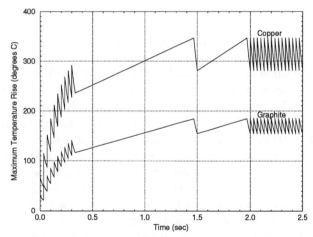

FIGURE 7. Maximum temperature rise ΔT relative to room temperature $T_0=27°C$ in 1 cm radius copper and graphite targets when irradiated by 8 GeV beam of 5×10^{13} protons at 30 Hz. ΔT for times between 0.3 and 2 sec is not shown in detail.

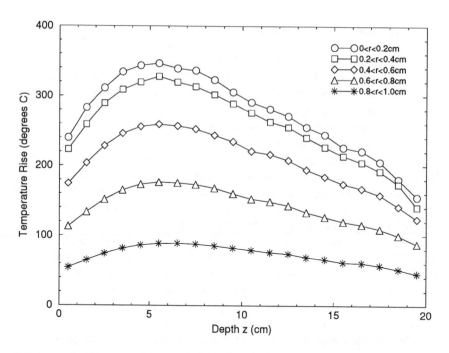

FIGURE 8. Equilibrium temperature rise ΔT relative to room temperature $T_0 = 27°C$ in 1 cm radius copper target after 60th pulse (t=2 sec). Target is irradiated by 8 GeV beam of 5×10^{13} protons at 30 Hz.

Solenoid Heating

Energy deposition in the primary 28 Tesla solenoid resulting from the intense radiation environment around the target might cause quenching. Based on hybrid designs reported in the literature, the solenoid is nominally assumed to consist of a normal-conducting 'insert', starting at 7.5 cm radius, and a superconducting 'outsert' starting at 30 cm. The latter will quench if the heat load becomes excessive. Fig. 9 shows average power density as a function of radius for 5×10^{13} protons at 30 Hz on a 1.5 λ_I Cu target. The end of the target coincides with the end of the primary solenoid. As expected, power density is highest at the downstream end of the solenoid (z=22.5 cm in this case). It is lower everywhere for lower Z targets. At 30 cm radius—where the superconducting solenoid starts—the power density is 4.5 mW/g which is below the experimentally determined quench limit of about 8 mW/g for Tevatron dipoles.

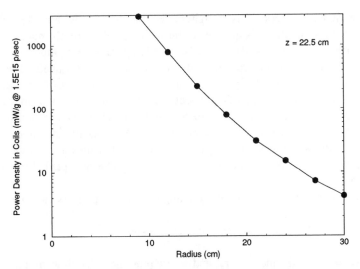

FIGURE 9. Peak power density in solenoid superconducting coils as a function of radius due to particle debris from 1.5 λ_I copper target irradiated by 8 GeV beam of 5×10^{13} protons at 30 Hz.

PION AND MUON COLLECTION

Particles produced in the target are transported along a beamline which forms the first stage of muon collection and acceleration en route to the collider. Attention must also be paid at this point to disposal of the other non-μ-producing particles, mostly nucleons, $e^{\pm}$, and γs. At a minimum such a beamline involves some focusing of the produced pions and their muon progeny just to keep them from being lost on the walls. Then, when an optimum population is reached, the muons are cooled and accelerated. More ambitious (and probably more realistic) schemes may begin cooling and/or acceleration earlier. But in this early phase of the study it is perhaps best to leave these more ambitious schemes for future consideration. This section thus concentrates only on the muon collection aspects of the post-target beamline. Even in this limited domain there appears a rich variety of strategies of which only a few are examined here.

Particle Decay

Pions and kaons immediately begin decaying into muons downstream of the target ($\lambda_\pi = 56\,p$, $\lambda_K = 7.5\,p$, and $\lambda_\mu = 6233\,p$ where λ is in meters and p in GeV/c). Particles that do not intercept the walls in their first Larmor gyration typically are transported down the entire 7 T channel. The vast majority of lost particles are wiped out in the first 15 meters. This straight collection channel (without RF debuncher cavities in the simulation) is quite efficient with only 40% of all muon-producing particles lost on the walls and close to 60% yielding transported muons.

While decay is fully incorporated into the simulations a few qualitative remarks may help interpret results. Only $\pi \to \mu\nu$ and $\mu \to e\nu\bar{\nu}$ decays are of real importance to this problem. Kaons are practically negligible as a source of muons in the present context: (1) their total yield is only about a tenth that of pions, (2) their branching ratios to muons are somewhat less favorable and (3) the decay kinematics produces muons typically with much larger $p_\perp$ than do pions. When they are included in a full simulation it is seen that only about 1% of all muons in the accepted phase space are due to kaons.

As a function of distance traversed along the pipe, z, pions decay to muons at a rate

$$dN_\pi/dz = \frac{1}{\lambda_\pi}e^{-z/\lambda_\pi} \tag{4}$$

where $\lambda_\pi = p_z^\pi \tau_\pi/m_\pi$ and m_π, τ_π, and p_z^π are pion mass, lifetime, and momentum along the pipe axis. There is a similar equation for muons. From the decay laws of radioactive chains, the fraction of muons at z is given by

$$N_\mu/N_\pi = \frac{\lambda_\mu}{\lambda_\mu - \lambda_\pi}\left(e^{-z/\lambda_\mu} - e^{-z/\lambda_\pi}\right). \tag{5}$$

From Eqn. (5) the maximum muon yield is realized at

$$z_{opt} = \frac{1}{\lambda_\pi - \lambda_\mu} \ln \frac{\lambda_\pi}{\lambda_\mu}. \tag{6}$$

To arrive at a more concrete (but approximate) estimate of z_{opt}, p_z^μ is replaced by its average value

$$\overline{p_z^\mu} = \frac{m_\pi^2 + m_\mu^2}{2m_\pi^2}p_z^\pi \simeq 0.785 p_z^\pi. \tag{7}$$

When inserted into Eqn. (6) this results in

$$z_{opt} \simeq 251 p_z^\pi \tag{8}$$

in meters with p_z^π in GeV/c. Eqn. (5) then indicates that at z_{opt} the number of muons per pion produced at the target is about 0.95.

For a *spectrum* of pions, optimization of z requires folding Eqn. (7) with the p_z^π of the spectrum. But even without such a folding, a rough knowledge of the spectrum establishes a distance scale for the decay channel. It also follows that for a spectrum 0.95 μ/π must be regarded as an upper limit. Since at distances of order z_{opt} and beyond most pions have decayed, the muon yield is governed by the muon decay length and one expects a broad maximum (at a z_{opt} corresponding roughly to the peak p_z^π of the spectrum produced at the target) where the theoretical maximum of 0.95 μ/π should be close to being realized. Taking 0.5 GeV/c as a characteristic pion momentum for the distribution, one expects the muon population to reach a maximum about 125 meters downstream of the target and fall off slowly after that. For a uniform distribution in the range 0.25–0.75 GeV/c a maximum 0.94 μ/π is attained at 130 m. Fig. 10 shows muon yield per proton versus distance from a simulation with a 22.5 cm long copper target. In this case the maximum yield is 0.52 μ^+ and 0.34 μ^- per 8 GeV proton.

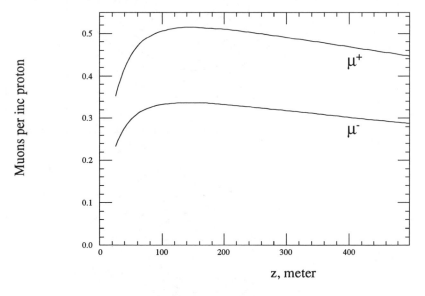

FIGURE 10. Muon yield vs distance from target for 22.5 cm copper target and standard straight decay solenoid.

Beam Collection in a Straight Solenoid

The transverse emittance of the nascent muon beam is of practical interest in the design of the downstream muon cooling channel and its required acceptance. Fig. 11 shows muon beam fractional contours in x or y transverse phase space 150 meters downstream of the target. The plot shows the fraction (in steps of 0.1) of the beam within the indicated x and $x' = p_x/p$ limits independent of y and y'. Plots for positive and negative muons are nearly identical. Note that the muon beam is well localized transversely in a channel with emittance $\epsilon_x(90\%) = 4.5 \times 10^{-2}$ meter, $|x| \le 10$ cm and $|x'| \le 0.45$. In the absence of x-y correlations, 81% of the muon beam is contained within both the $\epsilon_x(90\%)$ and $\epsilon_y(90\%)$ contours. For reference note that $\epsilon_x(30\%) = 5.4 \times 10^{-3}$ meter and $\epsilon_x(60\%) = 1.6 \times 10^{-2}$ meter.

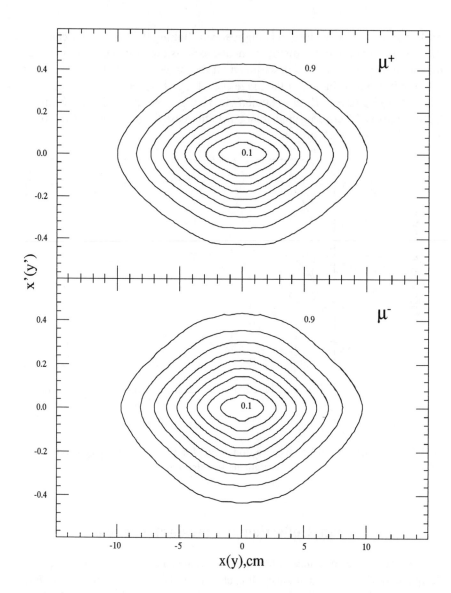

FIGURE 11. Contours of relative fractional transverse emittance for μ^+ (top) μ^- (bottom) in steps of 0.1, e. g., innermost contour contains 10 % of all muons.

Muon yield as a function of momentum is of particular interest in designing an RF system to reduce the momentum spread. Figs. 12(a–c) display momentum spectra of μ^+ from a copper target within the 0.3, 0.6, and 0.9 x and y emittance contours (cfr. Fig. 11). Figs. 13(a–c) present the corresponding μ^- spectra. Total number of muons per proton within the indicated contours are shown in parentheses. As expected, muons with small transverse emittance tend to have somewhat higher momenta. These plots quantify the trade-off in yield between momentum spread and transverse emittance. For example, the spectra of Figs. 12c or 13c each contain 81% of all μ^+ and μ^-, respectively. About 60% of this beam is contained in the momentum range 0.22 to 0.72 GeV/c corresponding to $\Delta p/p = \pm 0.53$ and $|\Delta v/c| = 0.075$. This constitues a relatively high density muon beam which contains a total of 0.26 μ^+/p and 0.17 μ^-/p. These plots characterize the raw muon beam and help provide insight for the design of a cavity system to debunch the parent pion beam. As stated earlier, this is not as yet included in the simulation.

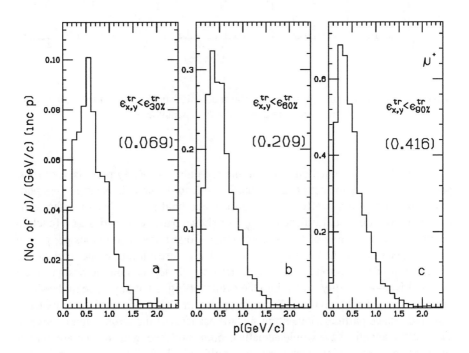

FIGURE 12. Momentum spectra of positive muons 150 m downstream within indicated transverse emittance. Totals in parentheses.

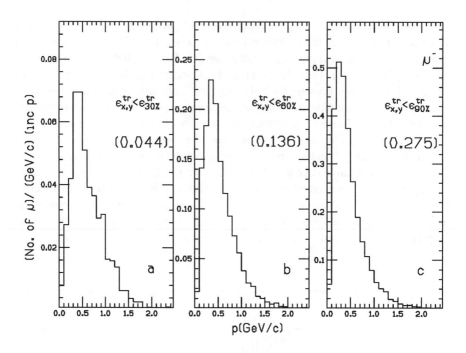

FIGURE 13. Momentum spectra of negative muons 150 m downstream within indicated transverse emittance. Totals in parentheses.

For each proton bunch on target an intense pulse of mostly protons, electrons, pions, kaons and muons starts down the 7 T solenoid channel. Neutrals like photons and neutrons are unaffected by the magnetic field and are lost onto the walls according to their initial trajectory. Fig. 14 shows particle densities as a function of time at the beginning of the decay channel. Total number of muons per proton of each species are indicated in parentheses. These distributions do not include the time spread of the proton beam. The latter—which depends on the design of the proton driver—is readily folded into the results of Fig. 14 at any stage in the simulation (prior to the RF cavities). When this pulse arrives at a debuncher cavity (proposed to reduce particle momentum spread) particles of the wrong sign relative to the electromagnetic wave would actually become bunched, with an increase in their momentum spread. Most such particles would quickly be lost downstream in any magnetic bend.

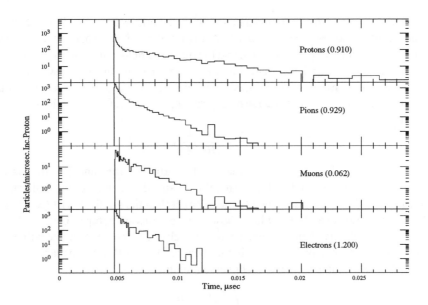

FIGURE 14. Particle densities as a function of time at end of matching region (1.15 m after end of target).

Curved Solenoid for Beam Separation

The proposed straight-solenoid plan uses two separate proton bunches to create separate positive and negative muon bunches accepting the loss of half the muons. In addition the debuncher cavities have to contend with a large population of protons, electrons and positrons that will tend to mask the desired π/μ bunches. Great advantage may be gained if the pions can be charge-separated as well as isolated from the bulk of protons and other charged debris before reaching the cavities. For the same number of proton bunches on target, a scheme which permits charge separation would produce a gain of a factor of two in luminosity. By coalescing the two proton bunches, this becomes a factor of four with little effect on target heating or integrity.

The solenoid causes all charged particles to execute Larmor gyrations as they travel down the decay line. As is well known from plasma physics, a gradient in the magnetic field or a curvature in the field produces drifts of the particle guide centers. Drift directions for this case are opposite for oppositely charged particles. Drift velocities depend quadratically on particle velocity components. This is exploited here by introducing a gentle curvature to the 7 T decay solenoid.

In the decay line, most particles moving in the curved solenoid field have a large velocity parallel to the magnetic field (v_s of order c) and a smaller perpendicular velocity ($v_\perp \simeq 0.3\,\mathrm{c}$ or less) associated with their Larmor gyration. In the curved solenoid the v_s motion gives rise to a centrifugal force and an associated 'curvature drift' perpendicular to both this force and the magnetic field. The field in the curved solenoid also has a gradient (field lines are closer near the inner radius than near the outer radius) resulting in an added 'gradient drift' in the *same* direction as the curvature drift. Averaged over a Larmor gyration, the combined drift velocities can be written as [8]

$$\vec{v}_R + \vec{v}_{\nabla B} = \frac{m\gamma}{q} \frac{\mathbf{R} \times \mathbf{B}}{R^2 B^2} (v_s^2 + \frac{1}{2}v_\perp^2),$$ (9)

where $m\gamma$ is the relativistic particle mass, q the particle charge, and R is the radius of curvature of the solenoid with central field B. Note that in the present application the curvature drift ($\propto v_s^2$) is typically much larger than the gradient drift ($\propto v_\perp^2/2$). This is in contrast to a plasma where these contributions are comparable.

The drift velocity changes sign according to charge so positive and negative pions become transversely separated. For unit charge and for $\mathbf{R} \perp \mathbf{B}$ the magnitude of the drift velocity can be written in convenient units as

$$\beta_d = \frac{E\left(\beta_s^2 + \frac{1}{2}\beta_\perp^2\right)}{0.3RB},$$ (10)

where E is particle energy in GeV, R is in meters and B in Tesla. The total drift displacement, D, experienced by a particle moving for a distance, s, along the field follows immediately from Eqn. (10)

$$D = \frac{1}{0.3B} \frac{s}{R} \frac{p_s^2 + \frac{1}{2}p_\perp^2}{p_s}$$ (11)

with D in meters, B in Tesla and momenta in GeV/c. Note that only the *ratio* s/R appears in Eqn. (11) which corresponds to the angle traversed along the curved solenoid. A typical 0.5 GeV pion ($p_\perp \ll p_s$) in a 7 T solenoid with R = 25 m has a drift velocity of about $10^{-2}c$. After moving 20 meters downstream in the solenoid, a 0.5 GeV positive and negative pion should be separated by about 35 cm.

The present study considers only *circularly* curved solenoids. Here the curvature and the $\nabla \times \boldsymbol{B} = 0$ condition requires the field, which is nonzero only along ϕ (i.e. along the axis of the curved beampipe) to have a $\frac{1}{R}$-dependence. This is readily incorporated in the detailed step-by-step simulations. Fig. 15a shows the pion distributions 20 m downstream of the target (which is in a 28 T field) calculated for a curved solenoid geometry. The centroid separation agrees well with what is expected from the drift formula. Also as expected, higher energy pions are shifted farther and low energy pions less. Decay muons created up to this point are separated by a comparable margin (Fig. 15b). At this point one could place a septum in the solenoid channel and send the two beams down separate lines to their own debunching cavities.

78

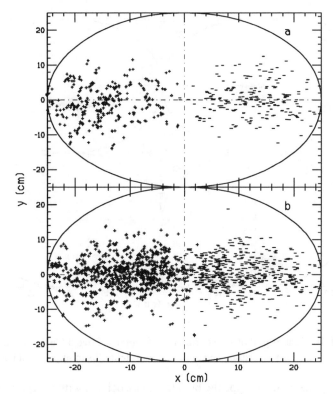

FIGURE 15. Position of (a) π^+ and π^-, and (b) μ^+ and μ^- 20 m downstream along curved solenoid (inner radius a=25 cm, R=25 m, B=7 T).

The curved solenoid also serves well to rid the beamline of neutral particles and most of the remnant protons after the target. Neutrals like photons and neutrons intercept the curved solenoid with their straight trajectories and deposit their energy over a large area. Beam protons which underwent little or no interaction in the target have such large forward momenta that they are unable to complete one full gyration before intercepting the curved wall downstream. Hence Larmor-averaged drift formulae cannot be applied. Fig. 16 provides some snapshots of this tight proton bunch moving away from the lower energy protons at successive downstream locations. At four meters downstream all beam-like protons have intercepted the wall. Protons remaining in the pipe for long distances have momenta similar to the positive pions and thus will accompany them downstream. Roughly 0.7 protons per positive pion/muon are still in the pipe at 10 meters which should not overburden the debuncher cavities with extraneous beam. Electrons and positrons have typically much lower momenta than pions and muons. Simulations indicate that they do not drift far from the curved solenoid axis and most would be lost at the septum.

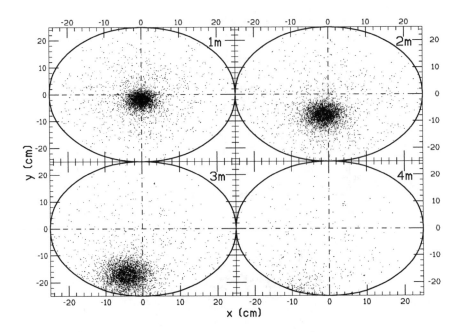

FIGURE 16. Scatter plot of x,y coordinates of protons above 5 GeV while traversing curved solenoid. Distance along center of curved solenoid is in upper right hand corner.

For the same diameter pipe the broadening associated with charge separation in the curved solenoid leads to increased particle losses on the walls compared with the straight case. An increase in pipe diameter is thus very desirable. To optimize the geometry with respect to yield would require many simulation runs. Since the curved regions may extend over long distances, this leads to much longer computation times for detailed step-by-step simulations to the point of becoming prohibitive when exploring a large parameter space. For survey type calculations a simplified procedure was therefore adopted.

The procedure adopted is then that for each pion encountered while reading a MARS file: (1) the position vector of the Larmor guide center is determined, to which (2) the drift displacement vector **D** is added, with appropriate sign, along the direction perpendicular to **B** and **R**. It is then determined whether (3) the entire Larmor circle fits inside the *half*-aperture appropriate to its charge, i.e., the side in the direction of the drift. More precisely, this last condition is $r_g < a - r_L$ and $\pm x_g > r_L$ where r_L is the Larmor gyration radius, a is the pipe radius and subscript g refers to the final guide center position with the sign of x_g dependent on particle charge. If a pion meets these criteria it is assumed to contribute to the yield. Decay of pions or muons is omitted from consideration. This was justified above (cfr. Particle Decay).

The study of pion/muon yield in a circularly curved beampipe (starting immediately after the target) with *constant* central field can thus be reduced to a problem with just three parameters B, s/R, and a. Some sensible range of values can
thus be readily explored over a reasonably dense grid. An optimization based on
yields alone is perhaps somewhat unrealistic. A measure of how effectively one can
separate the two components into different beamlines is provided by computing the
centroid of each distribution as well as its *rms* radius. Computation is very fast and
readily repeated for different sets of parameters to perform a more complete optimization.

A more realistic scenario starts with a 28 T field surrounding the target followed
by a *curved matching region* which accomplishes simultaneously both transition to
lower field and charge separation. The changing field causes an adiabatic decline in
$p_\perp$ according to

$$p_{\perp f}^2 = p_{\perp i}^2 \frac{B_f}{B_i} \tag{12}$$

along with a correponding change in p_s so as to conserve total p. Subscript i refers
to initial and f to final values of B and $p_\perp$, i.e., those prevailing at S, the total distance along the central field line. Ignoring the other field components—due to the
declining field, cfr. Eqn. (3)—the total drift becomes

$$D = \int v_D dt = \int v_D \frac{ds}{v_s} = \int_0^S \frac{1}{300BR} \frac{p_s^2 + \frac{1}{2}p_\perp^2}{p_s} ds, \tag{13}$$

where now B, p_s and $p_\perp$ all depend on s. Assuming a *linear* decline of the central
field $B = B_i(1 - as)$, and the dependence of p_s and $p_\perp$ on s this entails, one obtains

$$D = \frac{S}{300(B_i - B_f)R} p_0 \left[\ln \frac{(p_0 - p_{sf})(p_0 + p_{si})}{(p_0 + p_{sf})(p_0 - p_{si})} + p_{sf} - p_{si} \right], \tag{14}$$

where p_0 is the total momentum of the pion.

Thus for fixed B_i the problem remains confined to three parameters: B_f, S/R,
and a. Note also that the Larmor radius changes with s here. Fig. 17a shows $\pi^+ \mu^+$
yield in a curved solenoid with a constant 50 cm pipe radius for different values of
the final magnetic field at the end of the matching region for a 22.5 cm long copper target. Eqn. (13) can also be applied to a field having the $B_0/(1 + \alpha s)$ dependence. Again an expression for D, though somewhat lengthier than Eqn. (14), is
readily obtained and again the problem remains one of the same three parameters.
For comparison, Fig. 17b shows $\pi^+ \mu^+$ yield for the $1/(1 + \alpha s)$ field dependence
with everything else as in Fig. 17a. Note that the yields peak at somewhat smaller
s/R.

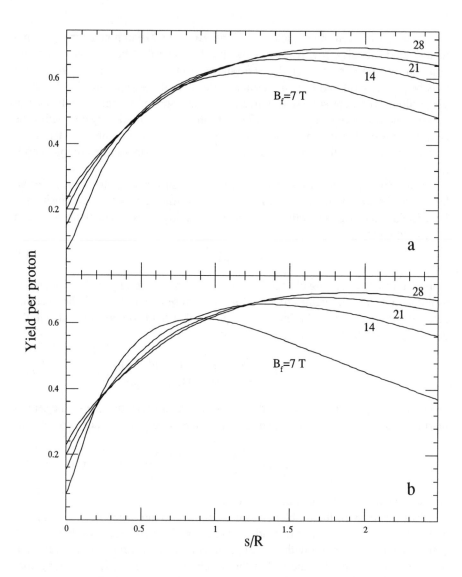

FIGURE 17. (a) Yield of positive pions and muons vs s/R for 22.5 cm copper target in straight solenoid with $B_0 = 28$ T followed by curved solenoid with $B = B_0(1 - \alpha s)$. Labels indicate final B reached at s/R. (b) Same for $B = B_0/(1 + \alpha s)$ in curved solenoid.

82

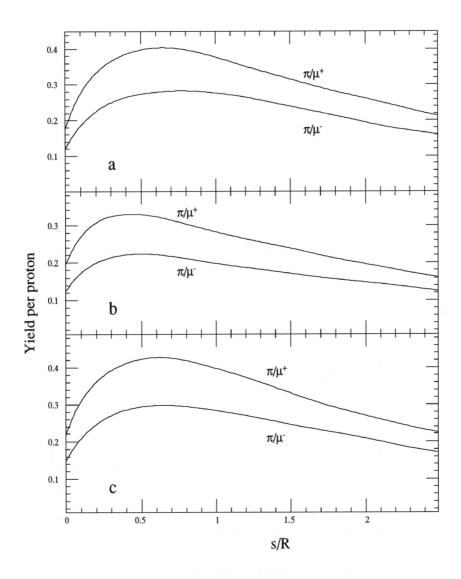

FIGURE 18. (a) Yield vs s/R for 22.5 cm copper target in solenoid of 7 T throughout. (b) Same for 57 cm carbon target. (c) Same for 30 cm copper target.

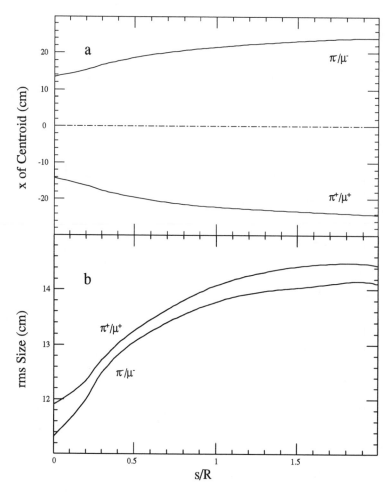

FIGURE 19. (a) Distance of centroid to magnet center vs s/R for 22.5 cm copper target, 7 T, r=50 cm solenoid throughout. Note beampipe extends beyond graph. (b) rms size of each beam vs s/R.

Because of the advanced magnet technology required for very high field (> 15T) solenoids, it is of interest to investigate yields obtained when lower magnetic fields prevail throughout the entire geometry. To keep matters simple a constant 7 T field and 50 cm solenoid radius is assumed—which might be considered state-of-the-art [9]. The solenoid is straight for the target portion, then curves to affect the desired charge separation, then straightens again to form the decay channel. In this last portion separation of plus and minus beams at a septum is to take place but details of this implementation are not considered here. Such a solenoid has a transverse momentum acceptance of 0.52 GeV/c and a normalized phase space acceptance of 1.87 m·rad. Fig. 18a,b show $\pi\mu$ yields for this type solenoid and for $1.5\lambda_I$ copper and carbon targets, respectively. Fig. 18c presents the yield curves for the copper target when the length is increased to $2\lambda_I$, suggesting longer targets to be better for this geometry. Yields are presented as a function of s/R and it is thus advantageous to begin the straight (decay) portion of the pipe in the region near the maxima. Fig. 19a,b show respectively the centroid position of the plus and minus beams within the beampipe and their *rms* size for the standard copper target. The latter refers only to the distribution of the guide centers and excludes the spread due to the Larmor motion. Recall that both centroid and *rms* size refer only to those particles for which the entire Larmor circle fits inside the proper half-aperture. This accounts for the non-zero centroid positions at z=0.

CONCLUSIONS AND OUTLOOK

The choice of target shows that while a 1–2 λ_I of copper is optimum for yield, lower-Z targets are not much worse—about 20% depending on the collection geometry. Hence lower-Z targets, because of the lower energy deposition associated with them, may still be the targets of choice. Microchannel cooling and/or elliptical beams and targets may be used to deal with target heating problems. Quenching due to energy deposition in superconducting solenoids is a problem only for high-field/small-diameter magnets. Lower field solenoids with larger diameter are much less likely to quench and also pose less technological difficulties. While the yields associated with them are somewhat lower there may exist some reasonable trade-off.

The simulations confirm the superiority of muon collection with the solenoid scheme in this energy regime. Total yields of about 0.5 muons per proton of either charge appear to be obtainable. Considerations of π/μ decay indicate a collection limit of about 0.95 muons per pion. Kaons appear to contribute far less than their numbers to the usable muon flux and are practically negligible in this application. The pion momentum spectrum generated by 8 GeV protons peaks around 0.35 GeV/c. The collection system tends favor the lower energies and most muons are expected to be in the 0.2 to 0.5 GeV/c range. Charge separation by *curved* solenoids practically doubles the number of muons collected and appears to be beneficial in disposing of the host of unwanted particles generated in the target along with the through-going beam protons. Yields are sensitive to field strengths and solenoid diameter as well as to the s/R parameter—which indicates where to end the curvature and send the beams their separate ways.

As is evident from Eqns. (11) and (14), there exists a strong correlation between drift distance and momentum with the larger momenta experiencing the larger drift. This could be exploited for cooling purposes, e.g., after each beam is traveling in its own channel and most pions have decayed, a (cylindrically symmetric) wedge would clearly introduce some longitudinal cooling. Even some 'pion cooling' might be beneficial during the separation stage. This would be accomplished by intermittently placing collars near the perimeter. Such a collar would only affect high momentum (large drift velocity), high $p_\perp$ (large Larmor radius) pions. It is clear that—unless a nuclear interaction occurs—some cooling results in both transverse and longitudinal space. Moreover the reduction in $p_\perp$ shrinks the pion's Larmor radius and tends to move its guide center in the general direction of the drift. The new Larmor circle is thus more likely to fit entirely into proper half-aperture of the beampipe with improved subsequent capture probability. The reduction in p_s reduces the drift velocity which might prevent some pions from being lost on the walls. Some of these benefits are offset by pions undergoing nuclear interaction or large angle scattering—although some salvagable pions might yet emerge from such events.

A definitive optimization of muon production and collection best awaits further studies of muon cooling and acceleration. Only when it is known—at least approximately—how to mesh these functions with the parts explored in this paper can one proceed reasonably efficiently towards this goal. But cooling and acceleration, in turn, depend strongly on the particulars of muon production and collection. This study thus provides a necessary step in the iteration which—it is hoped—eventually will lead to a realistic design of a muon collider.

References

[1] Palmer, R., Proc. 1st $\mu^+\mu^-$ Collider Workshop, Napa, CA (1992), in *Nucl. Inst. Meth.* **A350**, 24–56 (1994).

[2] Noble, R., Proc. 3rd Int. Workshop on Advanced Accelerator Concepts, Port Jefferson, NY, AIP Conf. Proc. **279**, p. 949 (1993).

[3] Palmer, R., et al., "Monte Carlo Simulation of Muon Production", BNL–61581 (1995).

[4] Mokhov, N. V., "The MARS Code System User's Guide, Version 13 (95)", Fermilab FN–628 (1995).

[5] See e.g., Montanet, L. et al., "Review of Particle Properties", *Phys. Rev.* **D50**, 1177 (1994).

[6] Jackson, J. D., *Classical Electrodynamics*, 2nd Ed., J. Wiley, New York, 1975.

[7] "ANSYS (rev. 5.1)", Swanson Analysis System, Inc., SASI/DN–P511:51, Houston (1994).

[8] Chen, F. F., *Introduction to Plasma Physics*, Plenum, New York, pp. 23–26 (1974).

[9] Green, M., "Pion Capture Magnet System", presented at 9th Adv. ICFA Beam Dynamics Workshop, Oct. 1995, Montauk, NY.

An Analysis of Pion Production Experiments

Hiroshi Takahashi
Xinyi Chen*

*Brookhaven National Laboratory
Upton, New York, 11973*

Abstract. To provide the design parameter of muon-muon collider, the production of positive pions in a proton-induced reaction at 14.6 GeV/c on Al and Cu, and the negative pions from the 6 inch thick target of Al and Cu measured by using the Alternating Gradient Synchrotron (AGS), were analyzed using the nuclear cascade model code Lahet.

INTRODUCTION

In recent years, there has been increasing interest in using high-intensity positive muon and negative muon collisions in muon-muon collider rings[1].
With the advent of high luminosity colliders, more stringent conditions are imposed on the intensity and emittance of the muon beam.

The muon beam's intensity and emittance are strongly related to the method of producing and collecting of secondary pions. Pions generated by injecting relativistic energy protons with high intensity into a target, such as solid tungsten, liquid mercury, or lead, have to be collected by a solenoid field. Due to the high-intensity proton beam, thermic processes and radiation damage in the target have to be considered, as is the case for an accelerator-driven reactor and spallation neutron source. In the latter case, we used the nuclear cascade code of LAHET[2] together with the MCNP code[3] which calculates the transport of neutrons less than 20 MeV energy, and also that of gamma-rays.

Positive and negative muons are produced from the decay of positive and negative pions, and the energy spectrum and angular distribution of the pions that are generated should be evaluated precisely so that these particle can be collected efficiently with a solenoid magnetic field. Currently, the pion productions has been studied theoretically by usng the Wang's model

*Permanent address: Department of Physics, Qinghua University, Beijing, People's Republic of China.

and ARC model.

In this paper, we analyze the production of positive pions in a proton-induced reaction at 14.6 GeV/c on Al and Cu, measured by using the E-802 spectrometer at Brookhaven National Laboratory's Alternate Gradient Synchrotron(AGS)[7]. The production of negative pions from the 6 inch thick target of Al and Cu[8] were analyzed using the Lahet code. The analyses indicated that the calculated yields are in reasonable agreement with the experimental findings for small angle production between 10 and 17 degrees azimuthal direction, but as the angle increases, the calculation gives a smaller value than the experimental one for thin target experiments. For the production of negative pions with low energy from thick targets, the theoretical and experimental values show reasonable agreement at small angles less than 20 degrees; however, the calculated production moving towards 30 degrees is much smaller than the experimental value. The reason for this difference might lie in the assumption used for the angular dependence of pion production which is adopted in the Lahet code. The Lahet code uses Sternheimer and Lindenbaum's isobaric model[7] for pion production, and a strongly anisotropic production of high energy pions is assumed in the single-pion production reaction.

LAHET CODE SYSTEM[2]

The Lahet code was developed by Los Alamos Scientific Laboratory (LASL) from the nucleon meson transport code (NMTC)[8] and its extension of the high energy transport code (HETC) code[9], which were originated at Oak Ridge National Laboratory (ORNL). It also incorporated the Isabel (Vegas) code [10] developed in Brookhaven National Laboratory (BNL). The Lahet code calculates an intra-nuclear cascade process that occurs in a high-energy nucleon-nucleus collision, and the inter-nuclear cascade of transport processes for nucleons, pions, muons, and high-energy photons in the assembly is expressed by general combinatorial geometry. This code system calculates heat generation, radiation damage, the production of charged particles, such as deuteron, He-3, tritium, and alpha particles, thus, it has been used extensively with Monte Carlo calculations for designing a spallation neutron source and accelerator-driven reactor, together with another Monte Carlo code, the MCNP code, which calculates the transport of low-energy neutrons of less than 20 MeV and also of gamma-rays.

As discussed, this code calculates pion production, and it also handles the decay of pions to muons, so it is a very useful tools for evaluating the design of the target for producing pions which can be captured by a solenoid magnetic field.

Meson production model

The Lahet code uses the Sternheimer-Lindenbaum's isobar model [SL,58,61] to assess the rpoduction of pions. Pionsare assumed to be produced through the decay of an isobar that is formed by a nucleon when it is excited in a collision. For example, $N + N \rightarrow N + N^*$,or $\rightarrow N^* + N^*$, with each $N^* \rightarrow \pi + N$. However, the final-state momentum distributions of the recoil and decay products are unspecified because neither the angular distributions of the decay pion nor the isobars themselves are specified.

Since this isobar model only accounts for single- and double-pion production in nucleon-nucleon collisions and single-pion production in pion-nucleon collisions, and because the practical thresholds for ternary-pion production by nucleons and double-pion production by pions are about 3.5 GeV and 2.5 GeV, respectively, the pion-production model used in these codes

limits the maximum nucleon and pion energies that may be treated using the intranuclear-cascade sub-program. To deal with the higher energy reactions, the extrapolation model of Garthier [11 Ga] was adopted.   We now consider a particle-nucleus collision by a particle (nucleon or charged pion) with energy E_0 and a collision with the same nucleus by the same type of particle but at some higher energy E_0', where E_0 and E_0' are kinetic energies in the laboratory system. The extrapolation model for relating the products from the " slow" collision at E_0 to the products from the " fast" collision at E_0' is based upon the following four assumptions:
(i) The total non-elastic cross section above E_0 is independent of the energy of the incident particle - i.e.: $\sigma(E_0') = \sigma(E_0)$.
(ii) The residual excitation energy after both the fast and slow collisions is the same.
(iii) The transverse momentum in the center of momentum (CM) system of each particle produced is assumed to be the same in the fast and slow collisions - i.e.

$$P_{ci}' \sin \theta_{ci} = P_{ci} \sin \theta_{ci} \qquad (1)$$

where c denotes CM quantities, i denotes the type of particle (neutron, proton, π^+, π^-, or π^0), P is the momentum, and θ is the polar angle with respect to the direction of the incident particle. To make this transformation unique, it is further assumed that the sign of $\cos\theta_{ci}$ is the same as the sign of $\cos\theta_{ci}$.
(iv) To relate the energies of the particles produced in the fast and slow collisions, the following scaling relation for kinetic energies is postulated :

$$\frac{E_{ci}'}{E_{c0}'} = \frac{E_{ci}}{E_{c0}} \qquad (2)$$

Using the above assumptions, we can determine the conservation of energy in the CM system for the fast and slow collisions and the momentum of the fast collision in the laboratory system; and the results of the intranuclear-cascade calculation at E_0, the energy, direction of each emitted particle and the excitation energy, the recoil energy, and the mass of the residual nucleus are determined for collisions at E_{c0}'.
In the Lahet code, E_{max} is fixed at 3.5 GeV for nucleon-nucleus collisions and at 2.5 GeV for pion - nucleus collisions since these are the maximum energies allowed by the intra-nuclear cascade routines.

ANALYSIS OF THE PION PRODUCTION EXPERIMENT USING THE LAHET CODE.

a. Analysis of the production of high-energy positive pions from thin targets of Al and Cu.

Particle production in proton-induced reactions at 14.6 GeV /c on Be, Al, Cu and Au targets was systematically measured using the E-802 spectrometer at the BNL's Alternating Gradient Synchrotron. Particles were measured in the angular range from 5 to 58 degrees. The calculation was encompassed a rather small sampling number of 10^4*10 batch protons.
Figure 1 compares the energy dependency of pion-production at angles at theta = 7.2 to 20.2 degrees for an aluminum target. At an angle of 7.2 the calculated production is little smaller that

the experimental value, although the calculated energy dependence is in good agreement. At a 10.6 degree angle, the calculation shows remarkably good agreement. As the angle increases to more than 16.5 degrees, the calculated production of 1 Mev energy pions agree well with experimental finding, but the values for assesed pions with energy above 1 MeV decrease very rapidly as pion energy increases.

Figure 2 compares the case of the Au target, where a similar trend is observed as in the case of aluminum target. At a 7.2 degree angle, the calculated value is slightly less than that in the experiment. At 10.6 degrees, the calculation shows good agreement with the experiment, but for the production of pions with 1 MeV energy, the code gives values that a factor two smaller than the experimental results. As the angle increases, the value for low-energy pion production are close, but the calculated production of high-energy pions becomes smaller than the experimental production as pion energy increases.

b. Analysis of the production of low-energy negative pion from thick targets of Al and Cu.

Another experiment with the AGS machine was carried out for using the thin and thick targets of Al, Cu and W to evaluate the production of negative pions for medical use.
The production cross section are measured for the negative pions with low momentum from 0.15 to 0.35 GeV/c momentum at four production angles from 0 to 30 degrees; the target materials were Al,and Cu with dimensions of 2"width and 2"height and thickness up to 6".
The protons were incident on the production target after passing through a Cerenkov counter, filled with Freon-12, which identified positive pions in the beam, and three beam counters with transverse dimensions smaller than the target, which therefore defined the target area through the trigger requirement. A momentum acceptance of proton $dp/p=12\%$ and negative secondary particles were identified by their time of flight for the 16 ft trajectory between the beam counter and the scintillation counter which was placed immediately behind each downstream hodoscope. The moments of the proton beam are 6 and 17 Gev/c, and its beam size was 1.25" wide and 1.5" high.
The total sampling number used in the Monte Carlo calculation is 10^4 x10 batches which is used in the same as the above case. The dependence of the yield of pions on target length can be investigated by analyzing these data.
Figure 3 compares the numbers of negative pions produced per unit incident proton in units of GeV/c momentum from 0 to 30 degrees at intervals of 10 degrees. At 0 degree, the calculated production of pion with up to 0.25 GeV/c momenta agrees excellently with the experimental values for both 6 GeV/c and 17 GeV/c incident momenta protons, but the agreement deteriorates for the production of pionswith higher momenta of 0.3 to 0.4 GeV/c. Similar trends are observed in pion production in both the 10 and 20 degree directions, although the agreements are not as good they are for 0 degree. For production in the 30 degree direction, the calculated values for 17 GeV proton incident momentum are much smaller than the experimental ones for high pion momentum of 0.25 to 0.4 MeV/c.
As shown in fig.3.d, pion production calculated for a proton incident momentum of 17 Gev/c is smaller than that for 6 GeV/c.
Similar trends are observed figure 4 for Copper target. The calculated production of high momentum pions is lower than the experiment.
Figure 5 shows negative pion production in the high momentum range up to 3 GeV/c for an

incident proton momentum of 6 GeV/c, and figure 6 shows the distribution of pion momentum up to 12 GeV/c for a 17 GeV/c incident proton momentum.

CONCLUSION

The analysis of the pion production experiments by the Lahet code system indicated that there is reasonable agreement between the calculations and the experimental data for small angular regions less than 20 degrees. The experiment shows the more isotropic angular distribution even for high momentum pions above 0.3 GeV/c. However, the calculated value for large angle production are far smaller than the experimental values. In the Lahet code, the angular distribution of single pion production is assumed to be composed of two components, of an isotropic part and a strongly anisotropic one, which is expressed as a delta function for forward and backward directions. The percentage (fan) of anisotropic components are shown in the table 1.

This strong anisotropic production for high-energy pions in the single pion production process might be the cause of the disagreement for large angle production. In the next study, we plan to correct this deficiency, and a revised code will be applied for designing the target assembly and the solenoid collecting coil for capturing pions and their decayed muons.

Table 1. The anisotropic factor for the angular distribution of single pion production.

proton energy range (GeV)	fan
0.-0.5	0.
0.5-1.0	0.25
1.0-1.3	0.5
1.3-2.5	0.75
2.5-3.5	1.0

ACKNOWLEGEMENT

The author would like to thank Drs. R.Palmer, J.Gallardo, R.Fernow, H.Kirk, Y.Torun, D.Kahana, T.Kycia, and F.Atchison for valuable discussions. And they also thank Dr. Woodhead for editing the manuscript.

References

[1] R. Palmer, "Muon collider review", in this proceeding.

[2] R. Prael and H. Lichtenstein, "User Guide to LCS: The LAHET Code System", LA-UR-89-3014, Los Alamos NAtional Laboratory, (Sep. 1989).

[3] MCNP, A General Monte Carlo Code for Neutron and Photon Tranasport (1986), Version 3A, J. F. Breismeister ed., LA-7396-MS, Rev. 2, Los Alamos National Laboratory, Los Alamos, NM.

[4] R. Fernow, BNL-61578 and the ARC model in Fermi Lab Work-shop.

[5-1] Y. Pang, T. J. Schlagel, and S. H. Kahana, BNL 1992.

[5-2] D. Kahana and Y. Torun, "Pion production studies", Transparencies presented at the 2+2 TeV μ^+-μ^- Collider Collaboration Meeting, Fermi National Accelerator Laboratory 1995.

[6] Y. Torun, "Pion production studies using ARC", Transparencies presented at the 2+2 TeV μ^+-μ^- Collider Collaboration Meeting, Fermi National Laboratory 1995.

[7] D. Berley et.al., "Production of negative pions of medical interest by high-enrgy proton", Proc 1973 particle acclerator conference, San Francisco, Clif., March 5-7, 1993.

[8] T. Abbot et.al., (E-802 collaboration), "Measurment of particle production in proton induced reactions at 14.6 GeV/c", Phys. Rev. D45, 3906 (1992).

[9] R. M. Sternheimer, and S. J. Lindenbaum, Phys. Rev. 123, 333 (1961); Phys. Rev. 109, 1723 (1958); Phys. Rev. 105, 1874 (1957).

[10-1] H. W. Bertini, "Monte Carlo Calculation on Intranuclear Cascade", ORNL-3383 (1963).

[10-2] W. A. Coleman and T. W. Armstrong, "The Nucleon-Meson Transport Code NMTC", ORNL-4606, Oak Ridge National Laboratory (1970).

[10-3] W. A. Coleman, and T. W. Armstrong, "NMTC - A Nucleon-Meson Transport Code", Nucl. Sci. and Engr. 43, 353 (1971).

[11] K. C. Chandler and T.W. Armstrong, "Operating Instructions for the High-Energy Nucleon-Meson Transport Code HETC," ORNL-4744, Oak Ridge National Laboratory (1972).

[12-1] M. R. Clover, R. M. DeVries, N. J. DiGiacomo, and Y. Yariv, Phys. Rev. C26, 2138 (1982).

[12-2] K. Chen, Z. Fraenkel, G. Friedlander, J. R. Grover, J. M. Miller, and Y. Shimamoto, Phys. Rev. 166, 949 (1968).

[12-3] K. Chen, G. Friedlander, and J. M. Miller, Phys. Rev. 176, 1208 (1968).

[13] T.A. Gabriel et.al., "An Extrapolation Method for Predicting Nucleon and Pion Differential Production Cross Sections from High Energy (>3 GeV) Nucleon-Nucleus Collisions" ORNL-4542 (1970).

9th Advance ICFA beam dynamics workshop:
Beam Dynamics and Technology Issue for mu^+ mu^- Colliders

Montauk Yacht Club, Montauk, Long Island, New York
October 15-20, 1995.

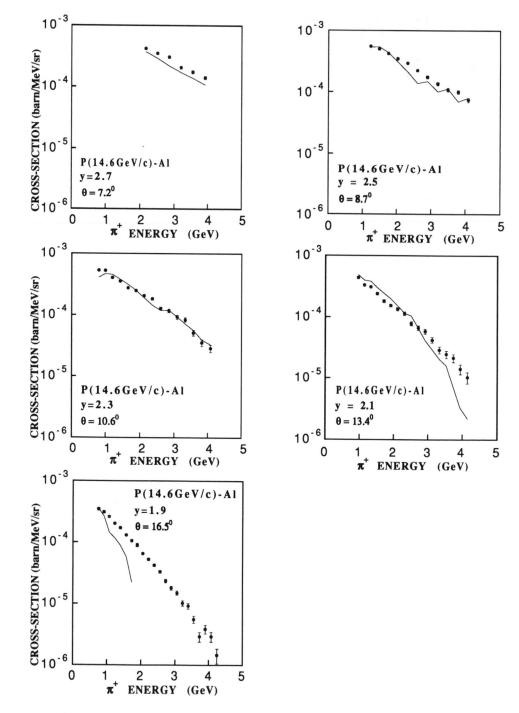

Fig.1 Cross section of positive pions production in a proton incuced reaction at 14.6 Gev/c on Al.

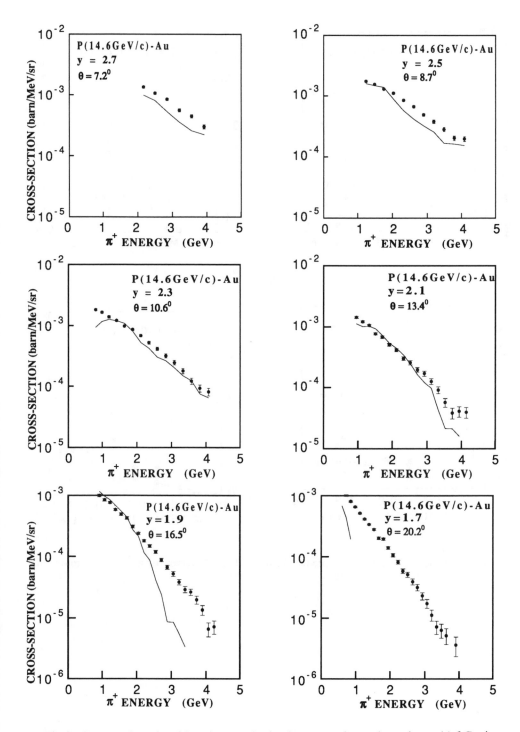

Fig.2 Cross section of positive pions production in a proton incuced reaction at 14.6 Gev/c on Cu.

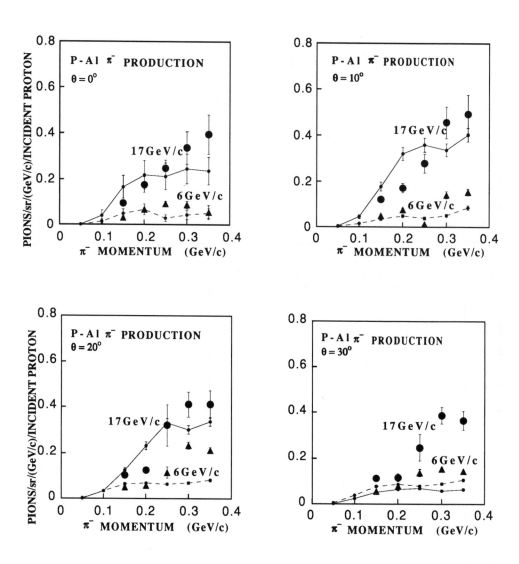

Fig.3 Number of negative pions produced from 6" thick Al target per an incident proton with
6 and 17 GeV/c momenta.

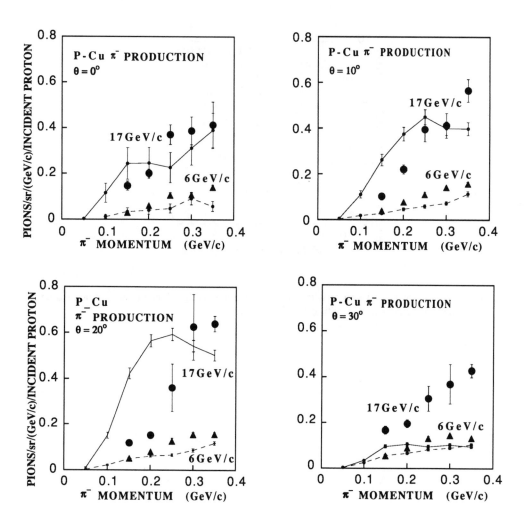

Fig.4   Number of negative pions produced from 6" thick Cu target per an incident proton with 6 and 17 GeV/c momenta.

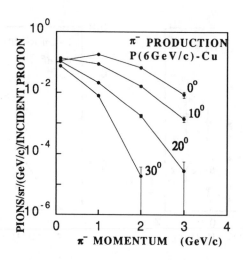

Fig.5 Number of negative pions with upto 3 GeV/c moment produced from 6" thick Cu target per an incident proton with 6 GeV/c momentum.

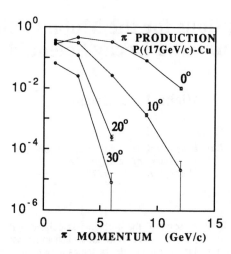

Fig.6 Number of negative pions with upto 13GeV/c moment produced from 6" thick Cu target per an incident proton with 17 GeV/c momentum.

99

Some Options for the Muon Collider Capture and Decay Solenoids

Michael A. Green

E. O. Lawrence Berkeley National Laboratory
University of California
Berkeley, CA. 94720

Abstract. This report discusses some of the problems associated with using solenoid magnets to capture the secondary particles that are created when an intense beam of 8 to 10 GeV protons interacts with the target at the center of the capture region. Hybrid capture solenoids with inductions of 28 T and a 22T are described. The first 14 to 15 T of the solenoid induction will be generated by a superconducting magnet. The remainder of the field will be generated by a Bitter type of water cooled solenoid. The capture solenoids include a transition section from the high field solenoid to a 7 T decay channel where pions and kaons that come off of the target decay into muons. A short 7 T solenoidal decay channel between the capture solenoid system and the phase rotation system is described. A concept for separation of negative and positive pions and kaons is briefly discussed.

BACKGROUND

A muon collider must have a number of different subsystems in order for the muons to collide at 2 TeV. The subsystems include: 1) a rapid cycling conventional accelerator that accelerates two bunches of about 5×10^{13} protons to 8 to 10 GeV at a repetition rate of 30 Hz., 2) a fixed target to produce pions and kaons from the protons, 3) a capture solenoid system to capture the particles that come from the target, 4) a decay and phase rotation channel that produces bunched muons, 5) a muon cooling channel that reduces the muon emittance from 0.025 m to 3×10^{-5} m at an energy of 0.2 GeV, 6) several stages of acceleration from 0.2 GeV to 2 TeV, 7) a single collider ring for both μ^+ and μ^- at an energy of 2 TeV, and 8) a single large detector to detect the products of μ^+ and μ^- collisions.

This report deals with the capture solenoid and the decay channel solenoids up to the first RF section used to bunch the pions and muons. The fixed target located inside of a capture solenoid converts 3×10^{15} protons per second at 8 to 10 GeV to pions and assorted other particles that will eventually decay to μ^+, μ^-, e^+, e^- and neutrinos There will also be left over protons from the original proton beam. The role of the capture and decay channel is to capture the pions and other particles and insure their decay to positive and negative muons. It may also be desirable to separate the muons from the left over protons and electrons and positrons produced by the target and the subsequent decay processes.

The transverse momentum that can be captured by a solenoidal capture field is proportional to the solenoid central induction and the solenoid radius. One can capture particles of a given transverse momentum at any field provided the solenoid is large enough. The advantages of using a high field capture solenoid are; 1) a small solenoid diameter and 2) a transfer of transverse momentum to longitudinal momentum when the field is reduced going into the decay solenoid. A transverse momentum of about 300 MeV/c can be captured by a 28 tesla solenoid with a radius of 75 mm. The process of reducing the induction from 28 to 7 tesla reduces the transverse momentum by about a factor of two. This means that a decay channel with 7 tesla solenoids will have a radius of about 150 mm.

High field solenoids (even those with a relatively small diameter) are expensive to build. For central inductions above 15 tesla, a pure superconducting solenoid is probably out of the question. For capture solenoids with a central induction above 15 tesla, the hybrid magnet with superconducting outer coils and water cooled on the inside becomes the design of choice. The water cooled solenoid inside of the superconducting solenoid uses a great deal of power. In addition water cooling has to be provided to absorb the energy from particles coming off of the target that are not captured.

If one tries to capture pions at a lower central induction, the capture efficiency theoretically goes down for a given solenoid diameter because the higher transverse momentum particles are lost. Unfortunately studies at Brookhaven and Fermi Lab have not yielded the same results in this regard(1,2). I suspect that one is comparing apples to oranges when the two studies are compared. The baseline study by Palmer uses a 28 tesla capture solenoid with a 7 tesla decay channel. Clearly other options are available for the capture and decay channels. The next section in this report describes a 28 tesla and a 22 tesla solenoid system for capturing the pions. In both cases the decay solenoids downstream are 7 tesla solenoids. The possibility of lower field capture solenoids and decay solenoids is discussed near the end of this report.

A COMPARISON OF A 28 TESLA AND A 22 TESLA CAPTURE SOLENOID MAGNETS

Capture solenoids with capture central inductions of 28 tesla and 22 tesla have been studied. Both of these capture solenoids are hybrid magnets with water cooled Bitter type solenoids inside of a 14 to 15 tesla superconducting solenoid. The superconducting solenoids are graded into low field niobium titanium outer coils and niobium tin high field inner coils. Both sets of superconducting capture solenoid coils would be cooled by a 1.8 K refrigerator. Both versions of the capture solenoid magnet system include a transition section where the central induction is reduced from the capture induction to a decay channel induction of 7.0 tesla. The warm bore diameter at 7 tesla is 300 mm in both magnet options. The final three meters of the capture solenoid is the start of the 7 tesla decay solenoid. The inner wall of this magnet section is heavy and water cooled, so the inner diameter of the niobium titanium coil is somewhat larger than it would be in subsequent decay channel magnets where the inner wall does not have to be as heavy. Table 1 compares the two capture solenoid designs that have been studied.

The superconducting coils for the 28 tesla capture solenoid are based on the cable in conduit superconducting coils that the National High Magnetic Field Laboratory is using for their 45 tesla hybrid magnet system(3). The water cooled Bitter magnet is longer and has a larger bore diameter than the Florida State magnet system. The projected power numbers in Table 1 reflect the longer water cooled solenoid length. It is assumed that a low Z temperature high melting point target will be used (for example, a beryllium target) A magnet inner bore of 150 mm was used for the 28 tesla capture solenoid. Figure 1 shows a cross-section of the 28 tesla solenoid.

The superconducting coils for the 22 tesla capture solenoid are based on the bath cooled superconducting coils that the Francis Bitter Magnet Laboratory is using for their 35 tesla hybrid magnet system(4). The water cooled Bitter magnet is longer and has a larger bore diameter than the MIT magnet system. The projected power numbers in Table 1 reflect this. It is assumed that a low Z target will be used A magnet inner bore of 180 mm was used for the 22 tesla capture solenoid. Figure 2 shows a cross-section of the 22 tesla capture solenoid system.

TABLE 1 Parameters for High Field and Low Field Capture Solenoids

Parameter	High Field Magnet	Low Field Magnet
Total Length of Solenoid System (mm)	5000	5000
Cryostat Outer Diameter (mm)	1850	1120
Niobium Titanium Capture Solenoid OD (mm)	1650	920
Niobium Titanium Capture Solenoid Length (mm)	1200	1000
Niobium Tin Capture Solenoid OD (mm)	1080	650
Niobium Tin Capture Solenoid Length (mm)	1000	1000
Water Cooled Bitter Coil OD (mm)	610	350
Water Cooled Bitter Coil Bore Diameter (mm)	150	180
Water Cooled Bitter Coil Length (mm)	900	900
Decay Channel Diameter (mm)	300	300
Transition Region Length (mm)	600	800
Decay Channel Length (mm)	3000	3000
Capture Solenoid Total Central Induction (T)	28.0	22.0
Superconducting Solenoid Central Induction (T)	14.6	14.2
Water Cooled Solenoid Central Induction (T)	13.4	7.8
Decay Solenoid Central Induction (T)	7.0	7.0
Total Magnet System Stored Energy (MJ)	150.7	51.1
Superconducting Magnet Stored Energy (MJ)	124.9	45.6
Power Required for the Water Cooled Solenoid (MW)	27.4	10.8
310 K Cooling Required for the Magnet System* (MW)	32.0	15.4
Total Mass of the Superconducting Magnets (Metric Tons)	26.0	16.6

* includes the water cooling required for the beam energy into the target

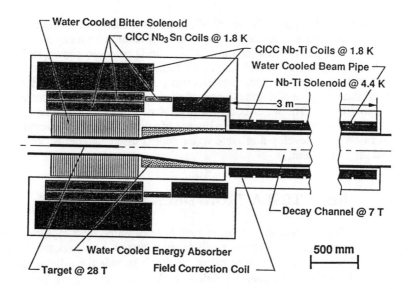

FIGURE 1 A Cross-section of a 28 T Capture Hybrid Solenoid based on the 45 T Florida State Hybrid Magnet System

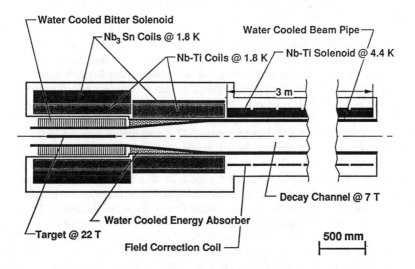

FIGURE 2 A Cross-section of a 22 T Capture Hybrid Solenoid based on the 35 T MIT Hybrid Magnet System

103

The theoretical capture efficiency of the 22 tesla solenoid is about 6 percent lower than the capture efficiency of the 28 tesla solenoid (from say 34 percent μ^- conversion to 32 percent conversion). The 22 tesla solenoid shown in Figure 2 should be less expensive to build than the 28 tesla solenoid shown in Figure 1. The 22 tesla capture solenoid uses less electric power and requires less cooling. However, there may be additional losses of particles in the transition section from 22 tesla to 7 tesla. The water cooled Bitter solenoid may be a problem in either case, because Bitter coils often have a limited life time. Clearly additional work is needed in order to develop a reliable water cooled insert magnet for the capture solenoid magnet system.

THE 7 TESLA PION DECAY SOLENOIDS

Once the fragments from the target have been captured by a capture solenoid system, many of the target fragments will decay to muons, electrons, positrons and neutrinos. The particles that come off of the target include the following: π^+, π^-, K^+, K^-, and protons. At energies below 1 GeV, the π^+, π^-, are produced in numbers that are similar to the number of low energy protons. The numbers of K^+ and K^- are about a factor of ten to thirty lower than the pions. There will be about 0.6 to 0.7 negative pions and kaons produced for every positive pion or kaon. The pions and kaon will decay to muons and electrons over a length of 150 meters or more. The electrons and positrons will form a tight spiral until they travel along the field lines with no transverse momentum.

The following conclusions result from computer simulations, done by Fermi Lab(5), of 8 GeV proton collisions with a target: 1) Copper targets about 1.5 interaction lengths long produced the highest numbers of muons. About ten percent of the beam power ends up in the target. 2) A water cooled Bitter solenoid that is 225 mm thick will reduce the power deposition into the superconducting magnet to around 5 mW cm^{-3}. This is below the quenching threshold in a well cooled magnet. 3) About 40 percent of the particles produced will be lost to the walls. Most will collide with the walls in the first 15 meters. 4) About ninety percent of the muons produced will have an energy in the range from 150 to 1200 MeV. 5) The maximum muon yield will be about 0.52 μ^+/p and 0.34 μ^-/p about 150 meters downstream from the target. The muon yield falls off slowly at longer distances.

From the Fermi Lab studies there is considerable latitude in the length of the phase rotation cavities and the distance to the first phase rotation cavity. The length of the decay channel should not be less than 75 meters. The issue of how long the decay channel should be before phase rotation is started was not addressed. Many think that phase rotation and pion decay can occur simultaneously. If there is no phase rotation in the first 150 meters and one neglects the particles with a momentum of 150 MeV/c, the bunch length will be about 30 meters. From the Fermi Lab studies it is clear that a range of capture and decay channel inductions are possible. In other words, one can capture at an induction below 28 tesla and the decay channel induction can be below 7 tesla. The minimum length of the decay channel before the first phase rotation cavity is probably about 30 meters.

There should be a short decay channel before phase rotation can take place. Table 2 presents the parameters for decay solenoid magnets that can be used in this channel. The central induction of these magnets assumed for this study is 7 tesla. The length of each magnet unit is determined by quench parameters for the magnets. The current density in the superconductor plus stabilizer matrix was nominally set to be 75 A mm^{-2} and the nominal magnet current should be 2000 A or larger. The bore tube wall (assumed to be water cooled copper) thickness was set at 25.4 mm. Figure 3 shows a schematic cross-section for a decay channel solenoid unit.

TABLE 2 Parameters for Pion Decay Channel Solenoids

Parameter

Total Length of the Channel (m)	>30
Number of Solenoid Magnet Units	>2
Nominal Length for Each Solenoid Unit (m)	15
Solenoid Cryostat Outer Diameter (mm)	~700
Niobium Titanium Coil Outer Diameters (mm)	530
Niobium Titanium Coil Inner Diameters (mm)	380
Solenoid Warm Bore Diameter (mm)	300
Nominal Bore tube Thickness (mm)	25.4
Design Central Induction of the Solenoid (T)	7.0
Nominal Solenoid Operating Temperature (K)	~4.4
Cold Mass per Solenoid Unit (metric tons)	17.8
Total Mass per Solenoid Unit (metric tons)	23.0
Stored Magnetic Energy per Solenoid Unit* (MJ)	47.5

*   at the design central induction

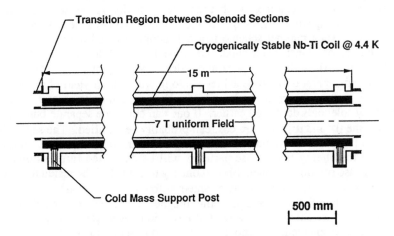

FIGURE 2   A Cross-section of a 7 T Decay Solenoid Section

DISCUSSION AND CONCLUSION

There are a number of issues that affect the efficiency and the cost of the muon production system for the muon collider. These include: 1) The capture induction affects both the cost and the efficiency of capture For constant capture efficiency, the bore of the capture solenoid must be increased as the induction of the capture solenoid decreases. 2) The type of target makes a difference. The Fermi Lab studies(1,5) suggest that the optimum material for the target is copper; the Brookhaven studies(1,2) suggest that a heavy material such as mercury is optimum. From the standpoint of power per unit mass and heat removal, a low Z material such as beryllium may be desirable. The target will define to some extent the yield and the amount of energy that is locally deposited in the superconducting magnet cryostat and the coils. Optimization of the target probably has to be done in combination with the optimization of the rest of the capture and decay system. 3) The thickness of the cryostat walls in the solenoid downstream from the target will affect the cost of the solenoids and the heating within them There is clearly an optimum wall thickness where solenoid capital cost is balanced with refrigeration cost. 4) The induction chosen for the decay channel and phase rotation channel clearly affects the cost of the magnet system. A lower central induction means larger diameter magnets, which in turn can affect the RF system for the phase rotation channel. Clearly the capture solenoid, the decay channel and the phase rotation system should be optimized as a unit.

The separation of π^+ and π^- from protons, electrons, positrons and other particles can have an effect on the overall efficiency of the injection system and the repetition rate of the collider. A method for separating π^+ and π^- was presented at the Montauk Workshop (5). This method involves the use of a curved solenoid. After the particles have been bent in the solenoid for about 10 meters, the π^+ and π^- are physically separated about 200 mm across the aperture of the solenoid. Once physical separation occurs, the bunches of π^+ and π^- have to be separated in time (and distance along the channel) so that the phase rotation RF system can act on both types of muons simultaneously. Some unusual magnet schemes have been proposed. One such magnet scheme is a cornucopia shaped magnet that has a central induction of 28 tesla at the small end of the horn and 7 to 10 tesla at the large end of the horn. All of the beam separation ideas involve larger bore solenoids if good π^+ and π^- separation is to be achieved along with a high yield of muons per proton. The cost of the cornucopia or larger bore solenoids is greater than the simple decay solenoids described earlier, If one can eliminate separate bunches of protons for the μ^+ and μ^-, the extra cost can probably be justified. The separation process for other particles coming off of the target appears to be imperfect. It is not clear where the energy from those particles will be deposited further down the decay and phase rotation channel. Should one try to get rid of the residual protons from the injector beam, and the decay e^+ and e^- before phase rotation?

From the standpoint of superconducting magnet technology, one would like to have a capture solenoid with a low central induction. A water cooled insert coil inside of the superconducting solenoid is still a good idea provided the water cooled solenoid can be made reliable enough to operate over the life of the machine. (To have to replace a radioactive water cooled solenoid during normal machine operation could be a problem.) The minimum thickness for the water cooled insert to a superconducting capture solenoid is about 250 mm (whether this insert generates part of the magnetic field or not) The optimum induction for the decay channel and the RF phase rotation channel is driven by the peak field in the phase rotation solenoids in the RF cavities and the physical size of the RF system. There is clearly a tradeoff between having to use niobium tin solenoids or a 1.8 K refrigeration system for the phase rotation magnets and having to build a larger diameter RF system. The minimum length of the drift space between the capture solenoid and the start of the phase rotation system appears to be driven by energy deposition into the walls of the decay channel. It appears that phase rotation should start as close to the capture solenoid as is practical (before the bunch length gets too large). The effect of the change in the particle gamma as it decays from a π to a μ on the phase rotation process is not clear. The separation of π^+ and π^- bunches appears to be desirable, but more study is needed.

ACKNOWLEDGMENTS

The author acknowledges the discussions with R. B. Palmer of the Brookhaven National Laboratory, and with R. Noble of Fermi National Laboratory concerning the capture of pions and other charged particles. This work was performed with the support of the Office of High Energy and Nuclear Physics, United States Department of Energy under contract number DE-AC03-76SF00098.

REFERENCES

1. Transparencies presented at the *A Second High Luminosity 2+2 TeV μ^+ μ^- Collider Collaboration Meeting,* Fermi National Laboratory 11-13 July 1995, compiled by Robert Noble of FNL
2. Transparencies presented at the *Ninth Advanced ICFA Beam Dynamics Workshop: Beam Dynamics and Technology Issues for μ^+ μ^- Colliders,* Montauk, New York, 15-20 October 1995, compiled by Juan C. Gallardo of BNL
3. Miller, J. R., Bird, M. D., Bole, S., et al, "An Overview of the 45T Hybrid Magnet System for the National High Field Magnet Laboratory," *IEEE Transactions on Magnetics* 30, No. 4, p 1563, (1994)
4. Iwasa,Y., Leupold, M. J., Weggel, R. J., and Williams J. E. C., "Hybrid III: The System, Test Results, the Next Step," IEEE Transactions on Applied Superconductivity 3, No. 1, p 58, (1993)
5. Transparencies of R. Noble, "Targeting and Collection Optimization Studies, "Ninth *Advanced ICFA Beam Dynamics Workshop: Beam Dynamics and Technology Issues for μ^+ μ^- Colliders,* Montauk, New York, 15-20 October 1995, compiled by Juan C. Gallardo of BNL

An Induction Linac Approach to Phase Rotation of a Muon Bunch in the Production Region of a μ⁺–μ⁻ Collider

William C. Turner

Lawrence Berkeley National Laboratory
1 Cyclotron Rd.
Berkeley, CA 94720

Abstract. The possibility of using an induction linac for phase rotation, or equivalently flattening the head to tail mean energy sweep, of a muon bunch in the production region of a μ⁺–μ⁻ collider is examined. Axial spreading of an accelerating bunch is analyzed and the form of appropriate induction cell voltage waveforms is derived. A set of parametric equations for the induction accelerator structure is given and specific solutions are presented which demonstrate the technological feasibility of the induction linac approach to phase rotation.

INTRODUCTION

In this paper we consider the possibility of using an induction linac for longitudinal phase rotation of the muon bunch in the production region of a μ⁺–μ⁻ collider(1). The induction linac is to be inserted in a region following the decay of pions into muons and prior to entering the muon phase space cooling region. The instantaneous muon energy spread is reduced from $\Delta E/E \sim \pm 100\%$ to $\sim \pm 10\%$ by allowing the bunch to drift and spread longitudinally. The instantaneous mean energy of muons arriving at a given axial location is highest at the head of a bunch and the lowest at the tail. An induction linac is then used to flatten the head to tail sweep of mean energy from $\sim 300\%$ to a few per cent by applying an appropriately shaped voltage waveform to the acceleration gaps. In order to analyze the feasibility of this idea it is necessary to characterize the muon pulse as a function of time. Figs. 1 and 2 show the results of typical Monte Carlo muon production calculations presented in this way(2). For these calculations the proton energy is 10 GeV, rms pulse width is 3 nsec and 0.2 muons per proton are captured in the decay channel at $z = 201$ m from the target. In Fig. 1 the mean instantaneous muon energy is plotted as a function of time of arrival at $z = 201$ m. (Throughout this paper "muon energy" is always total energy, including the rest mass.) At the head of the pulse the mean energy is 1.2 GeV, decreasing to 0.5 GeV at $T = 38.3$ nsec. With the exception of the Monte Carlo data point at the head of the pulse, the mean energy is reasonably well fit with an exponential function;

$$\langle E(GeV)\rangle = 0.21 + 0.79e^{-T(n\,sec)/38.3}.\tag{1}$$

Because the muons are relativistic ($\beta = 0.9945$ at 1 GeV) axial drift is only partially successful in reducing the instantaneous energy spread to the desired goal of $\pm 10\%$. In particular at the head of the pulse the muon spectrum extends above 2 GeV, compared to the mean 1.2 GeV. In practice it won't be possible to use muons above some cutoff energy and introducing such a cutoff will bring the Monte Carlo data at the head of the bunch into closer agreement with eqn. 1. For the feasibility discussion below we will assume eqn. 1 is a reasonable approximation of the mean muon energy.

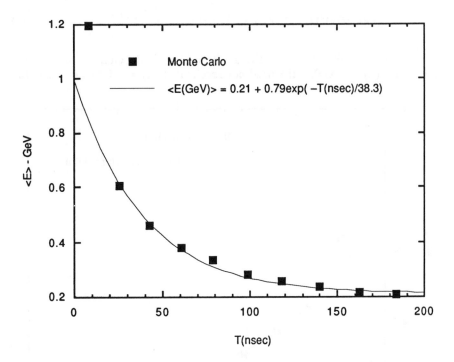

FIGURE 1. Mean muon energy versus time at z = 201 m from the proton target.

The instantaneous muon current is plotted versus arrival time in Fig. 2, again at z = 201 m from the production target. The Monte Carlo results have been normalized to 5×10^{13} protons on the target producing 2×10^{13} captured muons. At the head of the bunch the current is 76 A, decreasing to 21 A at T = 50 nsec. Except for the first two Monte Carlo data points, these data are also reasonably well fit by an exponential function;

$$I(A) = 57e^{-T(n\,sec)/\!50}. \tag{2}$$

Again cutting off high energy muons at the head of the bunch will tend to bring the Monte Carlo results into closer agreement with eqn. 2 and we will assume eqn. 2 is a reasonable approximation of the muon current. The integrated number of muons represented by eqn. 2 is 1.8×10^{13}.

At this point it is worth making a few observations based on Figs. 1 and 2. First, in order to flatten the mean energy to E_S a summation of voltage waveforms equal to the difference between E_S and eqn. 1 would be applied;

$$\sum_i eV_i(GeV) = E_s(GeV) - 0.21 - 0.79e^{-T(nsec)/38.3}. \qquad (3)$$

Depending on the choice of E_S, the head may be decelerated and the tail accelerated ($E_S > 1$ GeV), the head not accelerated ($E_S = 1$ GeV) or the entire

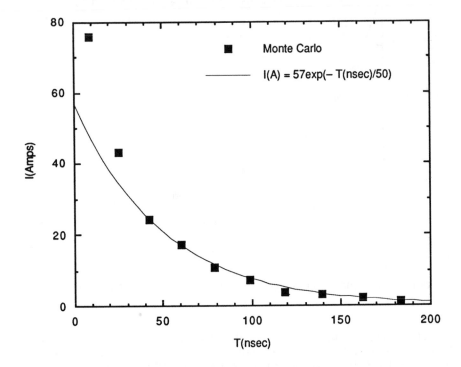

FIGURE 2. Muon current at z = 201 m.

pulse accelerated ($E_S > 1$ GeV). The first choice leads to bipolar waveforms which, while possible, presents a complication in the design of the pulsed power system feeding the induction cells. The third choice doesn't seem interesting in the absence of a rationale for accelerating muons to greater than 1 GeV before injection into the cooling section. Therefore we consider the second, $E_S = 1$ GeV.

For a reasonable accelerating gradient $V' \sim 1$ MV/m, the total length of induction accelerator would be ~ 0.5 - 1 km. The issue of magnitude of accelerating gradient and a consistent induction cell structure will be discussed in detail below. The summation of waveforms indicated by eqn. 3 will need to be applied to a large number of induction cells extending over hundreds of meters. The bunch will continue to spread as it propagates through the accelerator and, as a consequence, the detailed shape of the accelerating waveforms will change and the time scale will lengthen from beginning to end of the accelerator compared to what is indicated in eqn. 3. This will be discussed quantitatively in the following section. Nevertheless it is clear from Fig. 1 that the risetime of the voltage waveforms will be of the order of 40 – 50 nsec. This is slow enough to allow flexibility in the radial size and the capacitance of the induction cells. As we will see this is fortunate. The very large radius of the beam tube $R_w = 15$ cm and the 7 T superconducting focusing solenoids that are required by the large transverse momentum spread of the muons dictates that the induction cores be placed at a large radius $\sim 2 - 3$ R_w.

Referring to Fig. 2, the muon current is less than 100 A, rather low compared to the kiloampere beams usually encountered in electron induction accelerators. This together with the large beam tube radius implies that the beam breakup instability will have a growth length many times the length of the accelerator so this instability does not introduce significant design concerns. The muon beam current is also low compared to what we can expect for an induction core leakage current of $\sim$ few kA. Pulse distortion by beam current loading will therefore be negligible and this is a design advantage. We need only be concerned with the effective load resistance of the induction core, the cell capacitance and the pulse power impedance to determine the shape of the voltage waveform. The accelerator efficiency will be low $\sim 0.1\%$ but that is a minor concern for this application. Finally we note from Fig. 2 that the Monte Carlo data extend to 183 nsec, however only 18% of the muons arrive after 80 nsec. Since the power consumed by the induction accelerator increases linearly with pulse width and is nearly independent of the beam loading it becomes relatively inefficient and expensive to accelerate the last few trailing muons. Here we will somewhat arbitrarily cut off the accelerated muons at 80 nsec. The number of muons from zero to 80 nsec represented by eqn. 2 is 1.4×10^{13} muons per 5×10^{13} protons.

In Fig. 3 we have indicated in a schematic way the layout of induction cells and focusing solenoids for an assumed module length $L_m = 1$ m. The 7 T superconducting solenoids and cryostats are shown extending from an inside radius $R_w = 15$ cm to an outside radius 30 cm and extending over an axial length fraction 60%. The induction cells are represented by induction cores, vacuum insulators and vacuum acceleration gaps. The induction cores have been stacked radially outside the insulators rather than axially in series. This maximizes the axial length fraction for magnetic material and therefore maximizes the effective accelerating gradient. The acceleration gaps are assumed to extend over the 40% of axial length between solenoids and to extend inward to $R_w = 15$ cm. For the magnetic induction cores two conceptual possibilities are indicated. In the first case, Fig. 3(a), the cores are located entirely between the solenoids. This minimizes the interference between induction cells and solenoids and places the induction cores at the minimum possible radius. It is essential that negligible magnetic flux from the solenoids leak out of the beam channel and reach the induction cores so the induction cores are shown outside the solenoid structure in Fig. 3(a). In addition it will be necessary to divide the solenoid and induction cell structure more finely than indicated schematically in Fig. 3(a) to avoid flux leakage into the cores. In the second case, Fig. 3(b), the induction cores are allowed to extend over the outside of the

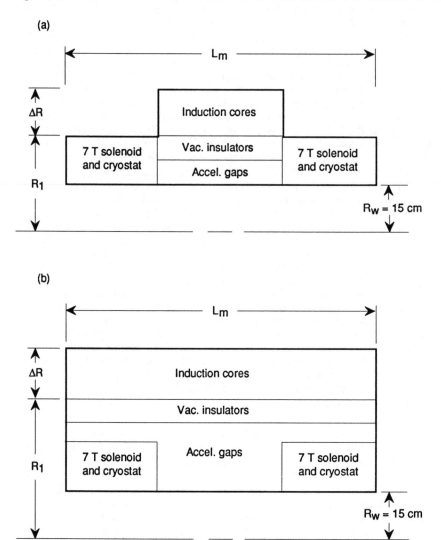

FIGURE 3. Schematics of induction cell components;(a) minimum interference between solenoids and induction cells and (b) maximum axial length for induction cores.

solenoids. Since more axial length is available for magnetic material, the accelerating gradient is increased up to a factor 2.5. However, now the vacuum insulators must be placed entirely outside the 7 T solenoid structure and the vacuum acceleration gaps must be mapped from outside the solenoids to the gaps between solenoids. All of this increases the radius, volume and weight of the cores as well as the cell capacitance. To see what is possible some detailed calculations have been done. The results are in the last section of the paper.

The remainder of the paper is divided into three sections. The first section describes the longitudinal dynamics of the muon bunch as it drifts and is accelerated. The second section presents an induction cell model and uses the model to analyze the induction core geometry and achievable accelerating gradient. The final section presents a table of parameters for four accelerator cases.

LONGITUDINAL DYNAMICS OF A MUON PULSE

We now derive the relationship between muon energy and time of arrival at a given axial position in the accelerator with the parametric dependence on accelerating gradient given explicitly. The time of arrival (T) is measured relative to an unaccelerated reference particle within the bunch labeled by the subscript "s". It turns out to be simplest to find $T(E,z)$ and then invert the result numerically to get $E(T,z)$. As a byproduct we also get the time dependence of the self consistent accelerator gap waveforms $V(T,z)$. As the muons propagate the change in their arrival time relative to the reference particle is given by;

$$\frac{\partial T}{\partial z} = \frac{1}{c}\left(\frac{1}{\beta_z} - \frac{1}{\beta_{s,z}}\right) \ . \tag{4}$$

where $c\beta_z$ is the velocity component parallel to the accelerator axis and $\beta_z^2 + \beta_\perp^2 = 1 - 1/\gamma^2$. In order to integrate eqn. 4 we imagine the bunch is subdivided into infinitesimal slices and apply an accelerating gradient $\gamma' = V'/mc^2$ to each slice. Space charge forces are unimportant since $\beta \sim 1$, $\gamma \gg 1$. The transverse velocity component is treated by writing $\beta_\perp^2$ in terms of transverse momentum and then taking the rms value within a slice: $\langle\beta_\perp^2\rangle = \langle p_\perp^2\rangle/(m\gamma c)^2 = \alpha/\gamma^2$. For numerical calculations we set the rms transverse momentum equal to the square of the cutoff momentum of a particle with an orbit passing through the axis of the solenoid focusing channel with $B = 7$ T and $R_W = 15$ cm; $\langle p_\perp^2\rangle = (158 MeV/c)^2$, $\alpha = 2.25$. γ' must vary from slice to slice in order to flatten the energy distribution but for each slice is conveniently chosen to be independent of z. This is a simple way of insuring that all slices reach the reference energy at the same axial location. We label each slice with a subscript "i" which can be thought of as its mean energy at $z = 0$. We also have in mind only cases where the mean energy is a monotonically decreasing function of arrival time and specifically exclude the possibility of one slice overtaking another. Owing to the spread of energies within a slice, the individual muons rearrange themselves somewhat amongst the slices, but we don't need to keep track of that. Then,

$$\gamma_i(z) = \gamma_i(z=0) + \gamma'_i z$$

$$= \frac{E_i(z)}{mc^2} \tag{5}$$

and

$$T_i(z) = \frac{1}{c\gamma'_i}\left\{\sqrt{(\beta\gamma)_i^2 - \alpha} - \sqrt{(\beta\gamma)_{i,z=0}^2 - \alpha}\right\} - \frac{z}{c\beta_{s,z}} + T_i(z=0). \tag{6}$$

If $\gamma'_i = 0$, eqn. 6 reduces to

$$T_i(z) = \frac{1}{c}\left(\frac{\gamma_i}{\sqrt{\gamma_i^2 - (1+\alpha^2)}} - \frac{\gamma_s}{\sqrt{\gamma_s^2 - (1+\alpha^2)}}\right)z + T_i(z=0). \tag{7}$$

Now we choose the magnitude of γ'_i to achieve the desired result of flattening the mean energy sweep;

$$\gamma'_i = \frac{V'_i}{mc^2}$$

$$= \frac{V'_m}{mc^2}\frac{E_s - E_i(z=0)}{\Delta E} \tag{8}$$

where ΔE is the magnitude of energy sweep that is to be reduced to zero and V'_m is the maximum accelerating gradient. The accelerator length is $z_m = \Delta E / V'_m$. We note that eqn. 5 maps the energy at $z = 0$ to the energy at z, eqn. 6 maps the energy at z to the arrival time at z and eqn. 8 maps the energy at $z = 0$ to the accelerating gradient. These may be used to obtain the mean muon energy and the accelerating gradient as functions of the arrival time within a bunch and the propagation distance z as shown in Figs. 4 and 5. For the initial condition we take eqn. 1 for $E_i(z=0)$ versus T. In Fig. 4a there is no acceleration so the muon bunch is simply continuing to spread out, noticeably more in the low energy tail than at the head of the bunch. At $z = 0$ km in Fig. 4a, the 0.3 GeV muons at the tail arrive 82.5 nsec after the 1 GeV muons at the head of the bunch; by the time the bunch reaches $z = 1$ km this time delay has grown to 1,000 nsec. In Fig. 4b we have added acceleration, chosen $E_s = 1$ GeV so the muons at the head of the bunch are not accelerated, $\Delta E = 0.7$ GeV and applied a maximum accelerating gradient $V_m' = 1$ MV/m to the tail of the pulse with energy 0.3 GeV at $z = 0$. The muon bunch reaches a flat energy distribution at $z_m = 700$ m. Acceleration quickly slows the

114

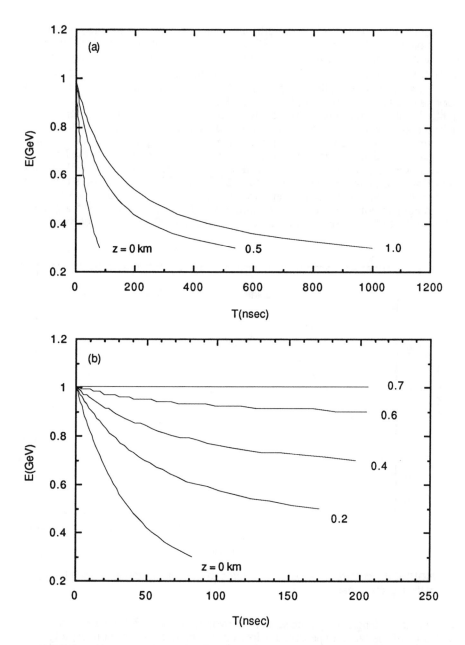

FIGURE 4. Muon pulse spreading and energy flattening with (a) drift only and (b) acceleration. For (b) the maximum acceleration gradient equals 1.0 MeV/m at the tail of the pulse. z = 0 is 201 m from the proton target.

spreading of the bunch so the width only reaches 205 nsec by the end of the induction linac. The decrease in spreading is most noticeable after the first 200 m so one should eventuallyconsider the possibility of increasing the accelerating gradient at the front of the accelerator. The condition of constant γ' can be met piecewise.

The accelerating gradient versus time that is consistent with Fig. 4b is shown for three axial locations $z = 0$, 0.2 and 0.7 km in Fig. 5. The time duration of the applied voltage gradient increases with distance because of the increase in differential arrival time of each muon slice relative to the head of the bunch (eqn. 4) and our stated condition of applying a constant accelerating gradient to each slice. One can check that the arrival times for the tail of the bunch relative to the head are consistent in Fig. 4b and Fig. 5. This is true for intermediate slices as well; muons with $E = 0.5$ GeV at $z = 0$ km arrive at $T = 38.2$ nsec, 0.2 km at $T = 66.8$ nsec and 0.7 km at $T = 85.2$ nsec and experience constant gradient $V' = 0.71$ MV/m etc.. The integrated volt-seconds/m for the three waveforms shown in Fig. 5 are .055, .122 and .134 volt-seconds/m at $z = 0$, 0.2 and 0.7 km respectively. In

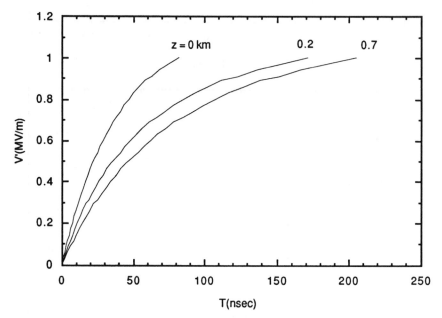

FIGURE 5. Acceleration waveforms at three axial locations from beginning to end of the induction linac.

addition to the lengthening time scale of the voltage pulse with distance there is also a change of the waveshape because low energy muons spread more rapidly than high energy ones. However the distortion is not too great and the simple exponential at $z = 0$ with $\tau_V = 38.3$ nsec maps reasonably accurately (i.e. within a couple of per cent) for our purposes to exponential waveforms at $z = 0.2$ and

0.7 km with $\tau_v = 66.8$ and 85.2 nsec. Eqn. 6 may be used to calculate the mapping of τ_v.

MODEL EQUATIONS

A schematic of an induction cell is shown in Fig. 6 which also defines some of the geometric quantities that will appear in the mathematical description. There are four components to the cell: (1) the high voltage pulsed power feed, (2) the magnetic core, (3) the vacuum insulator and (4) the acceleration gap. The induction core has axial length w and radial width ΔR. The inside and outside radii of the magnetic core are R_1 and R_2. The high voltage pulsed power lead enters along one side of the core, encircles it and returns to ground. The magnetic core volume behind the vacuum insulator is filled with insulating dielectric fluid. There are three insulating gap widths indicated: the dielectric gap g_d, the vacuum insulator gap g_s and the acceleration gap g_v. These must withstand dielectric breakdown, vacuum surface flashover and vacuum breakdown respectively. The angle between the insulator surface and the metal electrode surfaces on the vacuum side is shown to be ~ 25^0 to 45^0 to maximize the breakdown limiting field strength. The re-entrant acceleration gap prevents radiation from the beam channel reaching the insulator surface and possibly initiating breakdown. The primary concern here is shielding the synchrotron radiation coming from electrons and positrons from muon decay. The total length of the induction cell is L_c.

We now turn to the equations for determining the magnitudes of the geometric quantities in terms of materials properties and parameters of the accelerator. There are three primary equations for the induction core; (1) the cell axial length, (2) the volt-seconds of the magnetic core and (3) the voltage risetime. A fourth equation describes the axial length needed for voltage insulation and shielding the vacuum insulator. The cell axial length is the minimum required by either the magnetic core region or the vacuum insulator plus accelerator gap region. The requirements of voltage insulation and shielding the vacuum insulator will determine a maximum accelerating gradient. For gradients chosen to be at or below this maximum the cell dimensions are then determined by the magnetic core region and there is a leftover marginal length for voltage insulation and shielding. We begin by treating the magnetic core region and then turn to voltage insulation and shielding of the vacuum insulator.

A simple exponential waveform applied to the induction cells is adequate for flattening the energy sweep;

$$V(T) = V_0 \left(1 - e^{-\frac{T}{\tau_v}} \right) \ . \tag{9}$$

Then the length of the cell L_c is related to the voltage and gradient applied at the end of the pulse by;

$$L_c = (1 + f_s)w + g_d$$

$$= f_c \frac{V_0}{V_m'} \left(1 - e^{-\frac{T_{max}}{\tau_v}} \right) \quad , \qquad (10)$$

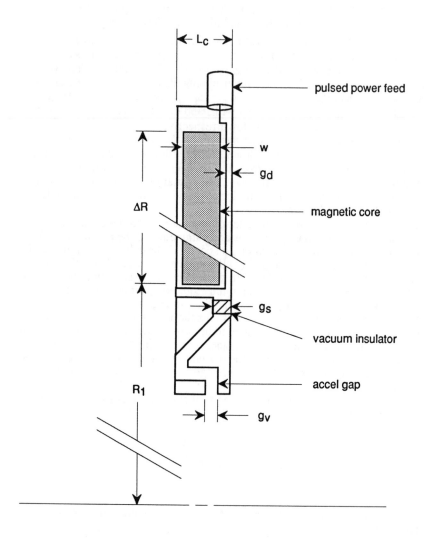

FIGURE 6. A schematic illustration of an induction cell showing the definition of some of the geometric quantities used in the model calculations.

where f_s is the fraction of cell length relative to the width of the induction core that must be allowed for mechanical structure, and f_c is the fraction of accelerator length occupied by induction cells. The accelerating gradient has been defined so it is the effective gradient over the entire length of accelerator and not over the reduced length occupied by induction cells.

The volt-seconds applied to the induction core is the integral of eqn. 9 and setting this equal to the magnetic flux swing of the induction core we obtain;

$$\frac{p_f \Delta B w \Delta R}{V_0 \tau_v} = \left(\frac{T_{max}}{\tau_v} - 1 \right) - e^{-\frac{T_{max}}{\tau_v}} \tag{11}$$

where ΔB is the change in magnetic induction and p_f is the effective packing fraction of magnetic core material.

Hysteresis losses of the magnetic core material (eddy currents and domain wall dissipation) dominate the impedance of the induction core viewed by the pulsed power circuitry. These losses may be represented by an effective load resistance R_L which will be discussed in a moment. The pulsed power impedance is then chosen to match R_L to minimize reflected energy and the voltage risetime may be written as;

$$\tau_v = \frac{R_L C}{2} + \tau_s \tag{12}$$

where C is the cell capacitance and τ_s is the intrinsic risetime of the pulsed power circuitry.

We now turn to expressions for C and R_L so that eqn. 12 can be written in terms of the cell geometry. The internal cell capacitance is dominated by the capacitance of the dielectric gap g_d. There are additional smaller contributions from the vacuum insulator and acceleration gap which we neglect. In addition it will be useful to add some external capacitance C_{ext} which is not shown in Fig. 6. The external capacitance has two functions: (1) it serves as a tuning element for adjusting τ_v and (2) it results in an increased dielectric gap g_d which, although not necessary for voltage holding in most cases we have examined, is useful for achieving gap dimensions that seem reasonable for mechanical fabrication tolerances. We then obtain;

$$C = \varepsilon \pi (R_1 + R_2) \frac{\Delta R}{g_d} + C_{ext} \tag{13}$$

where ε is the dielectric constant of the insulating fluid. In numerical calculations we will choose $C_{ext} = 0.5*C$.

The voltage applied to the induction cell is related to the core averaged time rate of change of magnetic induction $\dot{B}$ by;

$$V = p_f \dot{B} w \Delta R. \tag{14}$$

If the instantaneous core loss power per unit volume averaged over the core volume is denoted by $\dot{e}_L$ then the product of cell voltage and load current I_L is;

$$VI_L = p_f \pi (R_1 + R_2) w \Delta R * \dot{e}_L. \tag{15}$$

Taking the ratio of eqns. 15 and 14 we obtain the load current,

$$I_L = \pi (R_1 + R_2) \frac{\dot{e}_L}{\dot{B}} \tag{16}$$

and the ratio of eqns. 14 and 16 defines the load impedance,

$$R_L = \frac{p_f w \Delta R}{\pi (R_1 + R_2)} \frac{\dot{B}^2}{\dot{e}_L} . \tag{17}$$

To complete the equations for numerical analysis an expression is needed for the specific core loss $\dot{e}_L$. The particular material we have in mind is an amorphous ferromagnetic alloy, Metglas 2605SC, chosen for its very high change in magnetic induction from negative remanance to saturation $\Delta B > 2.5$ T(3). The core loss data that are available measure the dynamic hysteresis energy loss per unit volume per cycle for square wave excitation to saturation as a function of $\dot{B} = \Delta B / T_{sat}(4)$. For $\dot{B}$ exceeding 1 T/μsec the data are reasonably well approximated by a linear dependence of e_L on $\dot{B}$. At lower values of $\dot{B}$ the increase in hysteresis loss is not as steep but this is well below our range of interest. To further parameterize e_L we also factor out the magnitude of the flux swing since if the core is not driven to saturation we expect the loss to drop proportionally for a given $\dot{B}$. This is valid for material with reasonably square shaped hysteresis loops. For Metglas 2605SC we then obtain;

$$e_L = \alpha_L \Delta B \dot{B} \tag{18}$$

and,

$$\dot{e}_L = \alpha_L \dot{B}^2 \tag{19}$$

with $\alpha_L = 7.9 \times 10^{-5}$ J•sec/T^{2}m^3 and $\dot{B} > 1$ T/μsec. In our applications the excitation is not square wave but we will assume the validity of eqn. 19 for the instantaneous specific core loss appearing in eqns. 16 and 17. We then obtain for the load current and load resistance;

$$I_L = \pi(R_1 + R_2)\alpha_L \dot{B} \qquad (20)$$

and,

$$R_L = \frac{p_f w \Delta R}{\pi(R_1 + R_2)} \frac{1}{\alpha_L}. \qquad (21)$$

We notice that with the parameterizations of eqns. 18 and 19 the load current and applied voltage have the same time dependence and R_L is time independent.

Choosing $C_{ext} = 0.5*C$ and inserting eqns. 13 and 21 into eqn. 12, the dielectric gap width is related to core dimensions and risetime by;

$$g_d = \frac{p_f \varepsilon}{\alpha_L} \frac{w(\Delta R)^2}{\tau_v - \tau_s}. \qquad (22)$$

Eqns. 10, 11 and 22 can now be combined to solve for the dielectric gap width g_d and the core dimensions w and ΔR. First we eliminate g_d and ΔR to obtain a quadratic equation for w;

$$w^2 - \frac{f_c}{1+f_s} \frac{V_0}{V_m'}\left(1 - e^{-\frac{T_{max}}{\tau_v}}\right)w + \frac{A_1 A_2^2}{1+f_s} = 0, \qquad (23)$$

where

$$A_1 = \frac{p_f \varepsilon}{\alpha_L} \frac{1}{\tau_v - \tau_s}, \qquad (24)$$

and

$$A_2 = \frac{V_0 \tau_v}{p_f \Delta B}\left[\left(\frac{T_{max}}{\tau_v} - 1\right) - e^{-\frac{T_{max}}{\tau_v}}\right]. \qquad (25)$$

The solutions for w, ΔR and g_d are then;

$$w_\pm = \frac{1}{2}\frac{f_c}{1+f_s}\frac{V_0}{V_m'}\left(1 - e^{-\frac{T_{max}}{\tau_v}}\right) \pm \frac{1}{2}\sqrt{\left(\frac{f_c}{1+f_s}\frac{V_0}{V_m'}\left(1 - e^{-\frac{T_{max}}{\tau_v}}\right)\right)^2 - 4\frac{A_1 A_2^2}{1+f_s}}, \qquad (26)$$

$$\Delta R_\pm = \frac{A_2}{w_\pm} \qquad (27)$$

and

121

$$g_{\pm} = A_1 w_{\pm} (\Delta R_{\pm})^2. \tag{28}$$

From the form of eqn. 26 it is clear that the following inequality must be satisfied in order for solutions to exist;

$$\frac{f_c}{1+f_s} \frac{V_0}{V'_m} \left(1 - e^{-\frac{T_{max}}{\tau_v}} \right) > 2 \left(\frac{A_1 A_2^2}{1+f_s} \right)^{1/2} \tag{29}$$

and this implies an upper bound on the accelerating gradient,

$$V'_m < \frac{1}{2} \frac{f_c}{(1+f_s)^{1/2}} \frac{p_f \Delta B}{\tau_v} \left(\frac{\alpha}{p_f \varepsilon} (\tau_v - \tau_s) \right)^{1/2} \frac{1 - e^{-\frac{T_{max}}{\tau_v}}}{\left(\frac{T_{max}}{\tau_v} - 1 \right) - e^{-\frac{T_{max}}{\tau_v}}}. \tag{30}$$

Physically, as the accelerating gradient increases the cell length must decrease in order to fit more cells into a given space. This is done by decreasing w and increasing ΔR so the volt-seconds product per core stays the same. Meanwhile the dielectric gap g_d increases to keep the capacitance and risetime from increasing. However the overall cell length decreases because the increase in g_d is more than offset by the decrease in w. Eventually g_d becomes equal in magnitude to $(1+f_s)*w$ (see eqn. 10) and the increase in g_d can no longer be compensated by a decrease in w. At this point a limiting maximum accelerating gradient has been reached.

In Figs. 7, 8 and 9 we have plotted the induction cell geometric parameters $w_{\pm}$, $\Delta R_{\pm}$ and $g_{d\pm}$ versus accelerating gradient V'_m. Results are shown for induction cells occupying fractions $f_c = 0.4$ and 0.8 of the accelerator length. Assumed values for the parameters appearing in eqns. 23 to 28 are given Table 1. The cell voltage $V_0 = 50$ kV was chosen with consideration of how the pulse power system might be configured. In principle one could choose spark gaps, thyratrons or saturating magnetic cores for the final stage switch into the induction cores since they can all be configured to handle the required power level. Spark gaps could allow operation up to $V_0 = 250$ kV. However electrode erosion would require replacement after a few million pulses, or a day of operation at 30 Hz, so they are ruled out. Saturating magnetic cores have been shown to switch coaxial lines charged up to ~ 200 kV at kHz rep rates and high power thyratrons are commercially available for switching up to ~ 50 kV so either of these seems possible. Fifty kilovolts was chosen as a compromise between these two possibilities; a saturating magnetic core discharging a coaxial line charged to $2V_0 = 100$ kV or a thyratron discharging a Blumlein line charged to $V_0 = 50$ kV.

The voltage risetime τ_v and duration T_{max} appear in the equations for $w_{\pm}$, $\Delta R_{\pm}$ and $g_{d\pm}$ and as we have seen in Fig. 5, these vary along the length of the accelerator. To be specific we have presented the results for the beginning of the accelerator $z = 0$ ($\tau_v = 38.3$ nsec, $T_{max} = 82.5$ nsec) in Figs. 7a, 8a and 9a and at the end $z = z_m$ in Figs. 7b, 8b and 9b. Since the length of the accelerator depends

TABLE 1. Parameter values used for the evaluation of Figs. 7 to 10.

Parameter	Symbol	Value
Cell voltage, kV	V_0	50
Pulse power intrinsic risetime, nsec	τ_s	10
Induction core packing fraction	p_f	.75
Insulating fluid relative dielectric constant	$\varepsilon/\varepsilon_0$	3.4
Structure fraction	f_s	0.2
Magnetic induction flux swing, T	ΔB	2.5
Induction core loss parameter, J·sec/T^{2}m^3	α_L	7.9×10^5

on the accelerating gradient τ_v and T_{max} do also; for each value of V'_m, τ_v and T_{max} are calculated using eqn. 6.

In Figs. 7 to 9 the positive roots of eqn. 27 to 28 are plotted as the solid lines and the negative roots as dashed lines. The negative roots lead to core widths w that are less than the insulating gap width g_v and are impractically thin for fabrication, so only the positive roots are useful. Looking at the positive root solutions, the range of accelerating gradients that seem possible (1 to 2.5 MV/m for $w > 1$ cm, $g_v > 1$ mm) are several times larger than what one ordinarily encounters for electron induction accelerators. The reasons for this are: (1) choosing a core aspect ratio $\Delta R/w > 1$, (2) stacking the insulator and induction cores radially rather than axially and (3) using induction cores fabricated from Metglas with $\Delta B = 2.5$ T rather than Ni-Zn ferrite with $\Delta B = 0.6$ T. The comparatively slow risetime $\tau_v > 38$ nsec is the factor which allows this.

So far we have dealt with the induction core geometry and voltage waveshape requirements. We now turn to the requirements of high voltage insulation. Because of the light beam loading we do not have to worry about significant mismatch overvoltage if the beam is absent. The length of a cell cannot be less than what is required to prevent high voltage breakdown and therefore high voltage insulation imposes an upper bound on accelerating gradient V'_m that is independent of the

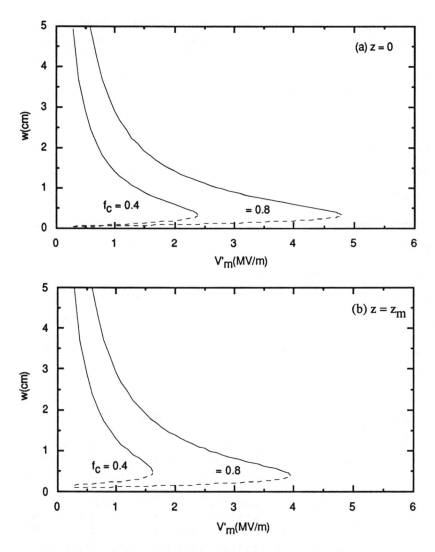

FIGURE 7. Axial width of the induction cell magnetic core versus accelerating gradient (a) at the beginning of the induction linac and (b) at the end. Induction cell voltage is $V_0 = 50$ kV. Results are shown for magnetic cores occupying fractions $f_c = 0.4$ and 0.8 of the length of the accelerator.

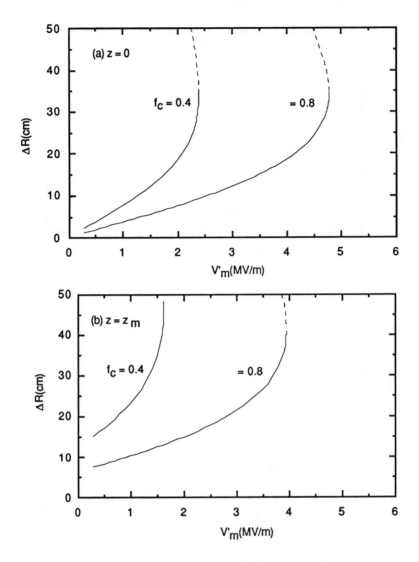

FIGURE 8. Radial width of the induction cell magnetic core versus accelerating gradient (a) at the beginning of the induction linac and (b) at the end. Conditions are the same as Fig. 7.

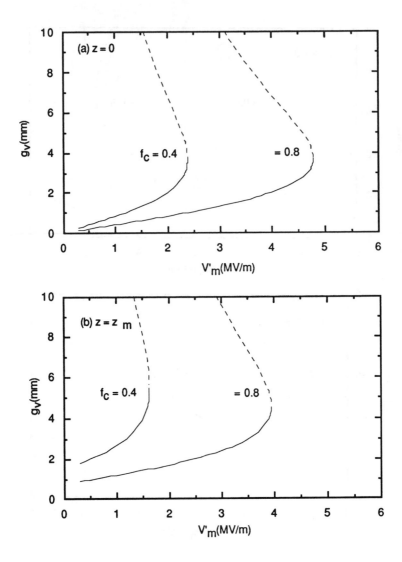

FIGURE 9. Dielectric insulation gap width versus accelerating gradient (a) at the beginning of the accelerator and (b) at the end. Conditions are the same as Fig. 7.

considerations of induction core size that have been discussed above. We first discuss the size of each insulation gap and then turn to the voltage breakdown limited accelerating gradient.

For the dielectric fluid we choose transformer oil and require

$$g_d > \frac{V_0}{E_{oil}}$$ (31)

with the breakdown electric field strength in oil taken to be (5);

$$E_{oil}(MV \,/\, cm) \approx \frac{0.5}{\tau_p^{1/3}(\mu \sec)A^{1/10}(cm^2)}.$$ (32)

The pulse width τ_p is the time the voltage waveform exceeds 63% of its maximum value (i.e. $T > \tau_v$ in eqn. 9) and A is the electrode area. The dependencies on τ_p and A are rather weak so we don't need precise values. Taking $\tau_p = .12$ μsec and $A \sim 3,000$ cm^2, $E_{oil} = 0.45$ MV/cm and for $V_0 = 50$ kV, we obtain $g_d > 1.1$ mm. The plots in Fig. 9 satisfy this constraint except at the lowest accelerating gradients. If more margin is needed it can easily be obtained by simply increasing g_d and at the same time increasing the external trim capacitance fraction C_{ext}/C so the risetime is unchanged. Another possibility would be to insert several layers of .025 mm Kapton in the dielectric gap.

The vacuum insulator is assumed to be Lexan (polycarbonate, $\varepsilon/\varepsilon_0 = 3.0$) or Rexolite (cross linked polystyrene, $\varepsilon/\varepsilon_0 = 2.5$). Care needs to be taken to avoid field enhancement at the triple points where vacuum, metal and insulator meet. Usually this is done by recessing the insulator slightly in the electrode surface. Scaling relations for surface flashover electric field strength as a function of pulse duration, gap width and surface area are not generally available. However for the size of insulators we are dealing with and for pulse duration ~110 nsec above 0.63 V_0 we take a limiting surface flashover field strength $E_s = 50$ kV/cm. The insulator height g_s should then satisfy

$$g_s > \frac{V_0}{E_s},$$ (33)

or for $V_0 = 50$ kV, $g_s > 1$ cm. Similarly for vacuum breakdown initiated on the metal electrode surfaces we take a limiting field strength $E_v = 100$ kV/cm and obtain for the acceleration gap,

$$g_v > \frac{V_0}{E_v},$$ (34)

or for $V_0 = 50$ kV, $g_v > 0.5$ cm. Care would need to be taken in the details of the shape of the electrodes between the vacuum insulator and acceleration gap, particularly at bends, to avoid excessive local enhancement of the field strength compared to eqn. 34.

We now turn to the voltage insulation limited accelerating gradient. In order to shield the insulator from the muon beam channel the sum of the vacuum insulator height and vacuum acceleration gap must be less than the cell length, or equivalently the following inequality must be obeyed,

$$\Delta_s = L_c - g_v - g_s - f_s w$$
$$= w + g_d - g_s - g_v \quad . \tag{35}$$
$$> 0$$

In eqn. 35 we have subtracted the allowance for structural width $f_s w$ supporting the induction core since this would presumably be extended to the insulation region. Δ_s has been plotted versus accelerating gradient in Fig. 10, again at the beginning (Fig. 10a) and end (Fig. 10b) of the accelerator. For these calculations the previously calculated w_+ and g_{d+} in Figs. 7 and 9 have been used. For the conditions we have assumed ($V_0 = 50$ kV, $g_s = 1.0$ cm and $g_v = 0.5$ cm) the voltage insulation limited accelerating gradient defined by $\Delta_s = 0$ is ~ 1 MV/m for f_c = 0.4 and ~ 2 MV/m for $f_c = 0.8$ and not significantly different at $z = 0$ and $z = z_m$.

SPECIFICATION OF ACCELERATOR PARAMETERS

In this final section we give some examples of accelerator parameters. Four cases are considered: (1) $V'_m = 1$ MV/m, $f_c = 0.4$, $\Delta B = 2.5$ T, (2) $V'_m = 2$ MV/m, $f_c = 0.8$, $\Delta B = 2.5$ T, (3) $V'_m = 1$ MV/m, $f_c = 0.8$, $\Delta B = 2.5$ T and (4) $V'_m = 1$ MV/m, $f_c = 0.8$, $\Delta B = 1.25$ T. The corresponding accelerator parameters are given in Table 2. The voltage gradients have been chosen to be consistent with the induction cell and voltage insulation requirements. An induction cell voltage V_0 = 50 kV has been assumed for all cases. The fractional width of an induction cell relative to the width of the induction core that is allowed for mechanical structure (f_s in eqn. 10) has been set equal to 0.2. The cell capacitances given in Table 2 are half due to internal structure and half external. The induction core magnetic material is assumed to be Metglas 2605SC with packing fraction $pf = 0.75$.

Where it makes a difference, the parameters tabulated in Table 2 are for the last cell which has the largest volt-second requirement (Fig. 5). For the $V'_m = 1$ MV/m cases the axially averaged volt-seconds per cell is reduced by 14% and for $V'_m = 2$ MV/m by 10.1%. If this reduction in volt-seconds were accomplished simply by reducing the radial width of the induction cores, keeping the axial length and voltage V_0 the same for all cells, then the total power and weight for the four cases in Table 2 are reduced by 18%, 11.6%, 15.6% and 17.2%. The numbers given for these totals in the last two lines of Table 2 reflect this reduction so the total power is not equal to the number of cells times the power per cell shown in Table 2 etc..

The first case in Table 2 was chosen as an example of minimum interference between focusing solenoids and induction cells with the cells only occupying a fraction $f_c = 0.4$ of the entire accelerator length (Fig. 3a). This case also puts the

TABLE 2. Induction linac parameters for four choices of accelerating gradient V'_m and axial length core fraction f_c: (1) $f_c = 0.4$, $V'_m = 1$ MV/m, $\Delta B = 2.5$ T, (2) $f_c = 0.8$, $V'_m = 2$ MV/m, $\Delta B = 2.5$ T, (3) $f_c = 0.8$, $V'_m = 1$ MV/m, $\Delta B = 2.5$ T and (4) $f_c = 0.8$, $V'_m = 1$ MV/m, $\Delta B = 1.25$ T.

Parameter	(1)	(2)	(3)	(4)
Voltage gradient, MV/m	1.0	2.0	1.0	1.0
Cell voltage, kV	50	50	50	50
Axial length fraction for cores, f_c	0.4	0.8	0.8	0.8
Cell length, cm	1.82	1.81	3.64	3.64
Accelerator length, m	700	350	700	700
Number of cells	15,383	15,506	15,383	15,383
Rep. rate, Hz	30	30	30	30
Pulse length, nsec[a]	205.34	143.89	205.34	205.34
Cell voltage risetime, nsec[a]	85.2	61.71	85.2	85.2
Volt seconds per cell[a]	6.39e-3	4.41e-3	6.39e-4	6.39e-3
Core flux swing, T	2.5	2.5	2.5	1.25
dB/dt, T/μsec[a]	12.2	17.4	12.2	6.09
Core loss per cycle, kJ/m^3 [a]	2.4	3.4	2.4	0.6
Inside magnetic core radius, cm	30	42	39	39
Core axial length, cm	1.30	1.37	2.94	2.60
Core radial width, cm[a]	23.1	14.9	10.2	23.1
Dielectric gap width, mm	2.63	1.67	1.16	5.26
Cell capacitance, nF[a]	13.8	16.7	14.7	8.40
Core leakage resistance, Ohms[a]	10.9	6.21	10.3	17.9
Core leakage current, peak, kA[a]	2.51	4.26	2.67	1.53
Core weight per cell, kg[a]	43.0	34.6	45.6	105
Hysteresis energy loss per cell per pulse, J[a]	0.424	0.487	0.450	0.258
Capacitance charging energy per cell per pulse, J[a]	0.429	0.509	0.455	0.261
Mismatch energy per cell per pulse, J[a]	0.560	0.741	0.595	0.341
Total magnetic core weight, tonne[b]	542	474	592	1,337
Total avg. power, MW[b]	17.8	23.9	19.5	10.9

[a] The numbers given for these quantities are for the last cell. Axially averaged values are less by 10 to 20%.
[b] Computed for axially averaged cell parameters.

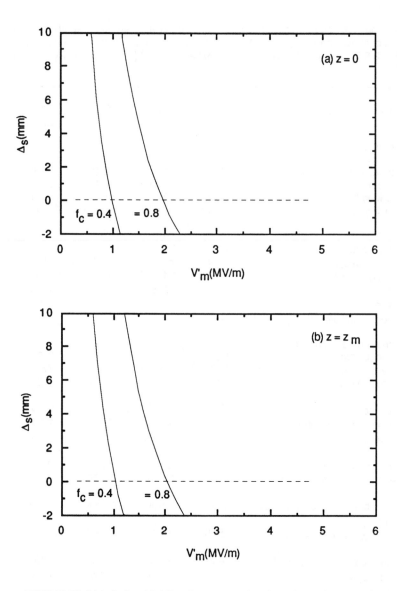

FIGURE 10. Margin for shielding the vacuum insulator from the muon beam channel; (a) at the beginning of the accelerator and (b) at the end. Conditions are the same as Fig. 7.

induction cores at the smallest possible radius $R_1 \sim 30$ cm and therefore minimizes their weight, 43 kg per cell) for given volt-seconds. The axial length of induction core for a single cell is 1.3 cm and the radial width 23 cm. The axial length is a factor of two or so less than one usually encounters with tape wound Metglas cores but seems doable. The length of the accelerator is 0.7 km, the total number of induction cells is ~ 15,400, the total weight of Metglas 2605SC is 542 tonne and the average pulsed power delivered to the induction cells is 17.8 MW. The energy delivered to a cell is approximately evenly divided between hysteresis loss (14.1 J), charging the cell capacitance (14.3 J) and mismatch power reflected from the cell (18.7 J). Unless the hysteresis loss is removed the induction core would reach the Curie temperature in a few hours so it is necessary to provide some cooling. A flow rate of the dielectric insulating fluid ~ 25 gm/sec would be sufficient to hold the maximum temperature rise inside the core to ~ 10°K. The core leakage current during a pulse ("peak current") is 2.5 kA, which is approximately three orders of magnitude greater than the muon current. The time average incremental power delivered to the muon beam is 25.9 kW, so dividing by the average pulsed power, the overall efficiency is 0.15 per cent.

For the second case the voltage gradient is doubled by increasing the fraction of accelerator length occupied by induction cells to $f_c = 0.8$ (Fig. 3b). This shortens the length of the accelerator to 0.35 km and reduces the pulse width from 205 nsec to 144 nsec. The reduced pulse width results in a decrease in volt-seconds required for the induction cores and a decrease in their weight. The decrease in weight is partially offset by the larger inside diameter since now the vacuum insulators must fit over the outside of the solenoid structure. Furthermore, the shorter pulse results in a higher hysteresis loss per unit volume and the average power increases 34% to 23.9 MW.

The third case is similar to the second with $f_c = 0.8$ but the accelerating gradient has been reduced to $V'_m = 1.0$ MV/m so the length is again 0.7 km. The main change here is that the axial length of the induction core is increased to 2.9 cm which is in the range of what is usually encountered for tape wound Metglas cores. Because the cores have larger radius their weight and hysteresis loss are slightly greater than the first case with the same gradient.

For the fourth case in Table 2, the total magnetic induction flux swing of the induction core material has been reduced by one half while $V'_m = 1$ MV/m and $f_c = 0.8$. The main motivation for this case is to reduce the magnitude of the hysteresis loss. The total average pulsed power is about a factor of two less than the third case, however the weight of Metglas has increased slightly more than a factor of two.

The cases presented in Table 2 have been chosen to give some idea of the tradeoffs that are possible. Ultimately the choice of parameters will depend on minimizing the cost and meeting the requirements imposed by the rest of the μ^+–μ^- accelerator complex. The induction accelerator described in this paper has mostly been thought of as being applied to separate μ^+ and μ^- production channels, in which case two of them are required. However, provided the μ^+ and μ^- pulses are separated in time it is conceivable that a single accelerator structure driven by a bipolar pulsed power system could be used for μ^+ and μ^-. For this approach the pulsed power system would necessarily be replicated for each polarity so the cost savings would only be in the accelerator structure.

ACKNOWLEDGMENTS

I would like to thank W. Barletta for encouraging me to have a look at the possibility of using an induction linac for phase rotation in a $\mu^+ -\mu^-$ collider. In addition I have benefited from helpful discussions with A. Faltens, E. Hoyer and L. Reginato on various aspects of this paper. J. Gallardo supplied the Monte Carlo muon production data.

REFERENCES

1. R.Palmer, D. Neuffer and J. Gallardo, *A Practical High-Energy High-Luminosity $\mu^+-\mu^-$ Collider*, Proc. of the Physics Potential and Development of a $\mu^+ -\mu^-$ Collider, (1995). W. Barletta is quoted in this paper as making the suggestion of using an induction linac rather than a low frequency rf linac for phase rotation.
2. J.G. Gallardo, private communication (1995).
3. Allied-Signal, Parsippany, New Jersey.
4. C.H. Smith, *Applications of Amorphous Magnetic Materials at Very-High Magnetization Rates*, Proc. of Magnetism and Magnetic Materials Conference, Boston (1989).
5. R.B. Miller, Intense Charged Particle Beams, Plenum, New York (1982).

IONIZATION COOLING AND MUON COLLIDER

A.N. Skrinsky
Budker Institute of Nuclear Physics, Novosibirsk

Abstract

Short review of ionization cooling features, important for preparation of muon beams for high luminosity collider, is presented. Some technical aspects to obtain ultimate efficiency of muons production, including polarized ones, are considered briefly.

The prospects to reach high energy high luminosity muon collider with the use of ionization cooling was considered at Novosibirsk INP still in 1960s, and mentioned in G.Budker's talks at the International Accelerator Conference (Erevan, 1969) and at the International High Energy Physics Conference (Kiev, 1970). Than, at the International Seminar on High Energy Physics Prospects (Morges, 1971) /1/ and at International Conference on High Energy Physics (Madison, 1980) /2/ we have presented a full list of necessary steps. In details, the cooling process we considered and main estimations for muon collider made in 1981 /3/.

Let us consider now some of the most important aspects of ionization cooling process with muon collider in mind, and first of all decrements, equilibrium emittances and muon losses in the process.

1. General formula for the 3-dimensional sum of emittance decrements δ_Σ, which results due to dissipative particle energy loss of power P_{fr}, is as following /3/:

$$\delta_{\Sigma 0} := \frac{2 \cdot P_{fr}}{P_\mu \cdot v_\mu} \cdot \left(1 - \frac{P_{\mu long}}{2 \cdot v_\mu} \frac{d}{dp_{\mu long}} v_\mu \right) + \frac{1}{v_\mu} \cdot \frac{d}{dp_{\mu long}} P_{fr}$$

where p_μ, $p_{\mu long}$ are particle momentum and its longitudinal projection, respectively, v_μ - velocity of particle.

For ionization energy losses the power (in CGSE) is

$$P_{fr}(\beta_\mu) := 4 \cdot \pi \cdot r_e^2 \cdot m_e \cdot c^2 \cdot N_e \cdot \frac{c}{\beta_\mu} \cdot \left[\ln \left[\frac{2 \cdot m_e \cdot c^2 \cdot \beta_\mu^2}{I \cdot \left[1 + \frac{2}{\sqrt{1 - \beta_\mu^2}} \cdot \frac{m_e}{M_\mu} + \left(\frac{m_e}{M_\mu}\right)^2 \right] \cdot \left(1 - \beta_\mu^2\right)} \right] - \beta_\mu^2 \right]$$

where N_e is the electron density in the target matter, $\beta_\mu c$ - the muon velocity.
This expression is valid for muon velocities much higher than mean electron velocities in target atoms.

For decrements sum expressed in cm^{-1} (assuming mean energy losses compensated by external electric field, the beam axially focused in transversal direction and autophased in longitudinal one) independently on the beam focusing structure at ionization channel, we obtain roughly:

$$\delta_{\Sigma 0}(\beta_\mu) := \frac{2 \cdot P_{fr}(\beta_\mu) \cdot \sqrt{1 - \beta_\mu^2}}{\left(M_\mu \cdot c^3\right) \cdot \beta_\mu^3} \cdot \left(\frac{1 + \beta_\mu^2}{2}\right) + \frac{\left(1 - \beta_\mu^2\right)^{\frac{3}{2}}}{\left(M_\mu \cdot c^3 \cdot \beta_\mu^2\right)} \cdot \left(\frac{d}{d\beta_\mu} P_{fr}(\beta_\mu)\right)$$

For muon beam, traveling in lithium, this formula gives dependence of the decrements sum inverse (the "cooling length of 6-dimensional emittance") on muon kinetic energy, presented at Fig. 1.

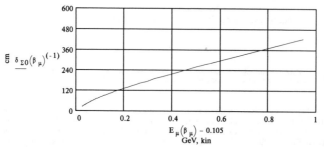

Fig. 1

2. Equilibrium transversal emittance of cooled muon beam directly influences the resulting luminosity of muon collider, which is inverse proportional to the emittance.

2.1. While traveling in matter, charged particle is influenced by Coulomb scattering, also. The joint action of energy losses and multiple scattering gives equilibrium angular spread. If decrement is of 1/3 of $\delta_{\Sigma 0}$, this spread will be

$$\theta_{mean}^{\ 2} := \frac{12 \cdot \pi \cdot e^4 \cdot N_e \cdot Z \cdot L_c \cdot \left(1 - \beta_\mu^{\ 2}\right)}{M_\mu^{\ 2} \cdot c^4 \cdot \beta_\mu^{\ 4}} \cdot \delta_{\Sigma 0}\left(\beta_\mu\right)^{(-1)}$$

where L_c is Coulomb logarithm, Z - the nuclei charge of the target.

If the beam travels in axially focusing channel with beta-function inside the matter equal to β_0, this equilibrium angular spread gives equilibrium (normalized) transversal emittance $\varepsilon_{neq}(=a^2/\beta_0)$:

$$\varepsilon_{neq}\left(\beta_\mu, \beta_0\right) := \frac{12 \cdot \pi \cdot e^4 \cdot N_e \cdot Z \cdot L_c \cdot \sqrt{1 - \beta_\mu^{\ 2}}}{M_\mu^{\ 2} \cdot c^3 \cdot \beta_\mu^{\ 2}} \cdot \delta_{\Sigma 0}\left(\beta_\mu\right)^{(-1)} \cdot \beta_0$$

For very low beta-value of 1 cm, the energy dependence of transversal emittance is given at Fig. 2

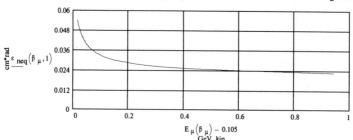

Fig. 2

We see, in case of constant beta-value, at higher energy the equilibrium normalized emittance does not depend on the cooling energy.

2.2. Let us evaluate the minimal transversal emittance reachable for muon beams via ionization cooling. Under conditions mentioned above, the best solution seems to be the use of lithium cylinder with focusing (pulsed) current. The beta-value is limited by magnetic field at the surface of the cylinder. For H_{max}=100 kGs and a_{li}=0.5 mm the dependence of ε_{neq} on muon kinetic energy is given at Fig. 3.

134

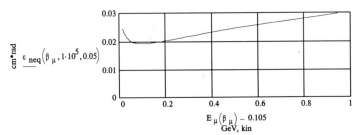

cm*rad $\underline{\varepsilon_{neq}\left(\beta_{\mu}, 1\cdot10^5, 0.05\right)}$

$\underset{\text{GeV, kin}}{E_{\mu}\left(\beta_{\mu}\right) - 0.105}$

Fig. 3

2.3. Let us mention /4/, that the using of ionization cooling in (continuous) high field solenoid with comparable beta-function (it corresponds to Larmor radius), gives the same equilibrium angular spread, but it does not lead to the shrinking of distances between spiral centers. Moreover, transversal size, and hence the muon beam emittance grows - slowly, but indefinitely - due to the diffusion of this centers because of multiple scattering.

3. While traveling in the cooling media, muons experience losses due to angular single and multiple scattering under presence of transversal acceptance limitation. Hence, some muons will be lost in the process, and the full path length in the matter is limited.

3.1. Assuming the acceptance angle equal to θ_A, we can evaluate the life-path in the target matter about

$$x_{life}\left(\beta_{\mu}, \theta_A\right) := \frac{1}{2\cdot\pi}\cdot\frac{1}{r_e^2\cdot N_e}\cdot\left(\frac{M_{\mu}}{m_e}\right)^2\cdot\frac{1}{Z}\cdot\frac{\beta_{\mu}^4}{1-\beta_{\mu}^2}\cdot\theta_A^2$$

or if normalized acceptance ε_{nA} of cooling channel is given:

$$x_{life}\left(\beta_{\mu}, \theta_A\right) := \frac{1}{2\cdot\pi}\cdot\frac{1}{r_e^2\cdot N_e}\cdot\left(\frac{M_{\mu}}{m_e}\right)^2\cdot\frac{1}{Z}\cdot\frac{\beta_{\mu}^3}{\sqrt{1-\beta_{\mu}^2}}\cdot\beta_0^{\varepsilon}\cdot nA$$

Such a value of life-path corresponds to number of coolings prior 1/2.7 fraction of muons is lost, as presented at Fig. 4 (for the current carrying lithium cylinder mentioned above):

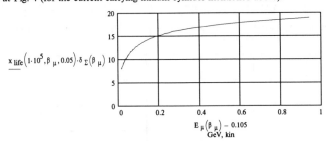

$\underline{x_{life}\left(1\cdot10^5, \beta_{\mu}, 0.05\right)\cdot\delta_{\Sigma}\left(\beta_{\mu}\right)}$

$\underset{\text{GeV, kin}}{E_{\mu}\left(\beta_{\mu}\right) - 0.105}$

Fig. 4

The acceptance needed to loose not more than κ_{lim} fraction of muons due to single scattering while performing n_{cool} coolings, corresponds to fulfilling of the inequality

$$n_{cool}{}^*x_{life}{}^*\delta_{transv} > 1/\kappa_{lim} .$$

135

3.2. The losses due to multiple scattering, very roughly, can be evaluated as following. Assuming beam distribution already equilibrium and thus Gaussian with mean θ^2 corresponding to normalized emittance ε_{neq}, we can evaluate the fraction of particles lost per 1 loss of full energy as fraction sitting in the distribution tail greater then acceptance angle, corresponding to the channel acceptance ε_{nA}.

Consequently, to lose less than fraction $\kappa_{lim} \ll 1$ of beam intensity in transferring of energy ΔE_Σ to ionization, the transversal normalized acceptance should be greater than

$$\varepsilon_{nA} := \varepsilon_{neq} \cdot \ln\left(\frac{1}{\kappa_{lim}} \cdot \frac{\Delta E_\Sigma}{E_\mu}\right)$$

For reference, the normalized acceptance of the current carrying lithium cylinder mentioned above is presented at Fig. 5.

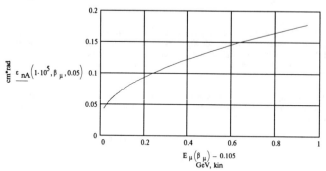

Fig. 5

4.1. The "natural" longitudinal cooling, generated by the dependence of ionization losses on muon energy, becomes positive above 200 MeV (kinetic) - below this level the dependence produces antidamping. But even there this cooling rate is around 7 times slower than 1/3 of sum of decrements.

To reach the faster cooling, we need to arrange the transversal energy dispersion and corresponding gradient in ionization losses (the higher energy muons shall lose more). In this case, muon beam can be cooled at energy down to 50 MeV (kinetic).

The longitudinal cooling length for these 2 cases is presented at Fig. 6 (pay attention to the factor 1/5 for the scale of the upper curve for the "natural" cooling!).

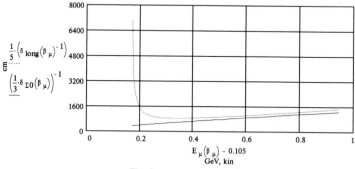

Fig. 6

4.2. The equilibrium energy spread, resulting due to balance of ionization cooling and fluctuations in ionization losses ("δ-electrons"), can be evaluated as

$$\Delta \, E\left(\beta_\mu\right) := \sqrt{2 \cdot \pi \cdot r_e^2 \cdot \left(\frac{m_e}{M_\mu}\right)^2 \cdot N_e \cdot \left(2 - \beta_\mu^2\right) \cdot \left(\delta_{\,\text{leff}}\left(\beta_\mu\right)\right)^{(-1)}}$$

where $\delta_{\text{leff}}(\beta_\mu)$ is longitudinal decrement arranged in the scheme used.

For both cases mentioned above, the energy dependence of this spread is presented at Fig. 7.

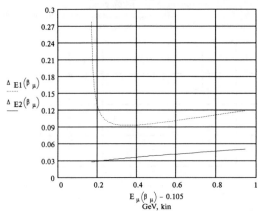

Fig. 7

4.3. Like for transversal motion, deviations of muons in longitudinal one - both in energy and in longitudinal position relative to the phase of autophasing oscillations - due to electron scattering can be larger then acceptance of the cooling installation. The domination of multiple or single scattering events depends mostly on maximal possible (relative) energy loss in a collision δ_{max}:

$$\delta_{\text{max}} = 2 \cdot \frac{m_e}{M_\mu} \cdot \frac{\beta_\mu^2}{\sqrt{1 - \beta_\mu^2}}$$

presented at Fig. 8.

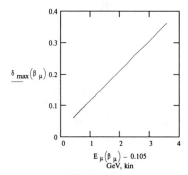

Fig. 8.

137

If δ_{max} is smaller than the channel acceptance δ_{acc}, the requirement for long enough life-path is the same as for the transversal case considered above: the equilibrium emittance should be several times smaller than the channel acceptance.

If δ_{max} is larger than δ_{acc}, the fraction κ of beam particles lost due to single scattering can be expressed as following:

$$\kappa = \frac{1}{2 \cdot L_i} \cdot \Delta_{ion} \cdot \left[\frac{1}{\delta_{acc}} - \frac{1}{\delta_{max}} \cdot \left(1 - \beta_\mu^2 \cdot \ln\left(\frac{\delta_{max}}{\delta_{acc}} \right) \right) \right]$$

where Δ_{ion} is the ratio of energy, transferred to ionization, to the muon energy.

As we can derive from Fig. 9 ($\delta_{acc}=0.1$, $\Delta_{ion}=3$), single scattering losses dominate above 1 GeV, and practically exclude possibility to use ionization cooling effectively at higher energies.

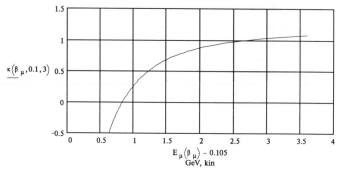

Fig. 9.

Longitudinal acceptance can be limited by geometric cuts of the channel or by chromatic aberration of its focusing system. The latter limitation seems to be the most serious one. The resulting energy spread by itself is not so important for further steps of acceleration and for collider operation - its absolute value, being (more or less) conserved, is quite small in the scale of the final energy.

5. The most difficult part of the cooling stages is the very last one, aimed to give the minimal transversal normalized emittance, which directly defines the collider luminosity. At this stage it is necessary not only to arrange the minimal beta-value - of the order of 1 cm - inside the matter giving ionization friction, of total length of more than 10 m. Simultaneously, we need to keep the (equilibrium) energy spread small enough, preventing the transversal emittance growth due to chromatic aberrations and the beam intensity losses due to the limited energy acceptance.

The better solution seems to be the focusing current carrying cylinder as ionizing target, as was mentioned above, but with special arrangement to redistribute of 1/3 of the decrement sum to the longitudinal direction. This aim can be reached by converting cylindrical target in to the spiral of few cm radius and introducing the high enough effective gradient of electron density along the dispersion induced by the spiral curvature.

For the preventing of losses in acceptances, it is worth to introduce along the spiral the additional quasi-uniform transversal magnetic field just giving proper curvature for the average energy muons inside the spiral target.

For the arranging of electron density gradient, it is worth to use as a target a bi-metallic cylinder (the outer half of the cylinder should be of beryllium, the inner half - of lithium). The lithium-beryllium boundary shall be made teeth-like, to make electron density gradient more or less constant. The important requirement is to arrange reasonably homogeneous conductivity across the cylinder.

138

6. The reaching of the ultimately low transversal emittance of muon beams is one part of the way to ultimate collider luminosity. The next part is the highest possible muon bunch intensity (the luminosity is proportional to the square of it). It is possible to use ionization cooling for rising number of muons effectively produced per incident proton - close to 1 muon/proton.

The idea is to arrange a multi-channel pion collection system.

Each channel can collect most efficiently pions, produced by a proton bunch, in some specific limited part of energy spectrum and transversal emittance. Then, pions enter separate decay channels with properly arranged and strong enough focusing. The out-coming bunches of muons are cooled preliminary; may be, it would be useful to equalize the mean energies of mouns in each channel by corresponding RF acceleration/deceleration.

Afterwards, all the muon bunches are summarized transversally (with the use of septum magnets) in a single bunch of proportionally larger emittance. Then, the resulting muon bunch is cooled up to final ultimately low emittance in a channel described above.

7. This approach brings not only practical possibility to reach ultimately high efficiency of proton/muon conversion. It gives also a natural way to produce polarized muon beams.

To reach this, in each channel of narrow enough pion energy spectrum and transversal acceptance, it is necessary to reject lower fraction of muon spectrum, using energy analizer at the exit of each channel, and then to repeat all the steps described in the previous section.

In this case, loosing half of the muon intensity, we shall obtain quite good muon polarization degree.

References.

1. A.Skrinsky
   Colliding Beams Program in Novosibirsk
   International Seminar on Prospects in High Energy Physics; Morges, 1971
2. A.Skrinsky
   Accelerator and instrumentation prospects of elementary particle physics.
   XX Intern. Conf. on High Energy Physics, Madison, 1980. New York, 1981, v.2, p.1056-1093;
   Uspekhi Fiz. Nauk, 1982, v.138, 1, pp.3-43
3. V. Parkhomchuk and A.Skrinsky
   Methods of cooling of charged particles beams.
   Phys. of Elementary Part. and Atomic Nucl., 1981, 12, pp. 557-613.
   Sov.J.Part.Nucl. 12(3), May-June 1981, pp. 223-232
4. T. Vsevolozhskaya
   Kinetics of ionization cooling of muons (Budker INP, Novosibirsk)
   The talk at this Workshop.

Achievable Transverse Emittance of Beam in Muon Collider

V. I. Balbekov

Insitute for High Energy Physics

Protvino, Moscow Region, 142284, Russia

Abstract

Ionization cooling of a muon beam by means of Li or Be lenses-absorbers is considered. It is shown that an achievable transverse emittance and radius of the beam depend on a current density of the lenses. The optimal scheme of the cooling is suggested.

Two factors are responsible for a final transverse emittance of muon beam subjected to ionization cooling: effective friction force due to energy losses, and diffusion of particles caused by the multiple Coulomb scattering [1,2]. The role of the second factor increases when β-function is large; therefore, it is necessary to use a lattice with an extremely small β-functions on both horizontal and vertical directions at final stages of the cooling process to achieve a minimal emittance of the cooled beam. The most suitable devices for this purpose are Li or Be lenses with high current density which can provide both a very strong focusing in transverse directions and ionization cooling of the beam with a minimal contribution from the scattering.

Therefore, let us consider a cell shown in the next picture as a period of a facility for final cooling of a muon beam.

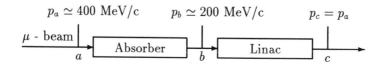

The beam passes over a Li or Be absorber of radius R along which a current J flows to produce a magnetic field with a gradient

$$G = \frac{2J}{cR^2}. \tag{1}$$

It provides a focusing both in horizontal and vertical directions. The average energy losses for ionization are compensated by re-acceleration in the linac. A few similar cells can be arranged in series to form a linear structure, or the beam may be forced to re-enter the cell repetitively with a special magnet system. If the latter is the case, we suppose this system to be dispersion-free through the section ab.

Without the scattering, unnormalized (to momentum) transverse emittance of the beam is kept constant within the absorber, while the normalized one is kept intact throughout the linac. Therefore, the total change of the normalized r.m.s. beam emittance after a single pass through the cell can be represented as

$$\epsilon_c = \epsilon_a \frac{x_b}{x_a} + \Delta\epsilon \tag{2}$$

where x_a and x_b are the normalized momenta of the particles ($x = p/mc$) before and after the absorber, respectively. A calculation gives the following formula for an increase of the r.m.s. beam emittance due to the Coulomb scattering:

$$\Delta\epsilon = \frac{mc^2 x_b}{2} \int_{x_b}^{x_a} \frac{\langle d\theta^2/ds \rangle}{\langle dE/ds \rangle} \frac{\beta(x)x}{\sqrt{1+x^2}} \, dx. \tag{3}$$

$\beta(x)$ is an amplitude function of transverse oscillations within the absorber-lens which, given gradient (1), reads:

$$\beta(x) = R\sqrt{\frac{xJ_0}{J}} \tag{4}$$

where $J_0 = mc^3/2e = 1762$ kA. The terms $\langle dE/ds \rangle$ and $\langle d\theta^2/ds \rangle$ are the average ionization energy losses and a mean square of the angle of the multiple Coulomb scattering per unit of length at momentum x. On extracting the most significant dependence of these functions on the momentum, we obtain [3]:

$$\left\langle \frac{dE}{ds} \right\rangle = \frac{mc^2(1+x^2)}{x^2 L_E} f(x), \tag{5}$$

$$\left\langle \frac{d\theta^2}{ds} \right\rangle = \frac{1+x^2}{x^4 L_\theta}, \tag{6}$$

Function $f(x) \simeq 1$ describes a logarithmic growth of the energy losses with momentum and effect of density; by definition, $f(3) = 1$ ($x = 3$ is approximately a point of minimum energy loss). Then the specific length L_E and L_θ are: $L_E = 1.4\text{m}$, $L_\theta = 88\text{m}$, $L_E/L_\theta = 0.016$ for Li; $L_E = 0.40\text{m}$, $L_\theta = 20\text{m}$, $L_E/L_\theta = 0.020$ for Be. It is easy to get from Eqs.(3)-(6) the following expression for the emittance increase:

$$\Delta\epsilon = R\frac{L_E}{L_\theta}\sqrt{\frac{J_0}{J}}\sqrt{\frac{x_b}{x_a}}(\sqrt{x_a} - \sqrt{x_b})F(x_a, x_b), \tag{7}$$

where

$$F(x_a, x_b) = \frac{\sqrt{x_a x_b}}{\sqrt{x_a} - \sqrt{x_b}} \int_{x_b}^{x_a} \frac{dx}{2f(x)\sqrt{x^3 + x}} \simeq 1 \tag{8}$$

Using Eq.(2) recurrently, one can find the r.m.s. beam emittance after n similar cells:

$$\epsilon_n = \Delta\epsilon + \lambda\epsilon_{n-1} = ... = \Delta\epsilon(1+\lambda+...+\lambda^{n-1}) + \epsilon_0\lambda^n = \epsilon_0\lambda^n + \Delta\epsilon\frac{1-\lambda^n}{1-\lambda}, \tag{9}$$

where $\lambda = x_b/x_a$. Therefore, the ultimate magnitude of the emittance at $n \to \infty$ is

$$\epsilon_\infty = \frac{x_a\Delta\epsilon}{x_a - x_b} = R\frac{L_E}{L_\theta}\sqrt{\frac{J_0}{J}}\frac{\sqrt{x_a x_b}}{\sqrt{x_a} + \sqrt{x_b}}F(x_a, x_b). \tag{10}$$

Corresponding r.m.s. beam radius which can be obtained with formula

$$\sigma = \sqrt{\frac{\epsilon\beta(x)}{x}} \tag{11}$$

has a maximal value in the point b, and with of Eqs.(4) and (10) taken into account can be written in the form

$$\sigma_\infty = k_\infty R, \qquad (12)$$

where k is a beam-to-absorber filling ratio:

$$k_\infty = \sqrt{\frac{L_E J_0}{L_\theta J} \frac{F(x_a, x_b)}{1 + \sqrt{x_b/x_a}}}. \qquad (13)$$

The most important conclusions following from Eqs.(10) and (13) are: the beam emittance depends only on the current density of the lens-absorber, whereas the filling ratio depends only on its total current.

Consider as an example the FNAL Li lens ($R = 1$ cm, $J = 500$ kA). Additionally, suppose $x_a = 4$, $x_b = 2$ which corresponds to deceleration of the particles from kinetic energy of 330 MeV to 130 MeV by absorber. In this case the factor $F = 0.94$, and Eqs.(10) and (13) give $\epsilon_\infty = 230$ mm·mrad (25% more for Be lens-absorber with the same current), and $k_\infty = 0.18$. Such an emittance exceeds essentially the required one for muon-muon collider, and should be decreased several times. It is seen from eq.(10) that an increase of the lens current is not very effective for this purpose, while a decrease of its diameter at the same current leads to the proportional decrease of the beam emittance and its radius.

Therefore, consider a modification of the cooling system which differs from above one only in that the lens radii R_n depend on cell number n whereas their other parameters are the same in all the cells. First of all, find the necessary rule to change the magnitudes R_n.

Suppose that the beam normalized emittance for the n-th cell is ϵ_{n-1} at the entrance and ϵ_n at the exit. It follows from Eqs.(4) and (11) that the r.m.s. beam radius at the entrance is

$$\sigma_{an} = \sqrt{\epsilon_{n-1} R_n \sqrt{J_0/J x_a}}. \qquad (14)$$

Requiring the filling ratio to have the same magnitude within all the lenses ($\sigma_{an} = k R_n$), we obtain:

$$R_n = \frac{\epsilon_{n-1}}{k^2} \sqrt{\frac{J_0}{J x_a}}. \qquad (15)$$

Substitution of this quantity into Eqs.(7) and (2) gives the emittance magnitude at the end of n-th cell:

$$\epsilon_n = \kappa \epsilon_{n-1}, \qquad (16)$$

where a damping coefficient κ does not depend on the cell number:

$$\kappa = \frac{x_b}{x_a} + \frac{L_E J_0}{L_\theta J k^2}\left(\sqrt{\frac{x_b}{x_a}} - \frac{x_b}{x_a}\right)F(x_a, x_b). \qquad (17)$$

It means that the beam emittance varies in such a system exponentially:

$$\epsilon_n = \epsilon_0 \kappa^n. \qquad (18)$$

Of course, one must require $\kappa < 1$ to provide a damping, setting a restriction

$$k > \sqrt{\frac{L_E J_0}{L_\theta J}\frac{F(x_a, x_b)}{1 + \sqrt{x_a/x_b}}}, \qquad (19)$$

As an example, consider lenses with current 500 kA at $x_a = 4$, $x_b = 2$. In this case the allowable filling ratio is 0.15, so a value of $k = 0.25$ is acceptable, providing $\kappa = 0.676$. Suppose the beam emittance is 10^4 mm·mrad at the entrance to first cell. According to Eq.(15), first lens radius must be 150 mm to accept such a beam. The following evolution of emittance and the lens radii required are shown in the next table. Magnitudes of β-functions at the end of each lens (where they are minimal) are presented as well. These are important parameters because it is necessary to afford a matching of entrance-exit of each lens with the linac optics which is not a simple matter at small β-functions.

Cell No	0	1	2	4	6	8	10	12	14
ϵ, mm·mr	10^4	6760	4570	2090	954	436	199	91.1	41.6
R, mm	-	150	101	46.3	21.2	9.68	4.42	2.02	0.92
β_b, mm	-	398	269	123	56.2	25.7	11.7	5.36	2.45

Thus, the muon beam emittance can be limited, finally, by technical possibilities to manufacture very thin Li or Be lenses with high current density. Take, for instance, the minimal radius of the lenses as 10 mm. Then one would obtain the beam emittance about 440 mm·mrad after 8-th cell. 5 - 6 additional cells like 8-th one allow to decrease the emittance to 230 mm·mrad, which is limiting value in this example. In any case, the final emittance depends only on the lens current density, and can not be less than the value defined by Eq.(10) where R and J are stand for the latest absorbers of the chain.

Certainly, the possibility to decrease ϵ_∞ by using a stronger deceleration of muons in absorber (to decrease x_b) is an attractive way, but it contradicts to a cooling of the longitudinal phase volume. An optimization of the momentum x_b should be made in the frames of a more general scheme including devices for a coupling of longitudinal and transverse motion to provide an optimal distribution of the decrements over them. It is not a goal of this work, but the next short remark seems to be important.

Because of statistical fluctuations, the absorber contributes significantly to the beam momentum spread. Each a passage of the absorber increases r.m.s. value of it by

$$\Delta \frac{\sigma_p^2}{p_b^2} = \frac{2.2 \cdot 10^{-4}}{x_b^2} \left(\frac{\gamma_a^3 - \gamma_b^3}{3} + \gamma_a - \gamma_b \right) \tag{20}$$

where $\gamma = \sqrt{1 + x^2}$ is normalized energy of muon. It gives 0.035^2 at $x_a = 4$ and $x_b = 2$, so a significant part of transverse decrements should be transformed to a longitudinal one to prevent an increase of the momentum spread, at least.

References

[1] Yu. M. Ado and V. I. Balbekov, Atomic Energy, V. 31, p. 40, 1971.

[2] D. Neuffer, Particle Accelerators, V. 14, p. 75, 1983.

[3] A. N. Kalinovskii, N. V. Mokhov, and Yu. P. Nikitin, "Passage of High-Energy Particles trough Matter". AIP, NY, 1989.

Possible demonstration of ionization cooling using absorbers in a solenoidal field

R.C. Fernow, J.C. Gallardo, H.G. Kirk, T. Kycia, Y.Y. Lee, L. Littenberg,
R.B. Palmer, V. Polychronakos & I. Stumer
Brookhaven National Laboratory, Upton, NY 11973

D.V. Neuffer
CEBAF, Newport News, VA 23606

D. R. Winn
Fairfield University, Fairfield, CT 06430

R. Thun & R. Ball
University of Michigan, Ann Arbor, MI 48109

M. Marx & Y. Torun
SUNY at Stony Brook, Stony Brook, NY 11794

M. Zeller
Yale University, New Haven, CT 06511

ABSTRACT

Ionization cooling may play an important role in reducing the phase space volume of muons for a future muon-muon collider. We describe a possible experiment to demonstrate transverse emittance cooling using a muon beam at the AGS at Brookhaven National Laboratory. The experiment uses device dimensions and parameters and beam conditions similar to what is expected in an actual muon-muon collider.

1 Introduction

There has been considerable recent interest in the development of high energy colliders using muon beams[1]. Most of the proposed designs require a high intensity pion beam resulting from protons interacting in a stationary target. The muons are collected from the decaying pions. However, the resulting muon beam typically has much too large an emittance to give high luminosity in a collider.

It is possible to reduce the transverse emittance of the muon beam through the process of ionization cooling[2]. Energy loss in a material reduces both the transverse and longitudinal components of the momentum. Multiple scattering in the material is a competing process that effectively heats the beam, resulting in an increase of the emittance. Subsequent acceleration cavities restore only the longitudinal component, thus leading to a reduction of the transverse emittance of the muon beam.

Transverse ionization cooling, although discussed by Skrinsky and others[3] in the 70's has, as far as we know, never been observed. Its demonstration is important because such cooling appears to allow the design of high luminosity muon-muon colliders that may provide the best possibility for physics at CM energies above 1 TeV.

We are considering the possibility of developing a prototype cooling system and testing its performance in a muon beam at the AGS at Brookhaven National Laboratory. A muon momentum of 210 MeV/c would be ideal for the experiment.

2 Ionization cooling

In order to generate sufficient muons it is necessary to capture a very large fraction of the pions generated at the target. These pions, and the muons into which they decay, are then necessarily very diffuse (i.e. they have a very large emittance). Since luminosity is inversely proportional to the final beam cross section, the beam must be small. Therefore, some form of cooling to lower the emittance is essential.

The large mass of the muon compared to that of the electron prevents cooling by radiation damping, while the short lifetime of the muon prevents conventional stochastic or electron cooling. Fortunately, the process of ionization cooling, which because of their long interaction length is possible only for muons, can be used. In this process the muon loses transverse and longitudinal momentum by dE/dx in a material, such as lithium or beryllium, and then has the longitudinal momentum (but not the transverse momentum) restored in a subsequent RF cavity.

The combined effect is to reduce the beam divergence and thus the emittance of the beam. The process is complicated by the simultaneous presence of multiple scattering in the material, which acts as a source of "heat" and increases the emittance. The cooling effect can dominate for low Z materials in the presence of strong focussing fields. One solution being considered for the collider is to use rods of lithium or beryllium inside a solenoid magnet. The rod provides the energy loss, while the large aperture solenoid provides the required focussing.

While the physics of the cooling process is rather well understood, the technology of a system that achieves it is not. As we mentioned in the Introduction, such cooling has never been convincingly observed. If a muon collider based on ionization cooling is to be proposed, then we believe that it is essential that such cooling must first be demonstrated.

The normalized transverse emittance ϵ_n can be reduced by ionization energy loss dE/dz in a material at the rate[4]

$$\frac{d\epsilon_n}{dz} = -\frac{1}{\beta^2}\frac{dE}{dz}\frac{\epsilon_n}{E} \tag{1}$$

where $E = \gamma m_\mu c^2$ is the total energy of the muon. At a momentum corresponding to minimum ionization, dE/dz is 0.88 MeV/cm for lithium and 2.95 MeV/cm for beryllium. In practical units we can write Eq. 1 as

$$\frac{d\epsilon_n}{dz}\left[\frac{mm-mrad}{cm}\right] = -\frac{1}{\beta^2}\frac{dE}{dz}\left[\frac{MeV}{cm}\right]\frac{\epsilon_n\,[mm-mrad]}{E\,[MeV]} \tag{2}$$

The rod also introduces heating of the normalized transverse emittance due to multiple scattering at a rate[4]

$$\frac{d\epsilon_n}{dz} = \frac{\beta_\perp}{2\,\beta^3}\frac{E_s^2}{E\,m_\mu c^2}\frac{1}{L_r} \tag{3}$$

where $\beta_\perp$ is betatron focusing parameter, β is the relativistic velocity of the muon, $E_s \approx 14$ MeV is the characteristic energy for the projected scattering angle from multiple scattering theory, $m_\mu c^2$ is the rest energy of the muon (105.6 MeV), and L_r is the radiation length of the material (155 cm for lithium and 35.3 cm for beryllium).

In practical units Eq. 3 can be written as

$$\frac{d\epsilon_n \, [mm\text{-}mrad]}{dz \, [cm]} = \frac{60 \, \beta_\perp \, [cm]}{\beta^3 \, E \, [MeV]} \tag{4}$$

for lithium. The minimum achievable emittance, which occurs when the cooling rate equals the heating rate, is given by

$$\min \epsilon_N = \frac{\beta_\perp \, E_S^2}{2 \, \beta \, m_\mu \, c^2 \, L_R \, \frac{dE}{dz}} \tag{5}$$

In practical units

$$\min \epsilon_N \, [mm\text{-}mrad] = \frac{9280}{\beta \, L_R \, [cm] \, \frac{dE}{dz} \left[\frac{MeV}{cm} \right]} \, \beta_\perp \, [cm] \tag{6}$$

For a relativistic particle the coefficient of $\beta_\perp$ is 68.0 for lithium and 89.1 for beryllium. The betatron focusing parameter for a solenoid is given by

$$\beta_\perp = \frac{2 \, p_z}{e \, B} \tag{7}$$

or in practical units

$$\beta_\perp \, [cm] = 0.667 \, \frac{p_z \, [MeV/c]}{B \, [T]} \tag{8}$$

A reduction in transverse emittance is accompanied by a corresponding increase in longitudinal emittance due to straggling. This introduces a spread in the mean energy loss in the material given by[4]

$$\frac{d(\sigma_E^2)}{dz} = 4 \, \pi \, (r_e m_e \, c^2)^2 \, n_e \, \gamma^2 \, (1 - \tfrac{1}{2}\beta^2) \tag{9}$$

where r_e is the classical radius of the electron, $m_e c^2$ is the rest energy of the electron, γ and β are the usual relativistic factors, and $n_e = ZN_A\rho/A$ is the number of electrons per unit volume of absorber material, where Z, A, ρ are the atomic number, atomic weight and density of the material.

3 Outline of the proposed experiment

We propose building a prototype cooling cell with properties as close as possible to those required in a muon collider. We would measure the performance of the device and compare it with predictions from the Monte Carlo programs we are using to design the collider. Since the cooling section may represent a large fraction of the total cost of a muon collider, it is essential that the design Monte Carlo accurately model the achievable cooling. We also need to check that muon losses in the focusing fields and absorbing material do not exceed the predicted amounts.

The proposed baseline experiment would use a cylindrical piece of lithium, 12 cm in radius and 60 cm in length as the cooling rod. The solenoid with a maximum central field strength of 7 T would focus the muons in the interior of the rod. The absorber material would be replaced at some point with different lengths of absorber and with Be or LiH absorbers in order to further check the accuracy of the Monte Carlo predictions. Beryllium would produce more efficient cooling than lithium for some regions of the muon parameters.

In order to simulate the conditions in the muon collider as closely as possible, the momentum of the incident muon should be around 210 MeV/c. For a 0.6 m lithium rod the final momentum would be about 140 MeV. The focusing $\beta_\perp$ in the rod should be as small as possible, since Eq. 5 shows that the minimum emittance that can be reached is proportional to $\beta_\perp$.

The layout of the proposed experiment is shown in Fig. 1. A large, diffuse muon beam passes through a hadron absorber and enters an approximately 6 m long experimental area. The dipole D1 together with the position measuring detectors P1-5 are used to measure the momentum of each incoming track. A fraction of the beam will be captured by the solenoid magnet S1 and focused into an absorber. The momentum of the muons leaving the solenoid is measured using dipole magnet D2 and detectors P6-10. Trigger counters would tag tracks entering the absorber. A time of flight system would be necessary to identify pion and electron background in the muon beam. The dipole D1 would have to be adjusted to examine a series of different central values for the incident momentum. The momentum spread could be determined by software selection of the tracks used in the analysis.

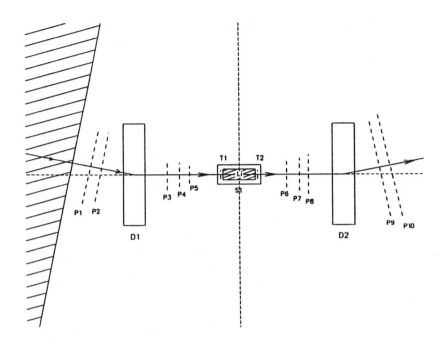

Fig.1 . Layout of the proposed experiment. D1,2 are dipole magnets, S1 is a solenoid magnet, P1-10 are position measuring detectors.

In the following we consider an example case with an initial normalized emittance of 4000 mm-mrad and a 7 T solenoid field. Some properties of the magnets are listed in Table 1. The dipole width and length are appropriate to existing 72D18 magnets at the AGS. We assume a bend angle in the dipoles of 0.2 rad.

Table 1 Magnet summary				
Magnet	length	width	gap	radius	B_o
	[cm]	[cm]	[cm]	[cm]	[T]
D1	46	183	109	-	0.30
S1	80	-	-	15	7.00
D2	46	183	148	-	0.20

a 1σ radius of about 3 cm inside the lithium. PARMELA results for the change in transverse normalized emitance as a function of axial distance z are shown in Fig. 2 and summarized in Table 2.

Table 2 Transverse cooling experiment			
	IN	OUT	
T	130	70	MeV
p	211	140	MeV/c
β	0.894	0.798	
γ	2.23	1.66	
ε_N	4000	3100	mm-mrad
ε_G	2006	2340	mm-mrad
$\beta_\perp$ (sol)	20.1	13.3	cm
σ_X	20	18	mm
$\sigma_{X'}$	100	133	mrad

Under the conditions listed in Table 2, the cooling rate is always larger than the heating rate and the normalized emittance of the beam leaving the rod is smaller than the normalized emittance entering it. However, the geometric emittance increases after traversing the rod. In an actual muon collider this stage must be followed by a reacceleration section that restores the starting momentum in order to get cooling of the geometric emittance.

One of the design requirements for the experiment is the ability to measure individual muon tracks. This will permit software "control" of the effective cooling rate. Software selection of initial beam tracks will allow us to reconstruct final beam emittance as a function of initial beam emittance.

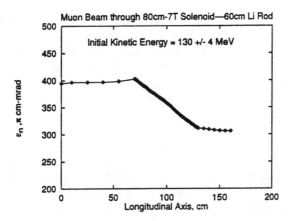

Fig. 2 PARMELA simulation of the normalized transverse emittance as a function of axial distance. The incident muon beam has a normalized emittance of 4000 mm-mrad.

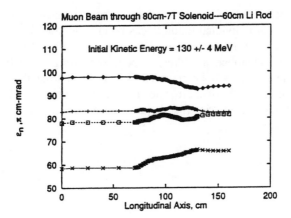

Fig. 3 PARMELA simulation of the change in normalized transverse emittance as a function of axial distance for smaller initial values of the emittance.

The experiment can be reanalyzed for smaller initial transverse emittances. As the initial emittances decrease, the cooling in the lithium becomes less effective relative to the heating and the net decrease in normalized emitance also drops. Fig. 3 shows that this cooling configuration should still produce net cooling until the initial emittance reaches about 820 mm-mrad. These measurements are important since they demonstrate the minimum emittance achievable for a given field strength and absorber length.

The region of the incident beam phase space for an initial emittance of 4000 mm-mrad that leads to cooling is shown in Fig. 4. The tilted ellipse represents a converging beam. Incident beams with smaller initial emittances are subsets of the data shown in the figure. The basic idea would be to use an incident beam phase space that completely covers the cooling region in Fig. 4. Then, since individual trajectories are measured, we would select by software those trajectories that lie in the cooling region. These trajectories should produce a known amount of normalized emittance reduction, as indicated in Fig. 2 and Table 2. Then subsets of these trajectories could be examined and the net cooling compared with the predictions shown in Fig. 3. Trajectories in the core area of Fig. 4 could also be examined to confirm that no cooling occurs.

The experimental test of cooling would require accurately measuring the position, angle, and momentum of individual muon tracks before and after the rod in order to measure the initial and final emittances. The positions have to be measured at 10 locations. Since the chambers are likely to be in the strong fringe field of the solenoid, a good field map will be required. We need at least three x and y measuring planes on each side of the rod to measure the curvature of the track. The six chambers closest to the rod require measurements with moderate resolution (≈ 300 µm).

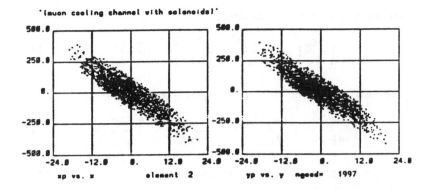

Fig. 4 PARMELA simulation of the 1σ geometric incident muon beam phase space that leads to the normalized emittance cooling shown in Fig. 2. (a) x vs x'; (b) y vs y'.

The transverse cooling experiments require measuring the normalized muon beam emittance before and after the rod. The normalized emittance at a beam waist is given by

$$\epsilon_N = \frac{\overline{p}}{m_\mu c} \, \sigma_X \, \sigma_\theta \tag{10}$$

where σ_X and σ_θ are the values for the entrance or the exit to the rod, p is the mean momentum for the distribution and m_μ is the mass of the muon.

155

The fractional error on the measurement of ϵ_N

$$\left(\frac{\delta\epsilon_N}{\epsilon_N}\right)^2 = \left(\frac{\delta\sigma_x}{\sigma_x}\right)^2 + \left(\frac{\delta\sigma_\theta}{\sigma_\theta}\right)^2 + \left(\frac{\delta p}{p}\right)^2 \qquad (11)$$

can be expressed entirely in terms of the position resolution as

$$\left(\frac{\delta\epsilon_N}{\epsilon_N}\right)^2 = \left(\frac{\delta\sigma_x}{\sigma_x}\right)^2 + \frac{2\,(\delta\sigma_x)^2}{D_\theta^2\,\sigma_\theta^2} + \frac{4\,(\delta\sigma_x)^2}{\alpha^2\,D_p^2} \qquad (12)$$

where D is the distance between chambers, and α is the dipole bend angle for the momentum measurement.

The expected fractional change in normalized emittance for the case described in Table 2 is about 23%. However, in order to measure the smaller effects expected with lower initial emittances, we would like to determine $\Delta\epsilon_N/\epsilon_N$ to an accuracy of 0.1%. Eq. 12 gives the accuracy required on the measurement of σ_x. This can in turn be used to determine the chamber resolution required once the muon flux is known.

The position chamber sizes are determined by the beam divergence angles. The detectors for the beam incident on the experiment (P1-5) cover a position range of up to $\pm 3\sigma$. In order to fully understand the performance of the cooling cell, we want to be sure that the chambers after the experiment can detect far into the tails of the angular distribution. For that reason these detectors (P6-10) cover a range of up to $\pm 4\sigma$. Detectors P3-8 must consist of an x and a y measuring plane. The other detectors only need to measure x.

4 Requirements for the AGS beam

In order to measure individual tracks we propose using the slow extracted beam from the AGS. The useful muon rate should be larger than $\approx 10^2$ / spill so that a complete measurement can be made in about an hour, in order to minimize the effects of systematic errors. The useful rate should be smaller than $\approx 10^4$ / spill in order to use a fairly simple data acquisition system. The mean muon beam momentum at the experiment should be centered around 210 MeV/c. A free region of about 6 m is required for the experiment following a hadron absorber-muon diffuser block.

Table 3 summarizes the muon beam requirements.

Table 3	Muon beam requirements	
1σ beam radius	24	cm
1σ beam divergence	100	mrad
μ mean momentum	211	MeV/c
useful μ rate	10^2 - 10^4	μ / spill

The momentum following the experiments is measured with a dipole in the spectrometer area shown in Fig. 1. In order to guarantee that all tracks leaving the rod are measured, no quadrupoles are used in the spectrometer.

An appropriate incoming muon beam might be formed from pions produced by sending an incident proton beam into a target. This could be followed by a free space, pion decay region, which is sufficiently long to allow the muon beam to grow to the desired radius. Then the beam could enter a thick block of iron, for example, which would (1) allow the muon beam divergence to grow to the desired amount and (2) absorb out most of the remaining pions in the beam.

Acknowledgements

This research was supported by the U.S. Department of Energy under Contract No. DE-AC02-76CH00016.

References

1. See, for example, Proc. of the Mini-workshop on $\mu^+\mu^-$ colliders: particle physics and design, Napa, CA, Nuc. Instr. Meth. in Phys. Res. A350: 24-56, 1994.

2. See, for example, D.V. Neuffer, $\mu^+\mu^-$ colliders: possibilities and challenges, Nuc. Instr. Meth. in Phys. Res. A350: 27-35, 1994; D. Neuffer, Principles and applications of muon cooling, Part. Accel. 14:75-90, 1983.

3. V.V. Parkhomchuk & A.N. Skrinsky, Ionization cooling: physics and applications, Proc. 12th Int. Conf. on High Energy Accel., p. 485-7, 1983.

4. D.V. Neuffer & R.B. Palmer, A high-energy high-luminosity $\mu^+\mu^-$ collider, Proc. 1994 European Particle Accelerator Conf; R.B. Palmer, D.V. Neuffer & J. Gallardo, A practical high-energy high-luminosity $\mu^+\mu^-$ collider, BNL report 61266.

KINETICS OF IONIZATION COOLING OF MUONS

Tatiana A. Vsevolozhskaya

Budker Institute for Nuclear Physics
630090 Novosibirsk, Russia

Abstract

The ionization cooling of muons is considered using the kinetic equations method. A comparison made between an efficiency of transverse cooling in the cases of attendant focus provided by a longitudinal current in cooling medium and by solenoid. The energy spread dynamics is analyzed by an account for straggling of ionization loss and dependence of its mean rate on particle energy.

1 INTRODUCTION

The ionization cooling of muons first discussed by Budker and Skrinsky in 1970–1971 [1] and analyzed by Skrinsky and Parkhomchuk in 1981 [2] is by now the most promising way for compression of transverse emitance of muon beams for projects of $\mu^+ - \mu^-$ colliders being now under wide discussion [3]. The cooling is based on the ionization loss of energy by particle passing through a medium and simultaneous acceleration compensating the longitudinal constituent of lost energy whereas the transverse one is decreasing down the equilibrium value determined by the multiple Coulomb scattering of particles. To reduce the effect of scattering a strong focus is to be applied to the beam to confine its transverse size and thus diminish an increase in beam emitance by particle scattering. Such a focus is created in the most efficient way by use of a strong longitudinal current driven through the cooling medium that can be practically fulfilled using the technique of lithium lenses [4]. Below we investigate the process of cooling with use of kinetic equations method.

2   KINETIC EQUATION

The general form of kinetic equation is (see [5]):

$$\frac{df}{dt} = \mathrm{St}f,$$

where f is a distribution function according to coordinates and momenta. In the left hand side here stands the derivative along the particle phase trajectory defined by the equations of motion whereas in the right – the collision integral making an account for all processes of particle interactions. For muons the only significant processes are the ionization loss of energy and the Coulomb scattering.

The multiple scattering contribution to collision integral was got in the theory of shower processes in cosmic rays (see [6]) in a form:

$$(\mathrm{St}f)_{scatt} = \frac{E_k^2}{X_0}\frac{cos\vartheta}{4p^2v}\Delta_\vartheta f,$$

where Δ_ϑ denotes the Laplacian over linear angle ϑ while E_k^2/X_0, with $E_k \cong 20 MeV$ being the characteristic energy of scattering and X_0 – the radiation length of slowing medium, is defined through the nuclear charge Z, atomic density n and Coulomb logarithm L_c as $E_k^2/X_0 = 4\pi Z^2 e^4 n L_c$. Left aside the expression is the single scattering by a large angle which will be considered separately.

The ionization contribution to $\mathrm{St}f$ is got if consider a variation of distribution function by ionization loss of energy, which is equal to a difference of numbers of particles entering an energy interval $(E, E + dE)$ from above and of leaving it down:

$$\delta f = \int\limits_{p'>p} f(\mathbf{p}')\frac{d^3\mathbf{p}'}{d^3\mathbf{p}}\omega(\delta s, E', E' - E)dE - f(\mathbf{p})\int\limits_{E'<E} \omega(\delta s, E, E - E')dE'.$$

Function $\omega(\delta s, E, \Delta E)$ here describes the probability for energy loss in an individual collision ΔE at a length segment $\delta s = v\delta t$, i.e. the straggling of ionization losses.

Having made an expansion of integrand in the first right hand side term into a power series over energy near E one gets:

$$\begin{aligned}
\delta f\, pE &= \frac{\partial}{\partial E}\left(pE\, f(\mathbf{p})\int \Delta E \omega(\delta s, E, \Delta E)d\Delta E\right) \\
&+ \frac{1}{2!}\frac{\partial^2}{\partial E^2}\left(pEf(\mathbf{p})\int \Delta E\,^2\omega(\delta s, E, \Delta E)d\Delta E\right) + \dots
\end{aligned}$$

160

The integrals in above equation are evidently equal to the mean values of energy loss $\overline{\Delta E}$ and of its square $\overline{\Delta E\,^2}$. Divided by $\delta s = v\delta t$ with $\delta t \to 0$ they define the mean values $-\frac{\overline{dE}}{ds}$ and $\frac{d\overline{\Delta E\,^2}}{ds}$. In result, the ionization contribution to Stf, got in the Fokker–Planck equation form, is:

$$(\mathrm{St}f)_{ion} = -\frac{1}{pE}\frac{\partial}{\partial E}\left(vpEf\,\frac{\overline{dE}}{ds}\right) + \frac{1}{2!pE}\frac{\partial^2}{\partial E^2}\left(vpEf\,\frac{d\overline{\Delta E\,^2}}{ds}\right) + \ldots$$

It seems reasonable to replace function $f(t,\mathbf{r},\mathbf{p})d^3\mathbf{p}$ with $P(t,\mathbf{r},\vartheta,E)d^2\vartheta\,dE$ defined as $P = pvEf$. With no apparent dependence on time the equation for P reads:

$$\frac{\partial P}{\partial z} + \frac{\partial P}{\partial \mathbf{r}}\frac{d\mathbf{r}}{dz} + \frac{\partial P}{\partial \vartheta}\frac{d\vartheta}{dz} + \left(\frac{\partial P}{\partial E} - \frac{2P}{pv} - \frac{\mathrm{tg}\vartheta}{pv}\frac{\partial P}{\partial \vartheta}\right)\left(\frac{dE}{dz}\right)_{acc} = \frac{E_k^2/X_0}{4(pv)^2}\Delta_\vartheta P$$

$$(1)$$

$$-\frac{1}{\cos\vartheta}\frac{\partial}{\partial E}\left[P\left(\frac{dE}{dz}\right)_{ion}\right] + \frac{1}{2!\cos\vartheta}\frac{\partial^2}{\partial E^2}\left[P\left(\frac{d\overline{\Delta E\,^2}}{dz}\right)_{ion}\right] + \ldots$$

where z is the longitudinal coordinate and $\mathbf{r}$ and ϑ – two-dimensional vectors of transverse coordinate and linear angle, $\left(\frac{dE}{dz}\right)_{acc}$ denotes an acceleration rate.

3   TRANSVERSE COOLING

The straggling of energy loss as well as dependence of its mean rate on particle energy do not influence much the transverse motion of particles. So we neglect them by consideration of transverse cooling. If as well we are bound with paraxial approximation $\sin\vartheta \cong \mathrm{tg}\vartheta \cong \vartheta$ only, the equation (1) simplifies to a form:

$$\frac{\partial P}{\partial z} + \vartheta\frac{\partial P}{\partial \mathbf{r}} - \left(\vartheta\frac{\xi_0}{pv} + k\mathbf{r}\right)\frac{\partial P}{\partial \vartheta} - 2\xi_0\frac{P}{pv} + (\xi_0 - \xi)\frac{\partial P}{\partial E} = \frac{E_k^2/X_0}{4(pv)^2}\Delta_\vartheta P \quad (2)$$

with substitutions $\frac{d\mathbf{r}}{dz} = \vartheta$ and $\frac{d\vartheta}{dz} = -k\mathbf{r}$, where k describes the attendant focus as $k = \frac{e}{pc}\frac{dH_\phi}{dr}$. Symbol ξ_0 here stands for $\left(\frac{dE}{dz}\right)_{acc}$ and ξ – for $\left(-\frac{\overline{dE}}{dz}\right)_{ion}$.

Multiplying (2) by ϑ^2, $\mathbf{r}\vartheta$ and $\mathbf{r}^2$ in turn and making integrations over all the transverse phase space under a condition of zero values of distribution function P and its derivatives at extreme r and ϑ one gets a system of differential equations for definition of $\langle\vartheta^2\rangle$, $\langle\mathbf{r}\vartheta\rangle$ and $\langle\mathbf{r}^2\rangle$ which at $\xi_0 = \xi$ reads:

$$\frac{\partial\langle\vartheta^2\rangle}{\partial z} + 2\frac{\xi}{pv}\langle\vartheta^2\rangle + 2k\langle\mathbf{r}\vartheta\rangle = \frac{E_k^2}{(pv)^2 X_0}$$

$$\frac{\partial \langle \mathbf{r}\vartheta \rangle}{\partial z} + \frac{\xi}{pv} \langle \mathbf{r}\vartheta \rangle + k \langle \mathbf{r}^2 \rangle - \langle \vartheta^2 \rangle = 0 \tag{3}$$

$$\frac{\partial \langle \mathbf{r}^2 \rangle}{\partial z} - 2 \langle \mathbf{r}\vartheta \rangle = 0$$

The differential equation for r. m. s. transverse emitance ϵ defined as: $\epsilon = \sqrt{\langle \mathbf{r}^2 \rangle \langle \vartheta^2 \rangle - \langle \mathbf{r}\vartheta \rangle^2}$, is got from this system in a form:

$$\frac{\partial \epsilon^2}{\partial z} + 2 \frac{\xi}{pv} \epsilon^2 = \frac{E_k^2}{(pv)^2 X_0} \langle \mathbf{r}^2 \rangle \tag{4}$$

For k independent on z, as in the case of focusing with magnetic field of longitudinal current, the general solution of system (3) is got in exponential form with exponents $-\delta z$, $-(\delta + i\omega)z$ and $-(\delta - i\omega)z$. Here $\delta = \frac{\xi}{pv}$ and $\omega = \sqrt{4k - \delta^2}$. The solution for $\langle \mathbf{r}^2 \rangle$ under condition $\omega \gg \delta$ reads:

$$\langle \mathbf{r}^2 \rangle = \frac{2E_k^2}{(pv)^2 X_0 \delta \omega^2}\left(1 - e^{-\delta z}\right) + \langle \mathbf{r}^2 \rangle_0 \frac{e^{-\delta z}}{2}(1 + \cos \omega z)$$

$$\tag{5}$$

$$+ \ \langle \mathbf{r}\vartheta \rangle_0 \frac{e^{-\delta z}}{\omega} 2 \sin \omega z + \langle \vartheta^2 \rangle_0 \frac{2e^{-\delta z}}{\omega^2}(1 - \cos \omega z).$$

Here $\langle \mathbf{r}^2 \rangle_0$, $\langle \mathbf{r}\vartheta \rangle_0$ and $\langle \vartheta^2 \rangle_0$ are the initial values of corresponding variables. By integration we neglected the logarithmic dependence of X_0 on z (through the Coulomb logarithm) using for L_c definition an effective thickness of cooling material z_{eff} which is defined from the equilibrium value of mean square angle at extreme z. By $\omega^2 \gg \delta^2$ this equilibrium value is: $\langle \vartheta^2 \rangle_{z \to \infty} = \frac{E_k^2}{(pv)^2 X_0} \frac{1}{2\delta}$ which defines the effective thickness for multiple scattering as $z_{\text{eff}} \cong \frac{1}{2\delta}$. The equilibrium value of mean square coordinate is: $\langle \mathbf{r}^2 \rangle_{z \to \infty} = \langle \vartheta^2 \rangle_{z \to \infty} / k$.

Solution for emitance in dependence on z is got in a form:

$$\epsilon(z) = \epsilon_0 e^{-\delta z} + \epsilon_{z \to \infty}(1 - e^{-\delta z}),$$

with equilibrium value at extreme z equal to:

$$\epsilon_{z \to \infty} = \frac{E_k^2}{(pv)^2 X_0 \delta \sqrt{\omega^2 + \delta^2}}, \tag{6}$$

The equation (4) can be simplified with use of substitution $\langle \mathbf{r}^2 \rangle = \beta \epsilon$, where β is an effective betatron function of particle beam, which, if matched, is equal to β–function of attendant focus. By that the equation is reduced by ϵ thus resulting in:

$$\frac{\partial \epsilon}{\partial z} + \frac{\xi}{pv}\epsilon = \frac{E_k^2}{2(pv)^2 X_0}\beta. \qquad (7)$$

The equilibrium value of emitance at extreme z in terms of β–function reads:

$$\epsilon_{z\to\infty} = \frac{E_k^2}{2\xi pv X_0}\beta, \qquad (8)$$

the condition of β smallness for efficient cooling of emitance is got from (7) in a form:

$$\beta << \epsilon \frac{2\xi pv}{E_k^2/X_0},$$

When β is a periodic function of z with periodicity length much shorter than cooling length $1/\delta$ the value of β in (7) and (8) is to be the averaged one.

If energy is not conserved constant during the cooling, that is $\xi \neq \xi_0$, the equation for emitance is:

$$\frac{\partial \epsilon}{\partial z} + \frac{\xi_0}{pv}\epsilon - (\xi - \xi_0)\frac{\partial \epsilon}{\partial E} = \frac{E_k^2 \beta}{2(pv)^2 X_0}. \qquad (9)$$

Its solution has a form:

$$\epsilon = \epsilon_1 \left(\frac{p}{p_1}\right)^\nu + \frac{E_k^2\, p^\nu}{2(\xi_0 - \xi)X_0 c^2} \int_{p_1}^{p} \frac{\beta(p')E'dp'}{p'^{\,3+\nu}} \qquad (10)$$

where $\nu = \frac{\xi_0}{\xi - \xi_0}$ and $E = E_1 - (\xi - \xi_0)z$. Index 1 denotes the initial values of particle energy, momentum and transverse emitance.

For longitudinal current focusing and relativistic energy the solution for ϵ simplifies to:

$$\epsilon = \epsilon_1 \left(\frac{E}{E_1}\right)^\nu + \frac{E_k^2}{E X_0 \sqrt{k}(\xi + \xi_0)}\left\{1 - \sqrt{\frac{E}{E_1}}\left(\frac{E}{E_1}\right)^\nu\right\}, \qquad (11)$$

in quadrupole FODO tract $(\overline{\beta} \sim \frac{1}{k})$ one gets:

$$\epsilon = \epsilon_1 \left(\frac{E}{E_1}\right)^\nu + \frac{E_k^2\overline{\beta}}{2\xi_0 X_0 E}\left\{1 - \left(\frac{E}{E_1}\right)^\nu\right\}$$

If focusing attendant the cooling is fulfilled with longitudinal magnetic field $H_\parallel$ of solenoid, vector $\frac{d\vartheta}{dz}$ in (1) has components $\frac{d\vartheta_x}{dz} = \vartheta_y \frac{eH_\parallel}{pc}$ and $\frac{d\vartheta_y}{dz} = -\vartheta_y \frac{eH_\parallel}{pc}$

with ϑ_x, ϑ_y being the components of vector ϑ as well as x, y – of vector $\mathbf{r}$, and instead of system (3) one gets for $\langle \mathbf{r}^2 \rangle$, $\langle \mathbf{r}\vartheta \rangle$ and $\langle \vartheta^2 \rangle$ definition:

$$\frac{\partial \langle \vartheta^2 \rangle}{\partial z} + 2\frac{\xi}{pv} \langle \vartheta^2 \rangle = \frac{E_k^2}{(pv)^2 X_0}$$

$$\frac{\partial \langle \mathbf{r}\vartheta \rangle}{\partial z} + \frac{\xi}{pv} \langle \mathbf{r}\vartheta \rangle - \omega \langle x\vartheta_y - y\vartheta_x \rangle - \langle \vartheta^2 \rangle = 0$$

$$\frac{\partial \langle x\vartheta_y - y\vartheta_x \rangle}{\partial z} + \frac{\xi}{pv} \langle x\vartheta_y - y\vartheta_x \rangle + \omega \langle \mathbf{r}\vartheta \rangle = 0$$

$$\frac{\partial \langle \mathbf{r}^2 \rangle}{\partial z} - 2 \langle \mathbf{r}\vartheta \rangle = 0,$$

where $\omega = \frac{eH_{\parallel}}{pc}$.

Solution of above system discovers a reduction of $\langle \vartheta^2 \rangle$ in proportionality with $e^{-2\delta z}$, where δ is still equal to $\frac{\xi}{pv}$, of $\langle \mathbf{r}\vartheta \rangle$ – with $e^{-\delta z}$, and no reduction of $\langle \mathbf{r}^2 \rangle$ at all. The former even grows linearly with z due to the multiple scattering. The emitance at $\delta z >> 1$ by $\omega >> \delta$ is defined as:

$$\epsilon_{z\to\infty}^2 \cong \frac{E_k^2}{2\xi X_0 pv} \left[\langle \mathbf{r}_0^2 \rangle + \frac{1}{\omega^2} \langle \vartheta_0^2 \rangle + \frac{1}{\omega^2} \frac{E_k^2}{2\xi X_0 pv}(1 + 2\delta z) \right],$$

where $\mathbf{r}_0$ and ϑ_0 denote the initial values of particle coordinate and angle.

Now let us evaluate an influence of single scattering by a large angle left aside the $(\mathrm{St}f)_{scatt}$ above. It can result in a loss of particles when they are scattered over the angular size of acceptance of focusing channel. The probability for single scattering over a fixed angle ϑ is inversely proportional to ϑ^2 and precisely equal to unity for an angle ϑ_1, used for definition of the mean square angle of multiple scattering: $\langle \vartheta^2 \rangle_{scatt} = \vartheta_1^2 L_c$, that is $\vartheta_1^2 = 4\pi Z^2 e^4 nt/(pv)^2$.

Let us conditionally consider the particle being lost when the single scattering angle exceeds the r.m.s. angular spread in the beam. For a beam cooled down to an equilibrium emitance in a length T of material the probability for particle to be scattered over the final r.m.s. angle is: $W(\langle \vartheta^2 \rangle_{z\to\infty}) = \vartheta_1^2(T)/\langle \vartheta^2 \rangle_{z\to\infty} = 2\delta T/L_c$. However, the probability for such a scattering is homogeneously distributed along the whole length of cooler where, at the most part, the r.m.s. angular spread is sufficiently larger than the final. Thus, the probability for particle to be lost in a sense defined above is found as $W_{lost} = \frac{\vartheta_1^2(T)}{T} \int_0^T \frac{dz}{\langle \vartheta^2(z) \rangle}$. If emitance is, for instance, cooled down by 100 times at 2 GeV energy, the probability for particle to get the single scattering over the final equilibrium angle is equal to 0.64, whereas the probability to be "lost" is $\sim 12\%$.

4 ENERGY SPREAD

The only terms in kinetic equation effecting the energy spread are those connected with straggling of ionization losses and dependence on energy of their mean rate.

Integrated over transverse coordinates and angles the distribution function $P(z, E)dE$ is defined by equation:

$$\frac{\partial P}{\partial z} + \xi_0 \frac{\partial P}{\partial E} = \frac{\partial}{\partial E}(\xi(E)\,P) + \frac{1}{2!}\frac{\partial^2}{\partial E^2}\left[P\left(\frac{\overline{d\Delta E}^{\,2}}{dz}\right)_{ion}\right] + \dots$$

where ξ_0 and ξ denote the rates of acceleration and ionization loss, as it was introduced in (2).

The mean energy of particles $\overline{E}$ varies as

$$\frac{\partial \overline{E}}{\partial z} = \xi_0 - \overline{\xi},$$

that is remains constant if acceleration rate is equal to the rate of ionization loss averaged over the particle spectrum. The mean square deviation of particle energy from the mean value by that is defined as:

$$\frac{\partial \left\langle E^2 - \overline{E}^2 \right\rangle}{\partial z} + 2\xi'\left(\overline{E}\right)\left\langle E^2 - \overline{E}^2 \right\rangle = \left\langle \left(\frac{\overline{d\Delta E}^{\,2}}{dz}\right)_{ion}\right\rangle \tag{12}$$

With probability distribution for energy transfer Q in an individual collision, defined for muon as ([7]):

$$\frac{dw(Q)}{dz}dQ = \frac{2\pi e^4 n Z}{mv^2}\frac{dQ}{Q^2}\left(1 - \beta^2\frac{Q}{Q_{max}} + \frac{Q^2}{2E^2}\right),$$

and the maximum energy transfer being $Q_{max} \cong \frac{\beta^2 E^2}{Mc^2\frac{M}{2m}}$ (in an energy range of several GeV or below), the mean square straggling of ionization losses is found to be (with the third term in brackets above neglected):

$$\left(\frac{\overline{d\Delta E}^{\,2}}{dz}\right)_{ion} \cong \frac{GE^2\left(1 - \frac{\beta^2}{2}\right)}{Mc^2\frac{M}{2m}}$$

where G stands for $\frac{2\pi e^4 n Z}{mc^2} = 0.035$ MeV/cm in lithium. By that (12) transforms into:

$$\frac{\partial \sigma_E^2}{\partial z} + \left[2\xi'\left(\overline{E}\right) - \frac{mG}{M^2 c^2}\right]\sigma_E^2 = \frac{mG\left(\overline{E}^2 + M^2 c^4\right)}{M^2 c^2}$$

with $\sigma_E^2 = \left\langle E^2 - \overline{E}^2 \right\rangle$.

Solution of this equation evidently is:

$$\sigma_E^2 = \sigma_{E,0}^2 \, e^{-\delta_E z} + \frac{mc^2 G \left(\overline{\gamma}^2 + 1\right)}{\delta_E} \left(1 - e^{-\delta_E z}\right), \qquad (13)$$

where $\delta_E = 2\xi'\left(\overline{E}\right) - \frac{mG}{M^2 c^2}$.

Dependence of mean rate of ionization loss $\xi = \frac{G}{\beta^2} L_i$, where L_i denotes the logarithmic factor $L_i = \ln \frac{2mc^2 \beta^2 \gamma^2 Q_{max}}{I^2} - 2\beta^2 - \delta$ (with δ making an account for density effect), on particle energy is defined by $\frac{1}{\beta^2}$ – in non-relativistic energy range, and by logarithmic grow of L_i – in relativistic. The former is proportional to $4 \ln \beta\gamma$ below an energy region where the density effect becomes significant, and to $2 \ln \beta\gamma$ at several GeV or higher where it achieves the full strength, while in an intermediate region there is a smooth transition between these two rates. The minimum mean rate of energy loss ($\xi' = 0$) for muons in lithium with account for density effect is found at $\beta\gamma \cong 3.9$. Below this point the decrement δ_E is negative, that means the heating of energy spread in a beam, whereas above the point the value of δ_E, defined in the main by ξ', soon becomes positive that means the energy spread cooling. Unfortunately this positive value is not large enough to provide with efficient cooling. The maximum value of δ_E, found near $\beta\gamma \cong 6.5$, is $\delta_{E,max} \cong 1.7 \ 10^{-2} m^{-1}$ being by about 10 times less than the transverse emitance decrement at this energy. At $E \sim 2$ GeV or more where $\delta_E \cong \frac{4G}{E}$ the ratio $\delta_\perp / \delta_E$ becomes ~ 7.

In the region of negative decrement its magnitude is growing fast with decrease of muon energy, achieving at kinetic energy $\sim 50 MeV$ the value $-\delta_E \cong 2 \ m^{-1}$, which is about equal to the transverse emitance decrement $\delta_\perp$ at this energy.

Heating of energy spread caused by the straggling of ionization losses gives at small enough z a linear grow of σ_E^2 with z: $\sigma_{E,z<<1/|\delta_E|}^2 \cong mc^2 Gz(\gamma^2 + 1)$, which later on is replaced or by exponential grow – in a region of negative decrement, or by tendency to an equilibrium value – in a region of positive. From (13) this equilibrium energy spread is found to be:

$$\sigma_{E,z\to\infty}^2 = \frac{mc^2 G \left(\gamma^2 + 1\right)}{\delta_E},$$

which in GeV's energy region simplifies to: $\sigma_{E,z\to\infty}^2 = \frac{mE^3}{4M^2 c^2}$.

To complete consideration of straggling effect in ionization loss of energy let us estimate a probability for particle to loose in an individual collision a fraction Q of energy, exceeding the r.m.s. energy spread σ_E. Such an estimation evidently makes sense for $\sigma_E < Q_{max}$ only because by the opposite correlation this probability is equal zero.

166

Integration of above presented expression for $w(Q)$ (with the third term in brackets neglected) from $Q = \sigma_E$ to $Q = Q_{max}$ gives:

$$W_{Q \geq \sigma_E} \cong \frac{Gz}{\beta^2 \sigma_E} \left[1 - \frac{\sigma_E}{Q_{max}} \left(1 + \beta^2 \ln \frac{Q_{max}}{\sigma_E} \right) \right].$$

For σ_E^2 linearly dependent on z and high enough energy this transforms into:

$$W_{Q \geq \sigma_E} \cong \frac{1}{\gamma} \sqrt{\frac{Gz}{mc^2}} \left[1 - \frac{1}{2\gamma} \sqrt{\frac{Gz}{mc^2}} \left(1 - \ln \frac{1}{2\gamma} \sqrt{\frac{Gz}{mc^2}} \right) \right].$$

The maximum value of $W_{Q \geq \sigma_E}$ in a range $0 \div 2$ of variable $\frac{1}{\gamma}\sqrt{\frac{Gz}{mc^2}} = \sqrt{\frac{M}{m \gamma L_i} \frac{\overline{\Delta E}}{E}}$, which corresponds to the range $0 \div Q_{max}$ of σ_E, appears at $\frac{1}{\gamma}\sqrt{\frac{Gz}{mc^2}} \cong 0.5$ and is equal to ~ 0.2.

Thus, the probability for particle to loose in an individual collision at a length z a fraction of energy exceeding the r. m. s. energy spread got at the same length does not exceed 20% . The probability to loose more than $2\sigma_E$ does not exceed 5% .

5   ACKNOWLEDGEMENT

The author is grateful to A. N. Skrinsky, V. V. Parkhomchuk and G. I. Silvestrov for fruitful discussions.

6   REFERENCES

1. G. I. Budker. Proc. XV Int. Conf. on High Energy Physics, Kiev, 1970.
A. N. Skrinsky, *Intersecting Storage Rings at Novosibirsk*, Morges Seminar 1971. CERN /D.Ph.II /YGC /mmg, 21.9.1971.
2. A. N. Skrinsky and V. V. Parkhomchuk, *Methods of Cooling of Charged Particle Beams*, Physics of Elementary Particles and Atomic Nuclei, v. $12, N3$, Moscow, Atomizdat,1981.
3. D. V. Neuffer, R. B. Palmer, Proc. European Particle Acc. Conf., London (1994)
4. B. F. Bayanov, Yu. N. Petrov, G. I. Silvestrov et al., *Large Cylindrical Lenses with Solid and Liquid Lithium.*, EPAC I, v. 1, Rome (1988).
5. L. D. Landau and E. M. Liphshitc, *Theoretical Physics, Volume X*, Moscow, NAUKA, 1979
6. S. Z. Belen'ky, *The Shower Processes in Cosmic Rays*, OGIZ, Moscow – Leningrad, 1948.
7. H. J. Bhabha, Proc. Roy. Soc., A164, 257 (1938)

LITHIUM LENSES FOR MUON COLLIDERS

Gregory I. Silvestrov

Budker Institute for Nuclear Physics
630090 Novosibirsk, Russia

Abstract

A possibility is considered for creation of new generation of lithium lenses to be used in muon collider projects being now under discussion. The lenses are to operate at 30 Hz repetition rate and have apertures from several centimeters down to one millimeter to solve the problems as of pion collection from target as of ionization cooling of muons. On the base of analyses of the main limitations to parameters and operational regimes for existing lenses with solid lithium it is shown that the most perspective way to create multi-hertz lenses consists in use of liquid lithium, pumped through the system. Several versions of liquid lithium lens design are presented and possible regimes of their operation are discussed.

1   INTRODUCTION

The projects of muon colliders being now under wide discussion make it vital to consider a possibility for creation of new generation of lithium lenses intended to be used as for pions collection from production target as for ionization cooling of muon beams. The lenses developed in 70 – 80 in BINP [1,2], FNAL [3] and CERN [4] were mainly intended for collection of antiprotons for proton–antiproton colliding beams. They had apertures from 2 to 3.8 cm, length of the order of 15 cm and repetition rate of operation about 0.5 Hz. By that with magnetic field at cylindrical surface being of the order of 10 Tesla the life time of lenses does not exceeded ∼ 1 million cycles. In muon collider projects (see [5]) the repetition rate is to be 10 – 30 Hz, the pion collection lens makes an interest by an aperture of 6 – 8 cm, and the current carrying lithium rods in muon cooling system are considered to be of 1 m length at least with apertures from 20 to 1 mm and magnetic field at the surface from 10 to 20 Tesla to provide at the final stage of cooling the field gradient up to 4 MOe/cm. The sum length of lithium rods in ionization cooling tract can achieve several hundreds meters. Evidently, the creation of lenses with such a parameters and by the orders longer life time requires the principally new technical and technological solutions as in lens design as in its power supply.

2 EXISTING LITHIUM LENSES DESIGN

Let us remind the main principles put in the base of existing lithium lenses design. The lenses are supplied by half–sine current pulse with a duration providing the homogeneous current density distribution over the lens cross–section to the moment of maximum current. As it was shown in [6], the sufficient homogeneity is got starting from the pulse duration corresponding to a ratio δ/r of skin depth δ to lens radius r equal to ~ 0.5. Thus at a temperature of lithium $T_{Li} \leq 100°C$ when the specific resistance ρ_{Li} is equal to $\sim 10^{-5}Ohm.cm.$, the lens of 1 cm radius is to be supplied by current pulse of duration $\tau \sim 0.5 \ 10^{-3}$. With change of lens radius the current pulse duration is to be changed quadratically because the skin depth depends on ρ and τ as $\delta = k\sqrt{\rho\tau}$ with $k = 7 \ 10^3 cm^{1/2}sec^{-1/2}$.

Typical design of lithium lens [2] is shown in fig. 1.

The operational part of lithium is contained inside a thin -wall titanium cylinder whose external surface is washed by water. This thin-wall envelope is inserted into another cylinder of larger diameter and connected to it at the ends thus providing a space for circulation of cooling water. The outer cylinder is cut in halves normally to the axis. Such a two-wall cylinder with a system for water circulation is placed in a rigid hermetic frame turning into the coaxial cylindrical current inputs. Whole free volume inside the case is filled by lithium under a pressure of 100 atmospheres.

During a current pulse the lithium inside the cylinder is heated and the pressure gets rise in proportionality with temperature. When $\delta/r \leq 0.7$ the temperature of pulsed heating of lithium cylinder does not depend on pulse duration and is determined only by square of field H_0 at the surface. With $H_0 = 10T$ it is $T_0 = 60°C$. In rigidly closed volume of lithium the pressure rise with temperature is defined as: $P = \frac{\alpha}{\chi}T_0$ where α is the coefficient of thermal expansion and χ – the compressibility. At $T_0 = 60°$ this gives $P = 1200$ atmospheres.

The pressure rise can be decreased by including in lens structure some elastic details.In the lens being considered such elastic details are the thin-wall envelope of heated lithium cylinder, and the lithium outside the operational volume which is not heated during the current pulse and accepts partly the pressure arising in operational volume. In result the pressure rise in former is defined as:

$$P = \frac{\alpha T_0}{\chi \left(1 + \frac{V_b}{V_0}\right) + \frac{2}{E_0}\frac{r}{\Delta}},$$

where V_0 denotes the operational lithium volume while V_b – the volume out of operation which we name a buffer volume, Δ stands for a wall thickness of inner titanium cylinder, E_0 – for elasticity module of titanium.

The main problem in increase of lens operation reliability consists in decrease of pressure rise in operational volume which defines the stresses arising in the system. An attractive way for this looks to consist in use of buffer volume in a region of current input which could be made large enough. The experiments made with model lenses of special geometry [7,8] have really shown that with large enough buffer volume the pulsed rise of pressure is sufficiently reduced. Unfortunately the use of this effect is strongly restricted because of very long time of stress relaxation in solid lithium. In acceptable design geometries the pressure transferred from heated to buffer volume within a time of sound propagation is relaxing during a time up to a minute or so.

The situation is fully changed when lithium filling the lens volume is melted. In this case the time of pressure relaxation is also defined by sound propagation only thus allowing to use efficiently the buffer volume. Moreover, the compressibility of melted lithium is by about 5 times higher than of solid which leads to further reduction of pulsed pressure rise. In result with preload fixed at a level of 20 atmospheres the maximum pressure in system can be reduced down to several tens of atmospheres by a field of 10 T.

The next problem in creation of lithium lenses consists in removal of heat released in a lens. With supply by half-sine current pulse the heat amount released per unit volume in the lens, $q_0 = 6 \ 10^{-7} H_0^2 \left(\frac{\delta}{r}\right)^2 \frac{cal}{cm^3}$ [9], is equal to $30 \frac{cal}{cm^3}$ by field at the surface $H_0 = 10T$ and $\frac{\delta}{r} = 0.7$. The mean power released per unit length, $w = \pi r^2 q_0 f$, with lens radius $r = 1cm$ and repetition rate $f = 1Hz$ is equal to about $400 \ W/cm$. By that the heat flux per unit square of cooled surface is equal to $62.5 \ W/cm^2$.

In existing lithium lenses the water cooling is carried out through thin envelope wall made of titanium alloy which has rather small thermal conductivity $\lambda_{Ti} = 0.02 cal cm^{-1} s^{-1} deg^{-1}$. The time of heat removal τ_r can be estimated as $\tau_{cool} \simeq \frac{r\Delta C_{Li}}{2\lambda_{Ti}}$ where C_{Li} is the heat capacity of lithium per unit volume equal to about $0.5 \frac{cal}{cm^3 deg}$ in a temperature range from $0°C$ to melting point $186°C$. For a lens with $r = 1cm$ and $\Delta = 0.15cm$ the time of heat removal is equal to about 2 s. With lens radius increase the envelope thickness is to be increased as well to get the same mechanical stress in the wall. So the time of cooling τ_{cool} is growing in proportionality with square radius. Simultaneously the heat flux density through the surface is growing in proportionality with lens radius.

The limit strong thermal regime was got in a lens produced in BINP in 1980 for super strong focusing of protons in the antiproton target station at FNAL [10]. The lens of radius 2.5 mm and wall thickness 0.7 mm was tested by field at the surface 10 T and repetition rate 13 Hz during 10 million pulses. By that the temperature drop on the wall was equal to $150°C$, mean temperature of lithium – to $170°C$ while the maximum temperature in lens center achieved the melting value $186°C$.

In lenses with radius more than 1 cm the cooling presents the main restriction by a choice of operational regime. With 10T field at the surface the repetition rate can not be higher than 1 Hz.

3 LIQUID LITHIUM LENSES

The principally new solution in lithium lens design consists in use of liquid lithium. It solves not only the problem of high pulsed pressure in system, as it was shown above, but the problem of heat removal, as well, if being pumped through the lens and heat exchanger simultaneously. By that the repetition rate is limited only with a pump power and productivity which can be made high enough with use of experience of cooling with liquid lithium got in the nuclear reactor technique.

The first liquid lithium lens was created in BINP for collection of polarized electrons and positrons for the linear collider project VLEPP [2,11]. The lenses with operational volume $\sim 1 cm^3$ were tested at 150 Hz repetition rate.

The next step in development of liquid lithium lens technology in BINP was the lens of 4 cm diameter and 15 cm length [12] shown in figure 2. Without the water cooling system the lens design becomes more simple that permits to make the structure whole-welded and titanium envelope – more thick and durable. Electromagnetic pump produced in BINP was used for pumping of lithium. The pump productivity was equal to several litters per minute so the repetition rate by tests was 0.1 Hz. The aim of tests consisted as in investigation of system behavior under the hydrodynamic shocks arising in lithium contour under an action of pulsed magnetic field and heating as in development of liquid lithium technology.

Because the specific resistance ρ in melted lithium is increased up to 2.5 $10^{-5}Ohm$ cm or more, the pulse duration may be decreased in inverse proportionality with ρ to keep constant the ratio δ/r. This makes another advantage in use of liquid lithium because decreases the losses in current contour, including the matching transformer, current inputs and pulse generator, by conserved power release in operational part of the lens.

Use of liquid lithium seems to be the only possible way for creation of lithium systems for muon colliders. In figure 3 there is presented a scheme draught of a version of large aperture lens for pion collection from production target. The liquid lithium is brought up through several tubes to a ring input collector in titanium body of lens, which is connected to operational lens volume through a number of radially drilled holes. After flow through the cylindrical part of lens the lithium returns to the output collector along a narrow cylindrical slot between two-wall titanium cylinder and made of copper coaxial current input. Because the main part of current runs in copper the energy release in this part of lens is sufficiently less than in the operational part.

With 5T field at lens surface and $\delta/r = 0.7$ the temperature rise in operational part of lens is 15°C per pulse. If temperature gradient from lens inlet to outlet is accepted to be 200°C (from 200°C to 400°C) the full exchange of lithium in operational part is to take place in about 14 cycles that is in 0.5 s at repetition rate 30 Hz. By that the necessary velocity of lithium in 15 cm length lens is to be 0.3 m/s and lithium expenditure – 1.2 litter/s.

In short lenses with length of the order of four radius, $l \sim 4r$, a sufficiently intensive cooling takes place at the lens ends, and thus the time of full exchange of lithium in operational part may be increased, so that the lithium expenditure becomes less than a litter per second. Modern electromagnetic pumps with such a productivity are simple enough devices with consumed power of several kW. However, with account for input tubes of large diameter, heat exchanger and operational volume of pump, the full volume of liquid lithium in the system will be several tens litters.

The full energy released in lens per one pulse is $\sim 20kJ$ and the power - $\sim 0.6MW$ at repetition rate 30 Hz. With average temperature of lithium being 300 °C its specific resistance will be $\sim 3~10^{-5}Ohm~cm$, and so the current pulse duration may be $\sim 3~10^{-3}s$.

Shown in figure 4 there is a scheme drought of lithium current carrying rod of 1m length and 2 cm diameter for a system of ionization cooling of muons. With $\delta/r \simeq 0.5$ and field 10T the temperature of lithium heating per pulse is 30 °C, and the full exchange of lithium is to be made for 6–7 cycles, that is for 0.2 second by 30 Hz repetition rate. Velocity of lithium flow is to be $\sim 1m/s$ by expenditure ~ 1.5 litter per second that requires a more powerful pump.

More exotic look the rods in final part of ionization cooling system, which are to be of $\sim 1mm$ diameter and $4Moe/cm$ field gradient. To provide the necessary velocity of lithium flow, being here of several meters per second, the required pressure is of several tens atmospheres, which can not be produced with electromagnetic pumps. The expenditure of lithium will, however, be as small as 20-30cm^3/s, which allows to use a method of mechanical pressing through the rod of lithium from a big bellows with welded bottom. This bellows-vessel is placed in a thick-wall hermetic cylinder where the air or liquid gallium-indium alloy is pumped in under a high pressure, which is transferred to the liquid lithium by compression of bellows. [1] By that the lithium passes

---

[1]Such a systems we use for production of pressure up to 2 thousands atmospheres in vessels filled with lithium.

from one bellows to another through the rod and heat exchanger with change of direction once per several minutes. The full volume of lithium in system can be several litters only.

With diameter 1 mm the rod is supplied by current pulse of $\sim 1 \mu s$ duration. By that the mechanical firmness will be provided by inertial properties of system and so the field up to 20T seems to be achievable [13].

4 REFERENCES

1. B. F. Bayanov et al., *Antiproton Target Station on Base of Lithium Lenses*, XI Int. Conf. on High Energy Acc., Geneva, CERN, 1980, p.362.

2. G. I. Silvestrov, *Problems of Intense Secondary Particle Beams Production*, XIII Int. Conf. on High Energy Acc., Novosibirsk, 1986, v.2, p. 258. (Preprint INP 86-163, Novosibirsk,1986).

3. G. Dugan, *P-bar Production and Collection of the FNAL Antiproton Source*, XIII Int. Conf. on High Energy Acc., Novosibirsk, 1986, v.2, p. 264.

4. R. Belone et al, *The Results of Prototype Test and Field Computations of CERN Lithium Lens*,XIII Int. Conf. on High Energy Acc., Novosibirsk, 1986, v.2, p. 272.

5. D. V. Neuffer, R. B. Palmer, Proc. European Particle Acc. Conf., London (1994)

6. T. A. Vsevolozhskaya et al, *Optical Properties of the Cylindrical Lenses*, ZhETP 45(1975), 2494.

7. B. F. Bayanov el al, *The Investigation and Design Development of Lithium Lenses with Large Operating Lithium Volume*, XII Int. Conf. on High Energy Acc., Fermilab, 1983, p. 587.

8. B. F. Bayanov, T. A. Vsevolozhskaya and G. I. Silvestrov, *Study of the Stress in and Design Development of Cylindrical Lithium Lenses*, Preprint INP 84-168, Novosibirsk, 1984.

9. T. A. Vsevolozhskaya and G. I. Silvestrov, *Thermal Regime of Cylindrical Lenses*, Preprint INP 84-100, Novosibirsk, 1984.

10. B. F. Bayanov el al, *A Lithium Lens for Axially Symmetric Focusing of High Energy Particles Beams*, NIM 190(1981), p. 9-14.

11. B. F. Bayanov el al, *Multi-Hertz Conic Lenses with Liquid Lithium for the Focusing of Low Energy Positrons*, IX National Accel.Conf., Dubna 1984, v. 1, p. 406.

12. B. F. Bayanov, Yu. N. Petrov, G. I. Silvestrov et al., *Large Cylindrical Lenses with Solid and Liquid Lithium.*, EPAC I, v. 1, Rome 1988, v. 1, p. 263.

13. B. F. Bayanov and G. I. Silvestrov, *The Possibility of Using of Lithium for Creation of Cylindrical Lenses with High Magnetic Fields*, ZhETP 48 (1978), 160.

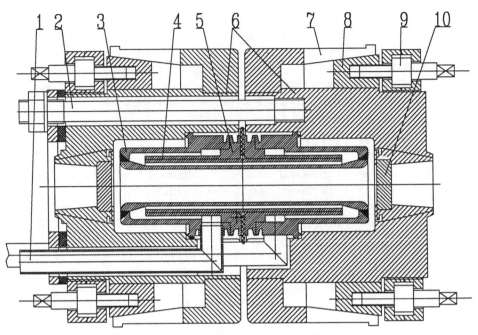

Fig.1. LENS WITH ELASTIC WALL

1—water supply;
2—retaining bolts;
3—titanium body of the lens;
4—distribution pipes of the water
   system;
5—flanges of the distribution pipes;
6—steel body of the lens;
7—collecting contact;
8—conic clamps;
9—bolts;
10—beryllium windows.

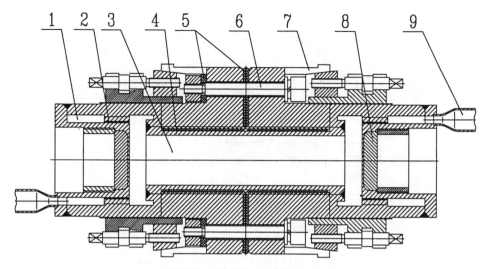

Fig.2. LENS WITH LIQUID LITHIUM

1—buffer volumes;
2—supply channels;
3—operating lithium volume;
4—thin—wall envelope of the operating
   lithium volume;
5—oxidized titanium insulators;
6—retaining bolts;
7—collecting contacts;
8—beryllium windows;
9—supply tubes for liquid lithium.

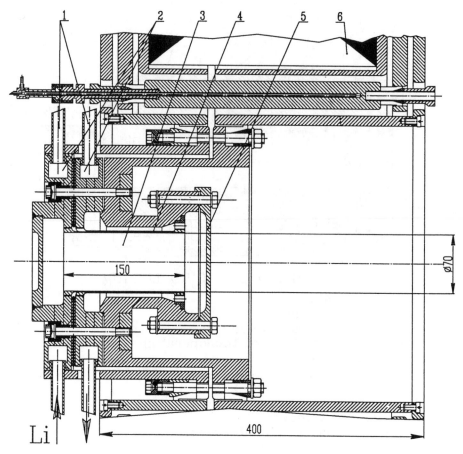

Fig.3. LIQUID LITHIUM PIONS COLLECTING LENS

1—tubes of liquid lithium supply;
2—inlet and outlet ring—bend collectors;
3—operating lithium volume; 4—two—wall insulated titanium cylinder;
5—titanium windows;6—matching transformer.

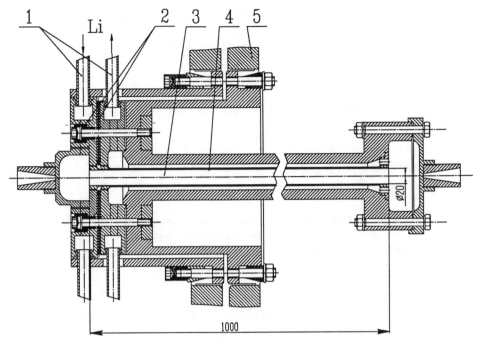

Fig.4. LITHIUM CURRENT CARRYING COOLING ROD

1 –tubes of liquid lithium supply;
2 –oxidized titanium insulators;
3 –operating lithium volume;
4 –two–wall insulated titanium
   cylinder;
5 –current input.

Interaction Regions with Increased Low-Betas for a 2-TeV Muon Collider

Carol Johnstone, King-Yuen Ng

Fermi National Accelerator Laboratory, P.O. Box 500, Batavia, IL 60510*

Dejan Trbojevic

Brookhaven National Laboratory, P.O. Box 5000, Upton, NY 11973*

Abstract

The difficulty encountered in designing an interaction region (IR) for a 2-TeV Muon Collider lies in the extreme constraints placed on beam parameters at the point of collision. This paper examines a relaxation of the interaction-point criterion insofar as it impacts luminosity, the design, and stability of the interaction region.

I. INTRODUCTION

It has been suggested for the muon collider that the transverse beam parameters at the interaction point (IP) have the values $\beta^*_{x,y}$ =3 mm (rms) for a beam with a normalized emittance of $\epsilon_N = 50\pi$ mm-mrad [1]. These stringent requirements were imposed in order to overcome partially the hour-glass effect and retain a luminosity of $\mathcal{L} = 3 \times 10^{33}$ cm^{-2} s^{-1} in the collision process; that is, assuming two 3 ps rms bunches of opposite charges populated by $N = 2 \times 10^{12}$ muons each. An additional requirement is that a ±150 mrad detector acceptance must be maintained about the collision point. This implies 2-6 m of separation must be allowed between the first IR quadrupole and the interaction point. The 3-mm low-beta and a long interaction drift produce a very sensitive and highly achromatic interaction region. The design is further hindered by the reduced gradient strengths of quadrupoles at 2-TeV.

Because the betatron amplitude scales as the square of the distance from the IP divided by β^*, the most direct approach to reduce chromaticities in the IR optics is to increase β^*. Moving quadrupoles closer to the interaction point in principle should also work, but in practice, unusually weak bending power of the quadrupoles reduce greatly any impact on the betatron functions. The purpose of this paper is to assess to what degree luminosity is affected by an increase in β^* and what gains are made insofar as the IR is concerned.

*Operated by the Universities Research Association, Inc., under contract with the U.S. Department of Energy.

II. LUMINOSITY CONSIDERATIONS

The interaction region of a muon collider is embedded in a storage ring, yet it must exhibit the tight final focus properties of a linear collider in order to achieve the projected luminosity. This represents the most critical issue in the design of the 2-TeV collider ring.

The actual or effective luminosity $\mathcal{L}_{\text{eff}}$ achievable in a 2-TeV muon collider ring is modified from the nominal luminosity, $\mathcal{L}_0$ which is given by the standard equation:

$$\mathcal{L}_0 = \frac{N^2 \gamma f}{4 \beta^* \epsilon_N}, \qquad (2.1)$$

where γ is the energy of the beam divided by the muon mass and f is the frequency of collision. At high muon intensities beam-beam interactions become important and enhance the effective luminosity due to the mutual focusing effect one colliding beam has on the other. A disruption parameter can be defined which characterizes the degree of beam-beam focusing. The other contribution that must be considered, the so-called hour-glass effect, arises from the rapidly increasing betatron functions, or transverse beam sizes, away from the collision point. The hour-glass effect is a measure of the depth of field and is strictly geometrical; it reduces the overall luminosity because of the dilution in particle density, and counteracts the disruption enhancement.

A prototype IR with a low-beta of 3 mm had been designed by Gallardo and Palmer [2]. Recently, Chen [3] performed a numerical simulation for the effective luminosity which included both the disruption and geometrical effects and obtained a luminosity enhancement factor of 0.87. Since the geometrical, or hour-glass, effect has been calculated independently to be 0.76, this implies that disruption causes a 14% increase in luminosity.

Relaxing the low-beta constraint for collision impacts the luminosity, but it facilitates achieving a workable IR design. The impact of β^* on luminosity can be estimated by noting that as β^* increases both the disruption-focusing and hour-glass effects disappear rapidly. The enhancement factor becomes one and effective luminosity, $\mathcal{L}_{\text{eff}}$, approaches the nominal, $\mathcal{L}_0$. According to Eq. (2.1), the effective luminosity then scales inversely to the ratio of β^*. With the enhancement factor at 3 mm taken into account, modified initial luminosity factors for β^*'s of 1 and 3 cm are given in the middle column of Table I.

The Gallardo-Palmer IR has a long length of about 947 m. The reason for this is that its high natural chromaticities of ~ -6000 have to be corrected locally. Either side of the IP is flanked by four high-beta, high-dispersion bumps. As will be shown below, IR's with $\beta^* = 3$ cm and 1 cm have chromaticities of only ~ -270 and ~ -500, respectively, making local correction at the IR unnecessary. As a result, their length can as short as ~ 200 m. For a collider ring of radius ~ 1 km with two IR's, this represents a 24% increase in revolution or collision rate and therefore luminosity. Taking into account path-length effects, the effective luminosity is recomputed and tabulated in the last column of Table I.

Disruption focusing also dilutes the emittances in both planes lowering the luminosity. Preliminary calculations by Chen [3] give a fractional increase of $\delta\epsilon_N/\epsilon_N = 6\%$ for the first interaction, but it is much less in subsequent collisions. Including emittance dilution may further reduce the discrepancy between 3 mm and centimeter IRs.

From the above analysis it is clear that the 3 mm β^* criteria could be relaxed to 1 or 3 cm without a severe sacrifice in luminosity.

Table I: Luminosity reduction versus β^*.

β^*	Reduction Ratio	
(cm)	H & D	Total
0.3	1.0	1.0
1.0	2.9	2.3
3.0	8.7	7.0

III. LOW-BETAS AND CHROMATICITIES

A relaxed β^* contributes enormously to the stability of the muon collider; in particular with respect to chromaticity, which has been a persistent problem in designing a 3 mm, 2-TeV IR. This can be seen from the simplified IR depicted in Fig. 1, where there are only two *thin* quadrupoles QF and QD to focus the beam horizontally and vertically, respectively.

The maximum of the vertical betatron function occurs at QD and is approximately equal to

$$\beta_y \gtrsim \frac{s_D^2}{\beta^*}, \tag{3.1}$$

where s_D is the distance measured from the collision point. The small beam size at the focus, or collision point, in the IR causes transverse sizes to increase rapidly as a function of distance from the IP. In order to reverse the rise of the β functions, the focal length of IR quadrupoles must be

$$f \lesssim s_F, \tag{3.2}$$

where s_F is the distance of the center of QF from the collision point. The vertical chromaticity is given very roughly by

$$\xi_y \sim -\frac{1}{4\pi}\frac{\beta_y}{f} \sim -\frac{s_D^2}{4\pi s_F \beta^*}. \tag{3.3}$$

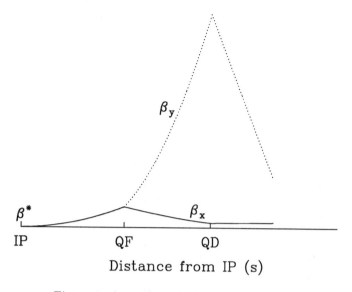

β_y

β^*

β_x

IP   QF   QD

Distance from IP (s)

Figure 1: A quadrupole doublet scheme at the IR.

Because of the high energy of this ring, the superconducting quadrupoles encounter gradient limitations, particularly in IR region. A pole-tip field of 9.5 T yields a quadrupole strength of only $k \lesssim .04$ m^{-2} at $\beta = 15$ km, or greater. Such weak gradients result in long quadrupoles, causing the amplitude in the defocusing direction to skyrocket inversely proportional to the value of β^*. (see Eq. (3.1)). For example, the quadrupole length is given by $\ell = 1/(fk)$. Therefore, although the distance of the interaction point to the first quadrupole is 6.5 m, the distance to the center of the first quadrupole s_F is 9 to 10 m, and s_D can be 20 to 40 m. The resulting chromaticity, as given by Eq. (3.3), is naturally large because of the distances and quadrupole lengths involved. In any case, very roughly, the natural chromaticities scale inversely with β^*.

Knowing the natural chromaticities of the Gallardo-Palmer IR, chromaticities of IR's with larger β^* can be estimated by scaling to the $\beta^* = 3$ mm case:

$$\beta^* = 3 \text{ mm}, \quad \xi_y \sim -6000$$
$$\beta^* = 1 \text{ cm}, \quad \xi_y \sim -1800$$
$$\beta^* = 3 \text{ cm}, \quad \xi_y \sim -600$$

One observes from these numbers that an order of magnitude improvement can be made in the chromaticity paying an equivalent price in β^*. Furthermore, scaling has overstimated actual values. This is because with larger β^*'s, transverse beam sizes in the IR will be much smaller. Smaller beam apertures mean quadrupoles with higher gradients can be used and they will be more effective in controlling the betatron functions. Table II displays the minimum aperture required as a the function of betatron amplitude for a 5-sigma beam width plus about 5 mm for shielding. Alongside these data are the maximum field gradient, B', and the strength, $k = B'/(B\rho)$, assuming a pole-tip field of 9.5 T. In Fig. 2, we plot dynamical parameters for a 5σ beam aperture and a beam energy of 2-TeV.

Table II: Maximum quadrupole gradient as a function of β for a 2-TeV collider.

β (km)	Aperture (cm)	Gradient (T/m)	k (m^{-2})
1	0.81	1170	0.175
3	0.41	675	0.101
5	1.82	522	0.078
7	2.15	442	0.066
9	2.44	390	0.058
$10-15$	~ 3.0	~ 325	0.049
$15-20$	~ 3.3	~ 375	0.043
$20-25$	~ 3.8	~ 245	0.037
$25-30$	~ 4.2	~ 220	0.034
$30-35$	~ 4.6	~ 205	0.030
$35-40$	~ 5.0	~ 190	0.028
$40-50$	~ 5.4	~ 175	0.026
$50-70$	~ 6.2	~ 150	0.023
$70-90$	~ 7.4	~ 130	0.020
$90-110$	~ 8.1	~ 115	0.017

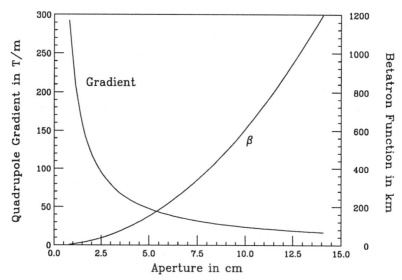

Figure 2. Dynamical parameters of collider at 2 TeV.

Here, we would like to point out that a $\mu^+\mu^-$ collider is very much different from an e^+e^- collider. Muons are decay products of pions which are in turn secondary particles produced by protons hitting a target, and cannot be cooled efficiently due to the short muon life time. As a result, the emittances for muons are, in general, much larger than those for the electrons. The transverse amplitudes of a muon bunch will also be much larger than for electron bunches. For this reason, the allowable quadrupole gradients will be more restricted in a muon collider than an e^+e^- collider.

IV. A 3-CM INTERACTION REGION

A. The IR Lattice

A stable, realistic IR with a 3-cm low-beta has been designed. It is comprised of essentially two quadrupoles; i.e., a doublet, as indicated in Fig. 1. Here, because of the much slower rise of the betatron functions as a function of distance from the point of collision, the first quadrupole gradient is strong enough to focus effectively before entering the second quadrupole. Consequently, a standard focusing doublet can be employed so that the β functions in both planes reach a minimum in approximately 50 m. In this design the first quadrupole is placed at the nominal 6.5 m from the IP. The maximum betatron function, as can be seen from the figure is 31.3 km, which is to be compared with the 400-km peak β in the 3 mm, Gallardo-Palmer IR. The 1-cm IR displayed in Figure 3 is an antisymmetric arrangement which allows the natural chromaticities to be minimized in both planes. The natural chromaticities have been optimized to values of $\xi_x = \xi_y = -270$.

Also investigated was the possibility of positioning the quadrupoles closer to the IP. We reduced the 6.5 m separation to 3 m. With lengths of approximately 6.5 m, the centers of the first and second quadrupoles are $s_F = 9.75$ m and $s_D = 16.25$ m away from the IP. Reducing the distance to the IP by 3 m shortens them to 6.25 and 12.75 m, respectively. Therefore, according to Eq. (3.3), we expect the chromaticity to decrease by approximately a factor of 1.04. In actual design, a reduction factor of 1.2 was found. It can be concluded that moving the quadrupoles nearer to the IP does not result in a large reduction in the chromaticity of the region.

In order to obtain some insight into the problem, we need to introduce the *bending power* or *bending efficiency* of the first IR quadrupole [4], for its effect of altering the slope of the betatron function. The change in Twiss parameter α across the quadrupole is given by

$$\Delta\alpha = \frac{\beta}{f}. \tag{4.1}$$

183

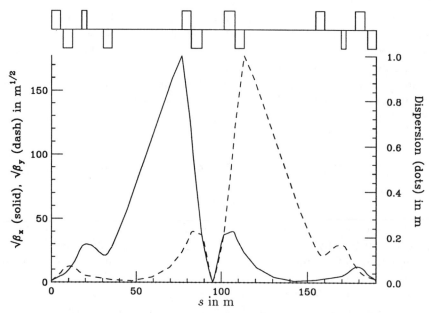

Dispersion max/min: 0.00000/−0.00000m, γ_t: (0.00, 0.00)

β_x max/min: 31433.13/ 0.03000m, ν_x: 1.56763, ξ_x : −270.32, Module length: 190.0354m

β_y max/min: 31433.13/ 0.03000m, ν_y: 1.58763, ξ_y : −270.32, Total bend angle: 0.00000 rad

Figure 3. The antisymmetric 3-cm β^* IR.

The maximum strength of the quadrupole depends on the transverse size of the beam, and therefore the betatron function; thus, $f \propto \sqrt{\beta}$. Therefore, the bending power of the first IR quadrupole is

$$\Delta\alpha \propto \sqrt{\beta} \propto s_F. \qquad (4.2)$$

As a result, when s_F is reduced, although the quadrupole strength is increased, its ability to bend the betatron function has, in fact, been lowered. This explains why the chromaticity will not decrease in the same way s_F decreases.

Since the outside of a superconducting quadrupole can be contained within a 40 cm radius, the clearance angle at the IP is $\theta \lesssim \pm 61$ mr when the IP-quadrupole separation is 6.5 m and $\lesssim \pm 134$ mr when it is 3.0 m. These angles are well below the detector acceptance criterion of $\theta \leq \pm150$ mr.

B. CHROMATICITY CANCELLATION

Because the inherent natural chromaticities of this IR are so small, local correction at the IR can be avoided. Sextupoles can be located in the arcs to do this job, reducing the length of the IR tremendously.

In lieu of a more complete design for the entire collider ring, we have used a quasi-isochronous $\beta^* = 3$ mm ring design from earlier work for computational purposes. Because of huge chromaticities arising from two $\beta^* = 3$ mm IR's, cancellation of chromaticities using only sextupoles in the arcs was never successful. In this work, we replace the IR's with 3-cm IR's and proceed to estimate the strengths of the correction sextupoles required. The constructed ring is two-fold symmetric. Half of it is composed of

Half Ring = 21 Normal Modules + IR + Zero-Dispersion Straight ,

where the *normal module* is a flexible momentum-compaction module [5] depicted in Fig. 4. With the previous lattice and this new IR, chromaticities were obtained which are listed in Table III.

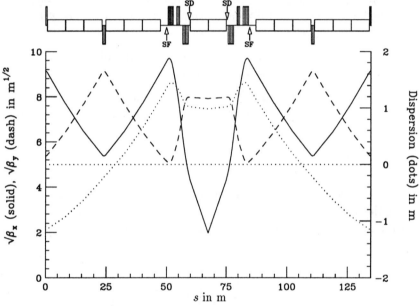

Dispersion max/min: 1.44677/−1.15250m, γ_t: (1066.87, 0.00)
β_x max/min: 94.11/ 3.90830m, ν_x: 0.75000, ξ_x : −1.32, Module length: 134.7766m
β_y max/min: 83.55/25.71391m, ν_y: 0.46953, ξ_y : −0.65, Total bend angle: 0.14096690 rad

Figure 4. The normal module.

For the correction scheme, two horizontal and two vertical sextupoles SF and SD can be placed in each basic module. In practice, sextupoles were placed only in 20 modules. This is because each module has a horizontal tune advance close to 270° and the nonlinearity due to sextupoles will be roughly canceled

Table III: Chromaticity for half a 3 cm β^* storage ring.

	IR	Arc	Total
ξ_x	-270	-176	-446
ξ_y	-270	-44	-314

every 4 consecutive modules. Using SYNCH [6], the integrated sextupole strengths required to reduced ξ_x by -446 and ξ_y by -314 were computed. The results are $S_F = 1.405$ m^{-2} and $S_D = -2.298$ m^{-2}.

If one assumes a maximum pole-tip field of 9.5 T for these sextupoles with a half aperture of 3 cm, we obtain $B'' = 2111$ T/m^2. For a muon energy of 2 TeV, the sextupole lengths are calculated to be 0.44 m for the horizontal and 0.73 m for the vertical. The space available in the normal module design is 3 m for SF and 0.82 m for SD. The betatron functions and dispersion at these points are listed in Table IV. Obviously, further optimization can be done; for example, increasing β_y and dispersion will shorten the length of SD.

Table IV: Lattice functions at sextupoles.

	β_x	β_y	D
S_F	95.3 m	26.0 m	1.295 m
S_D	18.5 m	64.2 m	1.038 m

C. NONLINEAR TUNE SPREADS

Chromaticity-correction sextupoles give rise to nonlinearities in the lattice optics. The most important nonlinear terms to evaluate are the second-order tune spreads as a function of transverse amplitudes. These are given by

$$\delta\nu_x = \alpha_{xx}\frac{\epsilon_x}{\pi} + \alpha_{xy}\frac{\epsilon_y}{\pi},$$

$$\delta\nu_y = \alpha_{yy}\frac{\epsilon_y}{\pi} + \alpha_{xy}\frac{\epsilon_x}{\pi}, \tag{4.3}$$

where ϵ_x and ϵ_y are the unnormalized transverse emittances of the muon beams. Calculations using SYNCH give for this arrangement of sextupoles for the *half* collider ring,

$$\begin{aligned}
\alpha_{xx} &= -0.263 \times 10^6 \text{ m}^{-1}, \\
\alpha_{xy} &= -0.566 \times 10^5 \text{ m}^{-1}, \\
\alpha_{yy} &= +0.492 \times 10^5 \text{ m}^{-1}.
\end{aligned} \tag{4.4}$$

Using 2σ for a transverse beam size, then $\epsilon_x = \epsilon_y = 1.056 \times 10^{-8}\pi$ m, and the maximum tune spreads are

$$\Delta\nu_x = -0.0034 \quad \text{and} \quad \Delta\nu_y = -0.0006, \tag{4.5}$$

which are clearly stable. For comparison, the Gallardo-Palmer [2] tune shift coefficients are

$$\begin{aligned}
\alpha_{xx} &= -0.564 \times 10^9 \ \text{m}^{-1}. \\
\alpha_{xy} &= -0.931 \times 10^8 \ \text{m}^{-1}, \\
\alpha_{yy} &= -0.315 \times 10^{10} \ \text{m}^{-1},
\end{aligned} \tag{4.6}$$

leading to large tune spreads for a single IR of:

$$\Delta\nu_x = -6.9 \quad \text{and} \quad \Delta\nu_y = -34. \tag{4.7}$$

We do learn that Gallardo and Palmer have recently improved the sextupole placements in their IR and brought the tune spreads down to an acceptable value of ~ 0.08. However, the correction scheme at the IR remains very complicated.

V. THE 1-CM INTERACTION REGION

A second IR region has also been attempted producing a β^* of 1 cm with a pair of doublets. Because of the larger horizontal betatron function and, therefore, larger aperture, the first quadrupole has a relatively weaker gradient than the one used in the 3-cm IR. It is just strong enough to damp the rise of β_x before entering the defocusing quadrupole. The relative strengths of the doublets have been adjusted to equalize the maximum betatron amplitudes in both planes. (They attain a maximum of approximately 28 km.) Equal maxima optimize the chromatic properties of the region and allow the IR to be symmetric about the collision point. Again, this is essentially a doublet scheme and provides the minimal chromaticity for a 1 cm low-beta region. In our design, the horizontal and vertical chromaticities are each about -500, independent of whether a symmetric or antisymmetric arrangement is constructed. A symmetric arrangement is shown in Fig. 5.

If we try to correct for the chromaticities as was done for the 3-cm IR, the corresponding sextupole strengths are $S_F = 2.187$ m^{-2} and $S_D = -3.890$ m^{-2} (implying sextupole lengths of 0.68 m and 1.24 m, respectively). Although the sextupole SD is too long to fit into the present normal module, small modifications will provide the required space. A previous design which eliminates one center dipole could be used to supply the extra space needed for sextupoles.

For the 1 cm IR, the nonlinear tune-spread coefficients are

$$\begin{aligned}
\alpha_{xx} &= -0.600 \times 10^6 \ \text{m}^{-1}, \\
\alpha_{xy} &= -0.170 \times 10^6 \ \text{m}^{-1}, \\
\alpha_{yy} &= +0.115 \times 10^6 \ \text{m}^{-1}.
\end{aligned} \tag{5.1}$$

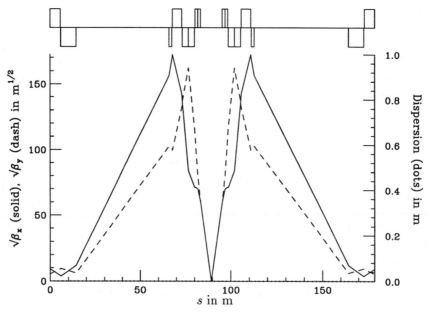

Dispersion max/min: 0.00000/−0.00000m, γ_t: (0.00, 0.00)

β_x max/min: 29739.78/ 0.01000m, ν_x: 1.45272, ξ_x : −501.21, Module length: 178.7508m

β_y max/min: 26296.47/ 0.01000m, ν_y: 1.54576, ξ_y : −544.96, Total bend angle: 0.00000 rad

Figure 5. The symmetric 1-cm β^* IR.

The maximum tune spreads at a 2σ amplitude in half of the collider ring are therefore

$$\Delta\nu_x = -0.0081 \quad \text{and} \quad \Delta\nu_y = 0.0018 \,, \tag{5.2}$$

which are still very small.

VI. CONCLUSION

In conclusion, we have analyzed both the advantages and the disadvantages of raising the low-beta specification for the IR of the muon collider ring. If β^* is raised from 3 mm to 1 cm, luminosity will be sacrificed by at most a factor of 2.3. In return, the structure of the IR becomes intrinsically simple. The proposed doublet scheme results in maximum betatron functions of only 30 km which represents a factor of 10 less than the 3 mm IR. At the same time, the natural chromaticities are about −500, which is more than a factor of 10 lower than a 3-mm IR. Also significant, local correction of chromaticity in the IR is eliminated. Chromaticity sextupoles can be placed in the arcs which correct the IR chromaticity, but contribute only minimally to the second-order tune spreads (as a function of amplitude). In constrast, an IR with high natural

188

chromaticities requires local correction; often giving rise to large nonlinearities. In this case, correction sextupoles flanking the IP must be carefully placed and the lattice carefully designed so that nonlinear terms cancel to a high degree of accuracy. For example, the betatron functions at a pair of consecutive sextupole locations must be equal to at least 3-4 significant figures and their phase difference has to be exactly equal to π rad to 3-4 significant figures [7]. This level of cancellation is very difficult to maintain. With the IR designs presented in this paper, no high-degree of cancellation is required, the ring lattice is tunable, and, operationally, it is more stable.

Our main conclusion is that relaxing the low-beta criterion from 3 mm to at least 1 cm dramatically decreases both the length and complexity of the IR. It results in a collider ring design which is inherently stable and with a tuning range. In summary, it is our view that a small sacrifice of luminosity is worth the price of an operable machine.

References

[1] See for example, R.B. Palmer, "Beam Dynamics Problems in a Muon Collider," Beam Dynamics Newsletter No. 8, 1995, p. 27, ed. K. Hirata, S.Y. Lee, and F. Willeke.

[2] J.C. Gallardo, private communication, 1995.

[3] Proc. 9th Advanced ICFA Beam Dynamics Workshop, Montauk, New York, Oct. 15-20, 1995, ed. J. Gallardo.

[4] K.Y. Ng, "A Study of the Autin-Wildner IR Scheme," these proceedings.

[5] S.Y. Lee, K.Y. Ng, and D. Trbojevic, Phys. Rev. **E48**, 3040 (1993).

[6] A.A. Garren, A.S. Kenney, E.D. Courant, and M.J. Syphers, "A User's Guide to SYNCH," Fermilab Report FN-420, 1985.

[7] R. Brinkmann, "Optimization of a Final Focus System for Large Momentum Bandwidth," DESY-M-90/14 (1990); B. Dunham and O. Napoly, "FFADA, Final Focus. Automatic Design and Analysis," CERN Report CLIC Note 222, (1994); O. Napoli, "CLIC Final Focus System: Upgraded Version with Increased Bandwidth and Error Analysis," CERN Report CLIC Note 227, (1994).

Doublet final focus

Bruno Autin

CERN, PS Division, 1211 Geneva 23 (Switzerland)

Abstract. A doublet scheme is designed in a self consistent analytical way for the low β insertion of a μ⁺-μ⁻ collider. It assumes focusing strengths of same magnitude and opposite signs in the quadrupoles. At the matching point, the β values are equal and the α values opposite. Two solutions using superconducting or permanent quadrupoles are discussed.

INTRODUCTION

The high luminosity of a muon collider requires a beta value as low as 3 mm at the interaction point (1). Several optical modules can be contemplated to achieve a round beam in the interaction region. The one discussed here is a doublet made of identical quadrupoles of opposite focusing strengths. Their thin lens parameters obey a very simple conjugation law and the generalization to finite length elements results from the solution of a transcendental equation. In order to reduce the chromaticity of the final focus, the quadrupoles have to be as close as possible to the interaction point yet satisfying the detector constraints. Permanent and superconducting quadrupoles are compared and the parameters of a superconducting scheme are given.

THIN LENS DOUBLET

The final focus of a round beam is a doublet made of two quadrupoles Q_1 and Q_2 of opposite focal lengths separated by a distance d. The theory of the doublet shows that the horizontal and vertical β-functions can be transported from the interaction point where they are both equal to $β^*$ to a matching point where the canonical conditions

$$\beta_h - \beta_v = 0$$
$$\alpha_h + \alpha_v = 0$$

are fulfilled if the distances from the interaction point to the next quadrupole Q_1 and from the matching point to the other quadrupole Q_2 are equal to the same value l^* and if the conjugation relation (2)

$$f = \sqrt{dl^*}$$

is satisfied. The β-value at the matching point is then the same as the value at Q_2 in the absence of doublet

$$\beta = \frac{(d + l^*)^2}{\beta^*}$$

and the absolute value of α

$$\alpha = \frac{d + l^*}{\beta^*} \sqrt{\frac{d}{l^*}}$$

is reduced with respect to the α–value at Q_2 by the factor $\sqrt{d/l^*}$ which is called k and plays a basic role in the discussion. All that expressions assume that β^* is small.

THICK LENS DOUBLET

The distance between quadrupoles is limited by the length l of the quadrupoles and its minimum value is precisely l. It will then be assumed that

$$d = l = k^2 l^*$$

and the focal length is re-written

$$f = kl^*$$

When the doublet is made of thick lenses, the global properties are maintained but the parameter is no longer a focal length but a phase u, solution of the transcendental equation

$$u(\cosh u \sin u - \cos u \sinh u) = a \cos u \cosh u ,$$

where a is the ratio between the length of the quadrupole and the length of the final drift space

$$a = \frac{l}{l_0},$$

a is related to k through the relation

$$a = \frac{2k^2}{2-k^2},$$

and the constant k must be smaller than $\sqrt{2}$. The numerical solution of the transcendental equation for various values of a shows that the fitting function

$$u_0 = 1.11205\sqrt{k} - 0.0108984\,k + 0.0905792\,k^3$$

is a very precise analytical solution in the interval $(0, \sqrt{2})$. The numerical solutions together with the fitting function are represented in Figure 1.

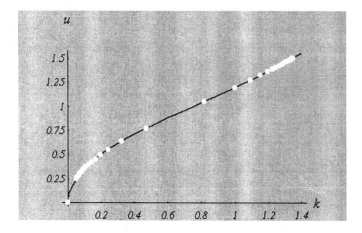

FIGURE 1. Variation of the phase u with the ratio k.

The phase u is related to the focusing strength K_L of a long quadrupole and to the length l through the relation

$$u = \sqrt{K_L}\,l$$

whereas the focal length of a thin lens depends on K_S and l according to the formula

$$f = \frac{1}{K_s l}$$

All the calculations of K in the thin lens approximation must then be corrected by the factor u_0^2/k to be valid for real quadrupoles. The focusing strength K is proportional to the gradient G of the magnetic field

$$K = \frac{e}{p} G$$

e and p being the charge and the momentum of the particle. The gradient can be expressed as the ratio B/r where B is the magnetic field at a beam radius r of sufficiently large emittance.

$$r = \sqrt{\frac{\varepsilon_N \beta}{\gamma}} \, ,$$

ε_N is the normalized emittance and γ the Lorentz factor. The β-function takes its maximum value β_2 at Q_2 and is equal to

$$\beta_2 = \frac{l^{*2}}{\beta^*} (1 + k + k^2)^2$$

By performing all the substitutions and equating the optical and magnetic expressions of the focal length, the final length is found to be

$$l^* = \frac{u_0^2 (1 + k + k^2)}{k^4 B} \frac{p}{e} \sqrt{\frac{\varepsilon_N}{\gamma \beta^*}}$$

The parameter k must be such that the contribution of Q_2 to the chromaticity

$$\frac{\beta_2}{4\pi f}$$

is as low as possible. It turns out that the function of k which enters the chromaticity is

$$\frac{u_0^2 (1 + k + k^2)^2}{k^5}$$

and it decreases with k. The quadrupoles must thus be as close as possible to the interaction point. The distance l_0 from the IP is given by the blind angle θ allowed by the detector:

$$l_0 = \frac{q\,r}{\theta}$$

where q is the ratio of the outer radius of the quadrupole to the beam radius. The previous relation can then be written as an equation in k:

$$1 - \frac{k^2}{2} = \frac{q}{\theta}\sqrt{\frac{\varepsilon_N}{\gamma\beta^*}}\left(1 + k + k^2\right)$$

DOUBLET PARAMETERS

The determination of the parameters starts with the evaluation of k from the above equation, then the lengths l^* and l are calculated and the properties of the doublet in the thin lens approximation are found with a dedicated function of the *BeamOptics* (3) program. Last, the real beam envelope in thick quadrupoles is derived by scaling the focusing strengths of the thin lens model.

The permanent and superconducting solutions will first be compared in the thin lens approximation for the general parameters

$$\varepsilon_N = 5\ 10^{-5}\ \text{m} \quad p = 2\ \text{TeV} \quad \beta^* = 3\ \text{mm} \quad \theta = .1\ \text{radian}$$

The normalized emittance corresponds to the r.m.s. value estimated after cooling. This is too small for a realistic design but too large emittances would lead to optics parameters very hard, if not impossible, to achieve at least for a doublet and for so small a β^* value. For permanent and superconducting quadrupoles, the field at the edge of the beam and the factor q are chosen as listed in the following table

	B [T]	q
permanent quadrupoles	1.2	3
superconducting quadrupoles	4	20

and the pattern of the β-functions is shown in Figure 2. Although the permanent magnets are closer to the interaction point, their relatively low field forces them to be much longer than the superconducting quadrupoles and the ratio of the peak β-values is almost 10. The chromaticity of the permanent magnet scheme is also higher in absolute value than in the super-conducting case: 5000 against 1500. It has nevertheless to be checked that a field of 4 Tesla is possible. As a comparison, the field in LHC low-β quadrupoles (4) is only 1.35 T at a radius equal to 10 times the beam r.m.s. value.

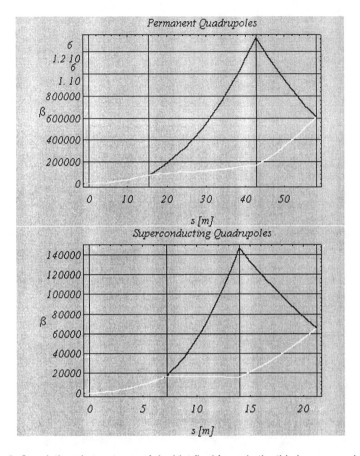

Figure 2. β-variations in two types of doublet final focus in the thin lens approximation.

The parameters of the super-conducting magnets are retained for the thick lens doublet (Figure 3). It turns out that the matching conditions are fulfilled but that the β-values are 10% higher about and that the doublet acceptance is reduced accordingly. Another aspect of this scheme is that the length of the matching straight section, which is equal to the interaction region length, is much longer than for permanent magnets and there is thus room to put other quadrupoles to complete the matching with the regular part of the lattice. The variations of the beam size are plotted in Figure 4, the outer radius of the quadrupole tank is slightly inferior to 400 mm and fits the solid angle of .1 radian since the interaction length is 3.83 m.

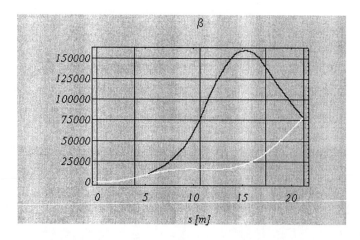

Figure 3. β-variations in a doublet made of thick super-conducting quadrupoles.

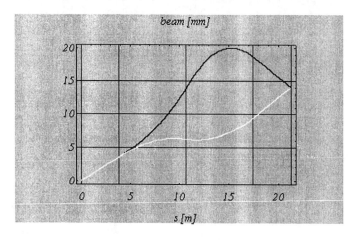

Figure 4. Beam size in a doublet made of thick super-conducting quadrupoles.

CONCLUSION

An analytical technique has been presented to derive the parameters of a doublet final focus. Its application to a muon collider shows that a doublet scheme is questionable in the present status of the parameters especially because the beam size has been taken to its r.m.s. value to maintain the maximum β-value within the already scary hundred kilometer range. Some improvement might be gained in reducing the aperture of the quadrupole next to the interaction point to shorten its length yet maintaining its focusing strength. If the doublet has the advantage to permit the optimum bunch length, it is severely limited by the defocusing effect in one plane. A triplet suffers from the same restriction. In contrast, a single lens avoids the dramatic de-focusing at the cost of a flat beam; the net consequence is either a loss of luminosity or a shorter bunch length. The machine performance however is not only determined by such basic machine parameters as β^*, it is also affected by its dynamic aperture and the quadrupole technology. It is thus necessary to carefully compare several schemes and, among them, the singlet scheme, to get a realistic global design.

REFERENCES

1. Palmer, R., Fernow R., Gallardo J., Lee Y., Torun, Y., Neuffer, D., Winn, D., "High-energy high-luminosity μ^+ μ^- collider design", Proceedings of 1995 Particle Accelerator Conference, Dallas, TX, May 1-5, 1995.
2. Autin, B., "Analytical model of final focus structures for flat and round beams at the interaction point", in *New Computing Techniques in Physics Research II*, World Scientific, 1992, pp. 763-771.
3. Autin, B., and Wildner E., "*BeamOptics*, a program for symbolic beam optics", PS/DI/Note 95-18.
4. The LHC study group, "LHC, the Large Hadron Collider Accelerator Project", CERN/AC/95-05

A Study of the Autin-Wildner IR Scheme

King-Yuen Ng

Fermi National Accelerator Laboratory, P.O. Box 500, Batavia, IL 60510

Abstract

The Autin-Wildner scheme [1] of implementing an interaction region (IR) with $\beta_x^* = \beta_y^*$ using two quadrupoles of equal but opposite strength is investigated. The impacts on chromaticities from quadrupole strength, low beta, and clearance from IP to first quadrupole are studied.

I. INTRODUCTION

Specifications for a 2 TeV muon collider call for a round beam at the point of collision [2]. Recently, Autin and Wildner [1] suggested a doublet scheme of quadrupoles to achieve the low-betatron functions for a round beam. The method is sketched in Fig. 1, where both the focusing and defocusing quadrupoles have the same focal length f, and are considered to be *thin*. The

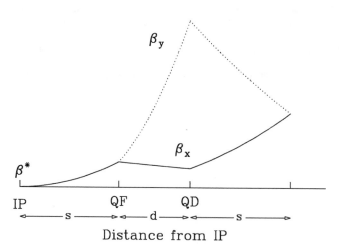

Figure 1: The Autin-Wildner doublet scheme at the IR.

---

*Operated by the Universities Research Association, Inc., under contracts with the U.S. Department of Energy.

first quadrupole, say the focusing one, QF, is at a distance s from the IP. It focuses the horizontal betatron function β_x, while it increases the rate of divergence of the vertical betatron function β_y. The second quadrupole QD is defocusing and positioned at a distance d from QF. It reverses the rise of β_y. The Autin-Wildner theorem says that, if the distance d between the two quadrupoles satisfies

$$f = \sqrt{sd}, \tag{1.1}$$

then, at a distance s downstream from QD, the two betatron functions are equal and their slopes obey the relation,

$$\alpha_x + \alpha_y = 0. \tag{1.2}$$

The proof of this theorem is straightforward but rather tedious. In this paper, we will study some general properties of the Autin-Wildner scheme: its feasibility and how the chromaticities are affected by the low-beta value at the IP, the quadrupole strength, and the clearance between the IP and first quadrupole. Although thin lenses are used in this model, the overall conclusions of our study should not be altered much when using a thick-lens calculation; therefore our results should serve as some guide lines in the actual design.

II. MINIMUM QUADRUPOLE LENGTH

The defocusing quadrupole will experience the largest betatron value and therefore have the largest aperture. If we require a 5-sigma aperture for the beam, then the half aperture of the quadrupole is given by

$$y = 5\sqrt{\frac{\beta_{yD}\epsilon}{\pi}}, \tag{2.1}$$

where the unnormalized rms emittance ϵ of the beam is related to the normalized emittance by

$$\epsilon = \frac{\epsilon_N}{\beta\gamma}, \tag{2.2}$$

with β and γ being the relativistic parameters of the beam. The vertical betatron function β_{yD} at QD can be derived using the relation of Eq. (1.1):

$$\beta_{yD} = \beta^* \left(1 + \frac{f}{s}\right)^2 + \frac{s^2}{\beta^*}\left(1 + \frac{f}{s} + \frac{f^2}{s^2}\right)^2. \tag{2.3}$$

With the pole-tip field B, the maximum field gradient of the quadrupole is therefore $G = B/y$, and the maximum quadrupole strength is

$$\frac{1}{f} = \frac{GL}{(B\rho)} = \frac{BL}{5(B\rho)}\sqrt{\frac{\pi}{\beta_{yD}\epsilon}}, \tag{2.4}$$

where $(B\rho)$ is rigidity of the beam and L the length of the *thin* quadrupoles.

For simplicity, let us define a parameter

$$\lambda = \frac{BL}{5(B\rho)}\sqrt{\frac{\pi}{\epsilon}} = \frac{\sqrt{\beta_{yD}}}{f} \qquad (2.5)$$

to denote the quadrupole length. Using Eqs. (2.3) and (2.4), one arrives at

$$\lambda = \frac{\sqrt{\beta^{*2}(f+s)^2 + (f^2 + fs + s^2)^2}}{\sqrt{\beta^*}fs} . \qquad (2.6)$$

In the case where the quadrupoles are extremely strong; i.e., $f \ll s$, the separation between the quadrupoles $d = f^2/s \to 0$, and the vertical betatron function at QD is just given by the simple quadratic expression

$$\beta_{yD} = \beta^* + \frac{s^2}{\beta^*} . \qquad (2.7)$$

Then we have

$$\lambda = \frac{\sqrt{\beta_{yD}}}{f} \to \frac{s}{f\sqrt{\beta^*}} . \qquad (2.8)$$

On the other extreme, when $f \gg s$, the focusing quadrupole is so weak that it does not significantly affect the vertical betatron function β_y, the defocusing quadrupole, being equal in strength to the focusing one, is also weak and it can only reverse the growth of β_y when the latter becomes very large. This is because the bending power of a quadrupole is given by $\Delta\alpha = \beta/f$. Here, the distance of QD from the IP is

$$s + d \approx d = \frac{f^2}{s} , \qquad (2.9)$$

so that β_{yD} can be computed according to Eq. (2.7), and we obtain

$$\lambda = \frac{\sqrt{\beta_{yD}}}{f} \to \frac{f}{s\sqrt{\beta^*}} . \qquad (2.10)$$

We learn from Eqs. (2.8) and (2.10) that $\lambda \to \infty$ as $f \to \infty$ or $f \to 0$. Thus, not all quadrupole lengths are possible and there is a minimum length.

In general, we do not want to keep s constant. Instead, we like to keep the distance s_0 from the IP to the *front end* of the first quadrupole constant. In other words,

$$s = s_0 + \frac{L}{2} , \qquad (2.11)$$

200

and Eq. (2.10) is an implicit function of the quadrupole length. To find the minimum quadrupole length, we equate $d\lambda/df$ to zero to obtain

$$f = (\beta^{*2}s + s^3)^{1/3} \approx s. \tag{2.12}$$

This implies

$$\lambda_{\min} \approx \frac{3}{\sqrt{\beta^*}}, \tag{2.13}$$

or

$$L_{\min} \approx \frac{15(B\rho)}{B}\sqrt{\frac{\epsilon}{\beta^*\pi}}. \tag{2.14}$$

It is important to note that the minimum allowable quadrupole length is independent of s_0.

Let us study the situation of a low-beta of $\beta^* = 3$ mm and a 2 TeV muon beam with normalized emittance $\epsilon_N = 50 \times 10^{-6}\pi$ m. If permanent magnets with a maximum pole-tip field of $B = 1$ T are used, we find immediately that the minimum quadrupole length is $L_{\min} = 93.9$ m, which implies that such magnets are too weak to be practical in building the IR. Even with conventional magnets and boosting the maximum field to $B = 2$ T, $L_{\min} = 47.0$ m is still too long to be practical.

Now, let us consider the strongest superconducting magnets which have a pole-tip field of 9.5 T. We obtain $L_{\min} = 9.88$ m. The distance d between the centers of the quadrupoles must be larger than L, or obviously the physical placement of the quadrupoles will not be possible. For a clearance of $s_0 = 6.5$ m between the IP and the first quadrupole, the quadrupole length L and the quadrupole separation d are plotted as functions of quadrupole focal length f in Fig. 2. We see that the physical allowable region starts at $f > 10.07$ m and the quadrupole length changes very slowly about its minimum value.

III. CHROMATICITY

Since the vertical betatron function will be much larger than the horizontal, the vertical chromaticity will be larger also. In this simple model, the vertical chromaticity receives its contribution from the defocusing quadrupole only and is given by

$$\xi_y = -\frac{\beta_{yD}}{4\pi f} = -\frac{\beta^{*2}(f + s)^2 + (f^2 + fs + s^2)^2}{4\pi\beta^* s^2 f}. \tag{3.1}$$

We can see with the help of Eq. (2.3) that $\xi_y \to -\infty$ regardless of whether $f \to 0$ or ∞. Thus, the chromaticity has a minimum also. In Fig. 2, the chromaticity is plotted as dots. It is clear that to minimize chromaticity, we must choose a quadrupole focal length of ~ 10 m and a quadrupole focal length as small as possible. In fact, it is true in general that, to minimize

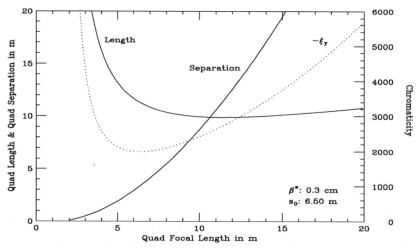

Figure 2: Quadrupole length and chromaticity versus quadrupole focal length when $\beta^* = 3$ mm and IP clearance $s_0 = 6.5$ m.

chromaticity, we should choose f at the position in the plot when L crosses d. In other words, the distance between the two quadrupoles must be made as short as possible.

IV. CLEARANCE BETWEEN IP AND QUADRUPOLE

For the detector acceptance, an open angle of $\theta \lesssim \pm 150$ mrad at the IP is required. Therefore, the clearance between the IP and the first quadrupole can actually be smaller. If the IR quadrupoles can be placed closer to the IP, the betatron functions at the quadrupoles will be smaller. However, for smaller β_x, the bending power of a quadrupole will not be so efficient. This can be seen as follows. At the center of the first IR quadrupole,

$$\beta_x \approx \frac{s^2}{\beta^*}. \tag{4.1}$$

The maximum strength of QF is

$$\frac{1}{f} = \frac{BL}{5(B\rho)} \sqrt{\frac{\epsilon}{\pi \beta_x}}. \tag{4.2}$$

The bending efficiency of the quadrupole can be defined as the change of the Twiss parameter α. Therefore, at the first IR quadrupole,

$$\Delta\alpha_x = \frac{\beta_x}{f} \approx \frac{BL}{5(B\rho)} \sqrt{\frac{\epsilon}{\pi \beta^*}} s. \tag{4.3}$$

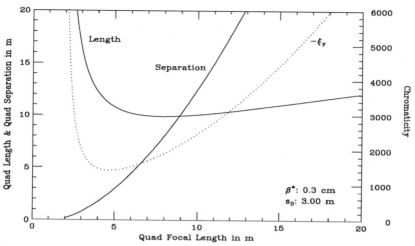

Figure 3: Quadrupole length and chromaticity versus quadrupole focal length
when $\beta^* = 3$ mm and IP clearance $s_0 = 3.0$ m.

Nevertheless, the vertical chromaticity which is given by β_y at QD is in general
somewhat smaller when s_0 is shortened. In Fig. 3, we make a similar plot with
s_0 shortened from 6.5 m to 3 m. We see that the maximum possible strength
of the quadrupoles is increased so that the focal length becomes 8.9 m. The
length of the quadrupoles is still around 10 m. The chromaticity is reduced
from $\xi_y = -2600$ to -2200. The reduction is only about 15%, although s_0 has
been reduced from 6.5 m to 3 m. However, if we use the distance from the IP to
the center of QF, this distance is actually reduced from $s = s_0 + L/2 = 11.5$ m
to 8.5 m only.

In passing, it is worth pointing out that the natural chromaticity of an IR
with $\beta^* = 3$ mm and $s_0 = 6.5$ is in practice of order -6000 [3], much larger
than the value of -2600 quoted above. This is because due to the limitation
of the quadrupole strengths, the doublet scheme is not able to lower the Twiss
parameters to reasonable values to match to a normal cell or module of the
collider ring. More quadrupoles are required and this raises the chromaticities.

V. LARGER LOW-BETAS

From Eq. (3.1), it is clear that chromaticity will be reduced if β^* is increased, the reason being that β_y which is inversely proportional to β^*, increases less rapidly. We perform the same plot of quadrupole length and
vertical chromaticity versus the quadrupole focal length in Fig. 4, for an IR

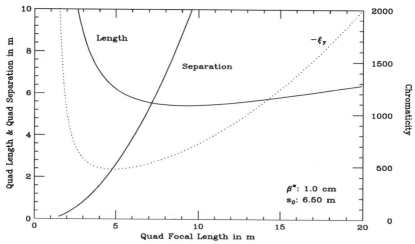

Figure 4: Quadrupole length and chromaticity versus quadrupole focal length when $\beta^* = 1$ cm and IP clearance $s_0 = 6.5$ m.

with $\beta^* = 1$ cm and IP clearance $s_0 = 6.5$ m. We see that the quadrupole length is around 5.5 m long and the focal length can be made as short as $f = 7.2$ m, which is shorter than the distance from the IP to the center of the first quadrupole, $s = s_0 + L/2 = 9.25$ m. This implies that the first quadrupole is capable of bending β_x by so much that α_x after passing through the quadrupole becomes positive, so that β_x starts dropping down after exiting this first quadrupole. In contrast, in the 3-mm low-beta IR described in Sec. II above, the first quadrupole has $f_{\min} = 10.07$ m, while $s = s_0 + L/2 = 11.4$ m, and β_x actually continues to increase after leaving the focusing quadrupole. Later, after passing through the defocusing quadrupole β_x will start to sky-rocket and another focusing quadrupole will be necessary to reverse its slope eventually. In other words, this Autin-Wildner doublet scheme is actually not feasible in the 3-mm low-beta IR.

According to Fig. 4, a vertical chromaticity as low as -500 can be achieved for the 1-cm IR. In fact, in an actual design [4], we did succeed in obtaining such an IR with $\xi_y \sim -500$.

Similarly, we plot in Fig. 5, the quadrupole length and vertical chromaticity as functions of quadrupole focal length when the low-beta is $\beta^* = 3$ cm. Now the quadrupole length can be made about 3.2 m and the quadrupole focal length $f_{\min} = 5.2$ m. Here, the first IR quadrupole is very strong. The horizontal betatron function β_x drops very fast after passing through the focusing quadrupole and does not rise significantly even after passing through the defocusing quadrupole, so that when it crosses the vertical betatron function at a distance $2s+d$ from the IP, where $\alpha_x+\alpha_y = 0$, both $|\alpha_x|$ and $|\alpha_y|$ are small and can be matched easily to ordinary cells or modules of the collider. In contrast, in the situation of the 3-mm low-beta IR, although we still have $\alpha_x + \alpha_y = 0$ when β_x and β_y cross according to Eq. (1.2), both $|\alpha_x|$ and $|\alpha_y|$ may be of order several thousands. Therefore, many more quadrupoles will be needed to control the betatron functions after this point, and chromaticities increase. Going back to the 3-cm low-beta IR, from Fig. 5 the vertical chromaticity can be as low as ~ -140. This prediction was also realized in an actual design [4]. There, we deviate from the Autin-Wildner model by employing a stronger focusing quadrupole and a weaker defocusing quadrupole, so that the horizontal betatron function continues to drop even after passing through the defocusing quadrupole. In this way, both β_x and β_y can be controlled more easily.

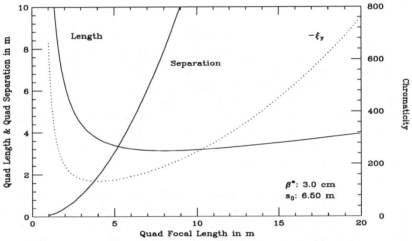

Figure 5: Quadrupole length and chromaticity versus quadrupole focal length when $\beta^* = 3$ cm and IP clearance $s_0 = 6.5$ m.

References

[1] B. Autin and E. Wildner, "BEAMOPTICS, a Program for Symbolic Beam Optics," CERN Internal Report PS/DI/Note 95-18, 1995.

[2] See for example, R.B. Palmer, "Beam Dynamics Problems in a Muon Collider," Beam Dynamics Newsletter No. 8, 1995, p. 27, ed. K. Hirata, S.Y. Lee, and F. Willeke.

[3] J. Gallardo, private communication.

[4] C. Johnstone, K.Y. Ng, and D. Trbojevic, "Interaction Regions with Increased Low-Betas for a 2-TeV Muon Collider," these proceedings.

STUDIES OF COLLECTIVE INSTABILITIES IN MUON COLLIDER RINGS

Wen-Hao Cheng, Andrew M. Sessler and Jonathan S. Wurtele
Lawrence Berkeley National Laboratory
Berkeley, California 94720

Abstract

Using a macroparticle model for both synchrotron motion and beta-tron motion, we provide formulae for the beam centroid shift, the tune shifts and the growth rates of the longitudinal and transverse head-tail instabilities. Both single turn and multi-turn effects are included. The transient amplitudes based on the two-particle model are estimated. We estimate the microwave instability for the muon collider, using three sets of design parameter, and three different impedance models: resonator, resistive wall and a SLAC-like or a CEBAF-like rf accelerating structure. For momentum compaction $\alpha \simeq 10^{-5}$, the synchrotron oscillation period is about a hundred turns, the impedance of the ring is dominated by the rf cavities, and the microwave instability is well beyond threshold. For $\alpha < 10^{-6}$, the synchrotron oscillation period is longer than the effective storage time, and the beam dynamics in the collider like that in a linac. For this case, the BBU instability is the most important collective effect. The implications of using BNS damping are briefly discussed. The growth rates of the head-tail instability due to longitudinal and transverse chromaticities, are presented. Finally we discuss briefly the beam dynamics issues related to the beam-beam interaction and a quasi-isochronous ring.

1 INTRODUCTION

The design of a muon collider ring, from the perspective of the physics of collective effects, has some unique features which need to be examined. (1) Muons have a very short life time: 41.6 ms at 2 Tev, corresponding to a thousand "effective" turns in a ring with the nominal design circumference of 7 kilometers. (2) The bunch is very short: $\sigma_z = 3$ mm. (3) The beam has a large charge: $N = 2 \times 10^{12}$. (4) The momentum compaction α

is very small: α is between 10^{-5} and 10^{-7}. These features lead us to some unusual aspects of the ring operation: (A) As the muon decays, the intensity decreases, and thus all the intensity-dependent instabilities would be most serious for the first few hundred turns. (B) The intense bunch required for the high luminosity of 2.5×10^{30}cm^{-2} per collision, makes instabilities likely. (C) Very small α requires careful estimations of nonlinear corrections to the particle orbit and to the collective dynamics.

In the case where the synchrotron tune ν_s is about $1/150$ (corresponding to $\alpha \simeq 10^{-5}$) [1], muons execute 6 synchrotron oscillations in 900 turns. For times much shorter than one-half of the synchrotron oscillation period ($T_s/2 \simeq 1.75$ ms) , particles are almost "frozen" in s, and act more like particles in a linac. A Beam-Break-Up (BBU) analysis treating the quasi-isochronous ring as a linac can thus give us the amplitude-doubling time and the BNS damping condition for the BBU instability [2].

Nevertheless, for times longer than the synchrotron oscillation period, the switching of head and tail particles stabilizes the BBU, but gives rise to the microwave instability with threshold. Evaluations of the thresholds for the impedances of a SLAC-like or a CEBAF-like rf structure show that the microwave instability is well beyond threshold.

We show that in preserving the bunch length and energy spread, the growth rate of the microwave instability per turn is scaled as the momentum compaction ($\alpha \simeq \eta$). A lattice with α of the order of magnitude 10^{-7} makes the synchrotron period about 10 times the storage time, and renders the beam dynamics almost linac-like during the effectve storage time.

Muon decay, which causes the beam intensity to decrease by about 40% in 900 turns, will be taken into account in future works.

2 MACROPARTICLE MODEL

2.1 Synchrotron Motion

The longitudinal equations of motion of a particle traveling in a circular machine can be written as [3]

$$z' = -\eta\delta, \tag{1}$$

$$\delta' = K(z), \tag{2}$$

where z is the oscillation amplitude with respect to the bunch center, $' = d/ds$, $\delta = dp/p$, $\eta = \alpha - 1/\gamma^2$, α has approximately equal contributions from pdC/Cdp and the variation of betatron amplitude at the interaction point. The force $K(z)$ that particle experiences has at least two parts, one is due to the radio frequency (rf) devices, and another the wake fields generated from the interaction between beam and cavities,

$$K(z) = K_{rf}(z) + K_{wake}(z), \tag{3}$$

where

$$K_{rf}(z) = \frac{eV_{rf}(z)}{CE} \approx \frac{\omega_{s0}^2 z}{\eta c^2}, \tag{4}$$

$$K_{wake}(z) = \frac{F_{\|wake}(z)}{E}$$

$$= -\frac{r_0}{\gamma C} \int_{-\infty}^{\infty} dz' \sum_{n=-\infty}^{\infty} \rho(z') e^{[-i\Omega(\frac{s}{c}+nT_0)]} W_0'(z - z' + ncT_0), \tag{5}$$

$T_0 = 2\pi/\omega_0 = C/c$, $\omega_{s0}^2 = \eta c \omega_{rf} \hat{V}/CE$, $E = \gamma m_0 c^2$, $r_0 = e^2/m_0 c^2$, $N = \int dz' \rho(z')$, and $C = 2\pi R$ is the circumference of the collider ring. In Eq. (5), the longitudinal monopole wakefield is integrated along the distribution in front of the considered particle, and is summed over all previous revolutions. Casuality means $W_0'(z) = 0$ if $z > 0$. In Eq. (4), we assumed the amplitude of synchrotron motion is small compared with the rf wavelength, and the rf voltage is linearized as $V_{rf}(z) = \hat{V} \sin(\omega_{rf} z/c) \approx \hat{V} \omega_{rf} z/c$. Similarly, the wake function can be approximated as

$$W_0'(z - z' + ncT_0) \simeq W_0'(ncT_0) + (z - z') W_0''(ncT_0). \tag{6}$$

The first term of Eq. (6) comes from the static effect of the parasitic loss of beam energy to the cavity. To offest the energy loss, the centroid of the particle bunch shifts,

$$z \rightarrow z - \triangle z, \tag{7}$$

where

$$\triangle z = \frac{N\eta r_0 c^2}{\gamma C \omega_s^2} \sum_n W_0'(ncT_0). \tag{8}$$

The second term of Eq. (6) creates the potential-well distortion, which is also a static effect, and gives the incoherent tune shift

$$\frac{\omega_s^2}{c^2} \rightarrow \frac{\omega_{s0}^2}{c^2} - \frac{N\eta r_0}{\gamma C} \sum_n W_0''(ncT_0). \tag{9}$$

The dynamics of the beam instability is described by the third term of Eq. (6). With the help of Eqs. (8) and (9), the equation of motion is simplified as

$$z'' + \frac{\omega_s^2}{c^2} z = -\frac{N\eta r_0}{\gamma C} \sum_n W_0''(ncT_0) e^{-i\Omega[(s/c)+nT_0]} \langle z \rangle, \tag{10}$$

where $\langle z \rangle = (1/N) \int dz' \rho(z') z'$. Fourier transforming Eq. (10) to frequency domain, the imaginary part of the mode frequency shift $(\Omega - \omega_s)$ gives the growth rate of Robinson instability. For the resonator model [cf. Appendix], in the limit $(\omega_s, \omega_R - h\omega_0) \ll \omega_R/2Q \ll \omega_0$, by tuning the machine so that $\eta(\omega_R - h\omega_0) < 0$, the longitudinal Robinson damping rate is

$$\tau_s^{-1} \simeq -\frac{4Nr_0R_s^{\|}Q^2}{\pi\gamma T_0 h}|\eta(\omega_R - h\omega_0)|, \qquad (11)$$

where h is the harmonic number, such that $h\omega_0$ close to the cavity resonance frequency ω_R. Robinson damping is a multiturn effect important for high Q structures, such as superconducting cavities.

The single turn wake gives the bunch centroid shift $\triangle z$, and the incoherent tune shift $\triangle\nu_s$, expressed explicitly via the resonator model, as

$$\triangle z/\sigma_z = 2\triangle\nu_s/\nu_s = \frac{N\eta r_0 R\omega_R R_s^{\|}}{4\pi Q\gamma\nu_s^2\sigma_z}, \qquad (12)$$

where $\nu_s = \omega_s/\omega_0$. Note that the tune shift is approximated, as $|\nu_s - \nu_{s0}| \ll \nu_{s0}$. For the resistive wall wake [cf. Appendix], we have

$$\triangle z/\sigma_z = \frac{4}{3}\triangle\nu_s/\nu_s = \frac{N\eta r_0 R^2}{2\pi\gamma\nu_s^2 b}\sqrt{\frac{c}{\sigma\sigma_z^5}}. \qquad (13)$$

The analysis above is for a rigid beam, i.e, we merely analyzed the beam as one macroparticle. For a two-particle model, we examine the longitudinal strong head-tail instability. We consider the two particles each carrying charge Ne/2 and executing synchrotron oscillations π out of phase. We further assume that only the wake field generated from the head particle can affect the tail particle, and not vice-versa. During the first half synchrotron period, i.e. $0 < s/c < T_s/2$, following Eq. (10), we have

$$z_1'' + \frac{\omega_s^2}{c^2}z_1 = 0, \quad z_2'' + \frac{\omega_s^2}{c^2}z_2 = -\frac{N\eta r_0}{2\gamma C}W_0'' z_1, \qquad (14)$$

and for the second half period, i.e. $T_s/2 < s/c < T_s$, $z_1 \leftrightarrow z_2$ in Eq. (14).

The solutions for the system after one synchrotron period can be written as

$$V_s(T_s) = M_2 \cdot M_1 \cdot V_s(0) = M_s \cdot V_s(0), \qquad (15)$$

where $V_s = [z_1, (c/\omega_s)z_1', z_2, (c/\omega_s)z_2']^T$,

$$M_2 = \begin{bmatrix} -1 & 0 & 0 & 2\Upsilon_s \\ 0 & -1 & -2\Upsilon_s & 0 \\ 0 & 0 & -1 & 0 \\ 0 & 0 & 0 & -1 \end{bmatrix}, M_1 = \begin{bmatrix} -1 & 0 & 0 & 0 \\ 0 & -1 & 0 & 0 \\ 0 & 2\Upsilon_s & -1 & 0 \\ -2\Upsilon_s & 0 & 0 & -1 \end{bmatrix}, \qquad (16)$$

and

$$\Upsilon_s \equiv -\frac{\pi N r_0 \eta c^2 W_0''}{8\gamma C\omega_s^2}. \qquad (17)$$

The eigenvalues of the transfer map M_s are

$$\lambda_{s\pm} = \begin{cases} 1 - 2\Upsilon_s^2 \pm i2\Upsilon_s\sqrt{1 - \Upsilon_s^2}, & \Upsilon_s < 1 \\ -1, & \Upsilon_s = 1 \\ 1 - 2\Upsilon_s^2 \pm 2\Upsilon_s\sqrt{\Upsilon_s^2 - 1}, & \Upsilon_s > 1. \end{cases} \qquad (18)$$

As the matrix M_s is simplectic, we have $\lambda_{s+} = 1/\lambda_{s-}$. The eigenvalues can be expressed as $\lambda_{s\pm} = |\lambda_{s\pm}| \exp[i\phi_{s\pm}]$, in which the phase of the strong head-tail oscillation is $\phi_{s\pm} = \tan^{-1}[Im(\lambda_{s\pm})/Re(\lambda_{s\pm})]$. When $\Upsilon_s < 1$, the modulus of the eigenvalue is one, and the system is stable. When the system is beyond the stability threshold, i.e. $\Upsilon_s > 1$, the growth rate in terms of synchrotron period is

$$\tau_{s\pm}^{-1} = \pm\log|\lambda_{s\pm}| = \pm\log\left|1 - 2\Upsilon_s^2 + 2\Upsilon_s\sqrt{\Upsilon_s^2 - 1}\right|. \qquad (19)$$

The synchrotron tune shift can be related to Υ_s, by $\Delta\nu_s/\nu_{s0} = \sqrt{1 + 2\Upsilon_s/\pi} - 1$. Furthermore, Υ_s can also be identified as the ratio of the voltage of the longitudinal wake field to the voltage of the rf field across the bunch length, therefore when Υ_s is not larger than one, we have the following approximate relation:

$$\Upsilon_s \propto \frac{\Delta\nu_s}{\nu_s} \propto \frac{\Delta z}{\sigma_z} \propto \frac{\hat{I}Z_0^{\parallel}/n}{\hat{V}}, \qquad (20)$$

where $\hat{I} = Nec/\sigma_z$, $n = \omega/\omega_0$ which is the mode index, and $Z_0^{\parallel}$ is evaluated at the beam spectral frequency $\omega = n\omega_0 \simeq c/\sigma_z$.

Even when the system is below the stability threshold, the amplitude of the macroparticle can correspond to a large excursion with respect to its initial value. A simple "transient" analysis can be found from the transfer matrix M_1 of Eq. (16). If we specify the initial condition as $V_s(0) = [\sigma_z/2, 0, -\sigma_z/2, 0]^T$, after the first half synchrotron period, we have

$$A_{s1}(\frac{T_s}{2}) = A_{s1}(0) = \frac{\sigma_z^2}{4}, \ A_{s2}(\frac{T_s}{2}) = (4\Upsilon_s^2 + 1)A_{s2}(0) = (4\Upsilon_s^2 + 1)\frac{\sigma_z^2}{4}, \ (21)$$

where the "longitudinal emittance" A_s is defined as $A_s \equiv z^2 + (c/\omega_s)^2 z'^2$. This implies that, near the threshold, although the system is still stable after a complete synchrotron oscillation, the amplitude of the tail particle can become as high as $\sqrt{5}$ of the initial amplitude. More generally, we will see in the next section that if no additional measures are taken, $\Upsilon_s > 1$, and the motions predicted by this model is unstable. Although the two-particle model might not accurately predict the amplitude excursion, the transient behavior could be unacceptably large.

Note that the formulae provided here are only valid for the case of zero order η, i.e. η_0. The first order of the momentum compaction factor will cause the longitudinal head-tail (LHT) instability [4,5] to occur, where $\eta \simeq \eta_0 + \eta_1 \delta$. Similar to the transverse head-tail instability due to the "transverse chromaticity", the LHT instability generated from the "longitudinal chromaticity" η_1, has no stability threshold. The growth rate of the LHT instability for dipole motion can be expressed as

$$\tau^{-1} = \frac{8\eta_1 N r_0 c^2}{3\pi \eta_0 \gamma C \sigma_z} \int \frac{d\phi}{\phi} Re Z_0^{\parallel}(\frac{2c\phi}{\sigma_z}) J_1(\phi) J_2(\phi). \tag{22}$$

2.2 Betatron Motion

Analgogus to Eq. (10), the transverse equation of motion is

$$y'' + \frac{\omega_\beta^2}{c^2} y = -\frac{N r_0}{\gamma C} \sum_n W_1(n c T_0) e^{-i\Omega[(s/c) + n T_0]} \langle y(s - n c T_0) \rangle. \tag{23}$$

The Fourier analysis gives us the transverse Robinson damping rate for the resonator model,

$$\tau_\beta^{-1} \simeq -\frac{2N r_0 c^2 R_s^\perp Q^2}{\pi^2 \gamma \omega_\beta h^3} |\triangle_\beta(\omega_R - h\omega_0)|, \tag{24}$$

where $-1/2 < \triangle_\beta < 1/2$, and $\triangle_\beta$ is the noninteger part of the betatron tune $\nu_\beta = \omega_\beta / \omega_0$. It is assumed that $(\triangle_\beta \omega_0, \omega_R - h\omega_0) \ll \omega_R/2Q \ll \omega_0$, and the machine is tuned in the condition $\triangle_\beta(\omega_R - h\omega_0) > 0$.

The two-particle analysis can be done in a fashion similar to that given previously. The dimensionless parameter for the transverse strong head-tail instability is

$$\Upsilon_\beta \equiv -\frac{\pi N r_0 c^2 W_1}{8\gamma C \omega_s \omega_\beta}. \tag{25}$$

Similar to the longitudinal case, Υ_β can be identified as the ratio of the betatron tune shift due to the transverse wake field to the synchrotron tune. For the resonator and resistive wall models, the head-tail parameter is shown, respectively, as

$$\Upsilon_\beta = \frac{N r_0 c R R_s^\perp}{16\gamma \nu_s \nu_\beta Q}, \quad \Upsilon_\beta = \frac{N r_0 R^2}{4\gamma b^3 \nu_s \nu_\beta} \sqrt{\frac{c}{\sigma \sigma_z}}. \tag{26}$$

The transverse amplitude excursion, just as in the longitudinal case, leads to $\sqrt{5}$ times of the initial amplitude when the parameter is near the threshold, i.e. $\Upsilon_\beta \sim 1$. For the collider parameters and the rf-structure wakes (see section 3), the beam is well above threshold and some BNS-like damping is desirable.

When the chromaticity ξ is not zero, the betatron frequency is modulated by δ, and leads to head-tail instability. In contrast to the strong

211

head-tail instability, the head-tail instability has no threshold. The growth rate is [3]

$$\tau_l^{-1} = -\frac{Nr_0c}{8\pi^2\gamma\nu_\beta} \int d\omega' Re Z_1^{\perp}(\omega') J_l^2(\omega'\frac{\hat{z}}{c} - \chi),$$ (27)

where l is the index for azimuthal mode and $\chi = \xi\omega_\beta\sigma_z/c\eta$ is the head-tail phase. Evaluation of Eq. (27) depends on the lattice design and will be examined in future papers.

2.3 Bousard Criteria

The dispersion relations can be obtained for the longitudinal and transverse motions, by a Fourier analysis and ensemble average on both sides of Eqs. (10) and (23). In doing so, the stability criteria are valid for the slow growing regime, i.e. when the growth rates $\ll \omega_s$. For the fast growing regime (growth rate $\gg \omega_s$), the Boussard criteria [6] for the longitudinal and transverse motion are

$$\left|\frac{Z_0^{\|}(n\omega_0)}{n}\right| < 0.6\frac{Z_0\nu_s^2\gamma\sigma_z^3}{N|\eta|r_0R^2}, \quad \left|Z_1^{\perp}(\omega_c)\right| < \frac{\pi Z_0\nu_s\nu_\beta\gamma\omega_c\sigma_z^2}{3Nr_0cR}.$$ (28)

If the cutoff frequency is $\omega_c = c/\sigma_z$, using the relations $Z_0^{\|} \sim W_0'/\omega$ and $Z_1^{\perp} \sim W_1/\omega$, the Bousard criteria are then consistent with the thresholds of head-tail instability with a geometric factor between one and two, dependent on the bunch shape. We will use the threshold parameters Υ_s and Υ_β as the parameters for estimations. Note that the formulae in Eq. (28) can be expressed in terms of the frequency spread or energy spread, since $\triangle\omega_{1/2} = \omega_0|\eta|\triangle\delta_{1/2}$ and $\triangle\delta_{1/2} = \omega_s\sigma_z/|\eta|c$.

3 ESTIMATES OF THRESHOLD AND GROWTH

The design parameters relevant to the collective effects in the main ring of a muon collider are given in Table 1. In Table 1, we show three sets of parameters. Since

$$\delta \propto \nu_s/\eta \propto \sqrt{V_{rf}\omega_{rf}/\eta},$$ (29)

as we keep both σ_z and δ as constants in preserving the bunch length and the rf phase space matching, V_{rf} is scaled as η/ω_{rf}, and ν_s is scaled as η.

With the numbers listed in Table 1, the threshold parameters using different impedance models are calculated in Table 2. A more accurate evaluation for the real machine should include a careful impedance budget. Note that we keep the broad band shunt impedances $R_s^{\|} = R_s[\Omega]$, $R_s^{\perp} = R_s/b^2[\Omega/\text{cm}^2]$ and Q as the variables for the resonator model, and $\sigma = \sigma_{aluminum} = 3.2 \times 10^{17}s^{-1}$ for the resistive wall impedance, where b is the pipe radius. The wake functions of a SLAC-like structure [7] are approximated as : $W_0' \sim 2$ cm^{-1}/cell and $W_1 \sim 1$ cm^{-2}/cell. The magnitude of W_0' and W_1 is decreased by about a factor of two, if the structure is L band, such as a CEBAF-like structure [8] (set II & III in Table 1), instead of S band (set I). We assume a superconducting structure with a gradient of $E_g = 10$ Mev/m, 3 cell/λ_{rf}. The total wake per turn is: W_0'(total) $\sim N_{cell} W_0'$ and W_1(total) $\sim N_{cell} W_1$, where $N_{cell} = 3 V_{rf}/\lambda_{rf} E_g$ is the total number of rf cells in the ring. The total rf length is $L_{rf} = V_{rf}/E_g$.

Beam energy E(Tev)	2		
Beam γ	18,929		
Muons per bunch N (10^{12})	2		
Bunch length σ_z (cm)	0.3		
Muon classical radius r_0 (cm)	1.36×10^{-15}		
Main ring radius R (km)	1.114		
Revolution frequency f_0 (KHz)	42.8		
Betatron tune ν_β	23.8		
	I	II	III
RF frequency f_{rf} (GHz)	3.0	1.5	1.5
Slippage factor η	1.2×10^{-5}	1.2×10^{-6}	1.2×10^{-7}
Synchrotron tune ν_s	0.01	0.001	0.0001
Harmonic number h	7.0×10^4	3.5×10^4	3.5×10^4
RF voltage V_{rf} (GV/turn)	3	0.3	0.03
Total number of rf cell N_{cell}	4500	450	45
Total RF length L_{rf} (m)	150	30	3

Table 1: Possible design parameters of the $\mu^+\mu^-$ collider ring.

The total parasitic loss $\triangle E$ due to the wake of a SLAC-like structure would be about 1.5 Gev/turn, the energy drop is roughly the energy gain from the rf. The energy loss can be scaled down by a L band structure.

For a system with different ω_{rf}, V_{rf} and η, keeping $V_{rf}\omega_{rf} \propto \eta$, the parameters are scaled as follows:

$$N_{cell} \propto \eta, \quad L_{rf} \propto \eta\lambda, \quad W_0'(\text{total}) \propto \eta W_0', \quad W_1(\text{total}) \propto \eta W_1, \quad (30)$$

The stability thresholds for the resonator and resistive wall impedances are scaled as

$$\Upsilon_s \propto 1/\eta, \quad \Upsilon_\beta \propto 1/\eta, \quad (31)$$

and for the SLAC/CEBAF-like rf structure,

$$\Upsilon_s \propto W_0'(\text{total})/\eta \propto W_0', \quad \Upsilon_\beta \propto W_1(\text{total})/\eta \propto W_1. \quad (32)$$

The percentage of energy loss can be scaled as

$$\triangle E/E_{rf} \propto W_0'(\text{total})/V_{rf} \propto \omega_{rf} W_0' \propto \omega_{rf}^2, \quad (33)$$

where we scaled $W_0' \propto \omega_{rf}$ approximately. The energy loss ratio decreases significantly, by a factor of 4, if L band is used rather than S band as the rf frequency. However, if $\alpha \sim 10^{-5}$ an L band structure would be excessively long.

		Resonator	Resistive Wall	SLAC/CEBAF-like
Υ_s	I	$1.3 \times 10^{-5} R_s/Qb$	$0.12/b$	1.08
	II	$1.3 \times 10^{-4} R_s/Qb$	$1.24/b$	0.54
	III	$1.3 \times 10^{-3} R_s/Qb$	$12.4/b$	0.54
Υ_β	I	$1.4 \times 10^{-4} R_s/Qb^2$	$1.05/b^3$	18.9
	II	$1.4 \times 10^{-3} R_s/Qb^2$	$10.5/b^3$	9.46
	III	$1.4 \times 10^{-2} R_s/Qb^2$	$105/b^3$	9.46

Table 2: Threshold parameters of the $\mu^+\mu^-$ collider ring for the parameters listed in Table 1. Note that the stability condition is $\Upsilon_s < 1 \,\&\, \Upsilon_\beta < 1$, where Υ_s and Υ_β are dimensionless, R_s is the shunt impedance in Ω, and b is the pipe radius in cm.

In case of when the system is beyond the stability threshold, the growth rates of the strong head-tail instability can be found from Eq. (19) for longitudinal motion. For the transverse growth rate, one replace Υ_s by Υ_β in Eq. (19). Note that the growth rate in terms of revolution is $\tau^{-1}[\text{turn}^{-1}] = \nu_s \tau^{-1}[T_s^{-1}]$. As Υ_s and Υ_β are independent of η, and $\nu_s \propto \eta$, the growth rate per turn scales as $\tau^{-1}[\text{turn}^{-1}] \propto \eta$, for both longitudinal and transverse motions. This indicates that the growth rates of strong head-tail instabilities per turn become smaller for a smaller η. If $\eta \lesssim 10^{-6}$, e.g. the parameters of set II and set III in Table 1, the crucial collective instability would be the BBU instability, as in a linac instead of the microwave instability in a ring. In this case, the microwave instability is immaterial, during the storage period, even the system is well beyond the threshold. However, when $\eta \gtrsim 10^{-6}$ such as for the parameters in set I, the beam makes one synchrotron oscillation in 100 turns, and the microwave instability then materializes.

For the multi-turn effects caused by high-Q cavities, the Robinson damping rates can be found from Eqs. (11) and (24), once the Q factor and the shunt impedances R_s are determined. We leave these estimates for a future paper.

Evaluation of the growth rate of the longitudinal head-tail instability from Eq. (22), requires an integral of the longitudinal impedance. Following Ref. 5, a rough estimate of the high frequency impedance based on the diffraction model [cf. Appendix] yields a growth rate $5.35 \eta_1 N_{cell}/\eta_0 b \times 10^{-3} \, [s]^{-1}$. If the impedances contribution to the LHT instability are dominated by the rf, then if $N_{cell} = 4500$, the growth time is about $0.125 b \eta_0/\eta_1 \, [s]$, which would be much longer than the storage time if $\eta_1 \sim \eta_0$. However, if $\eta_1 \sim \eta_0/\delta$, and as $\delta \sim 10^{-3}$, the growth time is then about $0.125b$ [ms], which is much shorter than the storage time. Using the scaling law in Eq. (30), where $N_{cell} \propto \eta_0$, the growth rate is scaled as η_1. As for the resistive wall impedance, the growth rate of the LHT instability is negligible.

4 DISCUSSION

In this paper, we discuss the growth rates of head-tail instabilities due to the chromaticities (η_1 for synchrotron motion, ξ for betatron motion), which are dependent on the lattice design. The evaluations of strong head-tail instability shows that a mechanism such as BNS damping is essential to stabilize the muon beam. Further studies, including analytical and numerical work, are in progress. From Table 2, one can see that the beam is unstable for the cases such as the SLAC/CEBAF-like structure, and the resistive wall impedance. The decrease of beam intensity due to muon decay by 5% in the first hundred turns does not alleviate the situation much.

However, the BBU-like instability can be suppressed by the BNS damping for the first few turns; after one synchrotron oscillation, the transverse strong head-tail instability is damped if the condition

$$\frac{\triangle \nu_\beta^{BNS}}{\nu_\beta} = -\frac{Nr_0W_1}{4k_\beta^2\gamma C} = \frac{2\Upsilon_\beta\omega_s}{\pi\omega_\beta}, \qquad (34)$$

is satisfied. The BNS "detuning" can be achieved by making the energy of the bunch's head higher than the energy of the bunch's tail, by $\triangle\delta = \triangle\nu_\beta/\nu_\beta\xi$. For the operation above transition, having the energy of bunch's head higher is also consistent with the rf bunching condition. For the muon collider, when $\nu_s = 0.01$, the BNS tune spread is $\triangle\nu_\beta^{BNS}/\nu_\beta = 2.7 \times 10^{-4}\Upsilon_\beta$, which can alternatively be achieved by introducing the radio frequency quadrupoles (RFQs). When $\nu_s = 0.001$ and $\nu_s = 0.0001$, the BNS tune spread is $2.7 \times 10^{-5}\Upsilon_\beta$ and $2.7 \times 10^{-6}\Upsilon_\beta$, respectively. If the design uses the parameters of set I, the total 150 m long rf section needs to be divided into many sections, with space in between for betatron phase advance, to perform BNS damping. By doing so, the amplitude growth due to the BBU effect is suppressed in between the rf subsections. Without BNS damping, the amplitude would grow as large as $\sqrt{4\Upsilon^2 + 1}$ [cf. Eq.(21)] in half of a synchrotron period. In reference to the parameters of sets I and II in Table 2, where $T_s/2$ is shorter than 1000 turns, the transient amplitude is very large even when the muon decay is considered. One needs to note that, taking into account the muon decay, the BNS tune spread should be controlled and decreased by the same decay rate as muons. The BNS damping scheme applied to a ring operating above threshold needs further study. Furthermore, the interplay of the bunch motion with the tight focus at the IP needs to be examined.

A preliminary simulation of the beam-beam effect for the muon collider has been done by Miguel Furman [9]. This was a "strong-strong" simulation carried out with the code TRS [10] in which the emittance of the two beams evolves dynamically from turn to turn under the mutual effect of the beam-beam interaction. The dynamics of the simulation is fully six dimensional, and includes finite bunch-length effects. The beam-beam kick is calculated in the "soft-Gaussian approximation," which effectively assumes that the bunch distribution remains Gaussian, albeit with time-dependent beam sizes. In this approximation, only incoherent effects show up. However, as the beam-beam parameter is only about 0.048, and the storage time is much shorter than the radiation damping time, coherent effects are unlikely to materialize. According to the simulation, the luminosity decreases roughly by 5% after 1000 turns due to the incoherent emittance blowup. The simulation does not account for the muon decay. Thus the beam-beam effect is overestimated because the muon decay leads to a significant decrease of the beam-beam interaction. The classical beam-beam effect on the luminosity performance is negligible compared with the degradation due to muon decay.

As a muon collider ring is operated very near transition, the effects of nonlinearities are very pronounced [11,12]. Issues such as dynamic aperture, synchro-betatron coupling, resonances, fringe fields, bunch lengthening, potential well distortion, matching and impedance above cutoff [13] are yet to be discussed in detail.

5 ACKNOWLEDGMENT

We wish to acknowledge support from Department of Energy under Contract No. DE-AC03-76SF00098. And we are also grateful to Miguel Furman for his simulation works and many useful discussions.

6 APPENDIX: Impedance Models

The impedances and the wake functions used in this paper are, for the resonator model,

$$Z_m^{\parallel}(\omega) = \frac{R_s^{\parallel}}{1 + iQ\left(\frac{\omega_R}{\omega} - \frac{\omega}{\omega_R}\right)}, \quad Z_m^{\perp}(\omega) = \frac{c}{\omega}\frac{R_s^{\perp}}{1 + iQ\left(\frac{\omega_R}{\omega} - \frac{\omega}{\omega_R}\right)},$$

$$W_0' \simeq \frac{\omega_R R_s^{\parallel}}{Q}, \quad W_1 \simeq -\frac{cR_s^{\perp}}{Q};$$

and for the resistive wall,

$$Z_m^{\parallel}(\omega) = \frac{\omega Z_m^{\perp}(\omega)}{c} = \frac{[1 - sgn(\omega)i]}{(1 + \delta_{m0})b^{2m+1}c}\sqrt{\frac{2|\omega|}{\pi\sigma}},$$

$$W_0'(z < 0) = \frac{C}{2\pi b}\sqrt{\frac{c}{\sigma|z|^3}}, \quad W_1(z < 0) = -\frac{2C}{\pi b^3}\sqrt{\frac{c}{\sigma|z|}},$$

where b is the pipe radius, σ is the conductivity. Note that, for the convenience of estimation of the stability threshold , we neglect the z dependent parts of the wake functions in the resonator model. And the wake functions are evaluated at $|z| = \sigma_z$ in the two-particle model, for the wakes of resonator and resistive wall. The impedance of the diffraction model used in evaluating the growth rate of LHT instability, is

$$Z_0^{\|}(\omega) = \frac{Z_0[1 + sgn(\omega)i]}{2\pi b}\sqrt{\frac{cg}{\pi|\omega|}},$$

where the gap width g can be approximated as $g \simeq c/2f_{rf}$, if the rf cavities dominate the structure. Note that the resonator model gives a description for the gross features of the cavity impedance. The resistive wall model is used for the coupling impedance caused by the beam pipe which is not perfectly conducting. The applicable range of the resistive wall model for a aluminum pipe is: $2 \times 10^{-3}b^{2/3}[cm] \ll |z| \ll 10^8 b^2[cm]$. And the diffraction model is for the cavity impedance at high frequency $\omega \geq c/b$.

7 REFERENCES

[1] R.B. Palmer, in this proceeding.

[2] K.Y. Ng, in this proceeding.

[3] A. W. Chao, *Physics of Collective Beam Instabilities in High Energy Accelerators*, (John Wiley & Sons, 1993).

[4] G. Hereward,*Rutherford Laboratory Report RL-74-062*,1972 and *RL-75-021*, 1975.

[5] B. Chen and A. W. Chao, Particle Accelerators, vol. 43(1-2), pp77-91,1993.

[6] D. Boussard, CERN Lab II/RF/Int 75-2, 1975.

[7] K. Bane and P. B. Wilson, *Proc. 11th Int. Conf. High Energy Accel.*, Geneva, 1980, p.592.

[8] Z. Li and J.J. Bisognano, *Proc. IEEE Par. Acc. Conf.*, Dallas, 1995.

[9] M. Furman, private communications.

[10] J. Tennyson, "The Two-Ring Simulation code TRS", undocumented, 1989.

[11] D. Robin *et al.*, Physical Review E, vol. 48, no. 3, 1993.

[12] A. Amiry *et al.*, Particle Accelerators, vol. 44, pp57-76, 1994.

[13] S. Chattopadhyay ed.,*Impedance Beyond Cutoff*, Special Issue of Particle Accelerators, vol. 25, no.2-4 , 1994.

Incoherent and Coherent Tune Shifts*

S.Y. Zhang , AGS Department, BNL, Upton, NY 11973, USA

Abstract

The longitudinal and transverse microwave instabilities are the two important limiting factors for the performance of an accelerator. Comparing with the fairly unified approach for the longitudinal microwave instabilities, different approaches have been used to define the transverse microwave instabilities. One reason of this is related to the role played by the space charge incoherent and coherent tune shifts. In this article, the transverse microwave instabilities will be discussed by defining separately the roles of the space charge incoherent and coherent tunes, which are reprersented by the space charge transverse impedances. Preliminary results for the AGS as proton driver are presented by using this approach.

1   Introduction

The longitudinal and transverse microwave instabilities are the two fundamental performance limiting factors for an accelerator. The longitudinal microwave instability threshold is usually obtained by using the Keil-Schnell and the Boussard criteria. The approach is fairly unified, and the obtained results are quite consistent. The approaches to the transverse microwave instability are usually not that unified, and the obtained results are somewhat diversified. One of the reasons is that the role played by the space charge incoherent and coherent tune shifts are not clearly defined, which can be shown by the conventionally defined transverse space charge impedance. This problem is probably not very important for higher energy machines, because of the relatively small space charge effects. It is, however, important for the machines operated at low or medium energy, such as the AGS. In this article, we separately define the incoherent and coherent space charge transverse impedances. The coherent space charge impedance, together with the broad band impedances, is responsible for the beam coherent motion, and the incoherent impedance, together with the chromatic and octupolar tune spread, etc., contributes to the transverse Landau damping. A preliminary study for the AGS as proton driver will be presented by using this approach.

2   Transverse Impedance

Derived from the Vlasov equation, taking $m = 0$ mode and also the first orthogonal polynomial in the expansion of radial modes, the transverse instability is determined by [1],

*Work performed under the auspices of the U.S. Department of Energy

$$\omega - \omega_\beta = \frac{jeI_0}{2Rm_0\gamma\nu_0\omega_0} \sum_{p=-\infty}^{\infty} Z_T(p)(\Lambda_0^{(0)}(p))^2 \tag{1}$$

where ω_β and ω_0 are the betatron and revolution frequencies, respectively, I_0 is the average beam current, and m_0 is the rest mass of proton. Also R is the machine radius and ν_0 is the betatron tune. The transverse impedance $Z_T(p)$ is sampled at the harmonic number p, and $\Lambda_0^{(0)}(p)$ is the bunch spectrum of the first orthogonal polynomial for $m = 0$ mode.

For microwave instability, we consider only the space charge and broad band impedances. The transverse broad band impedance is obtained by using a rough estimate from the longitudinal counterpart $Z_{LBB}(p)/p$,

$$Z_{TBB}(p) = \frac{2c}{b^2\omega_0} \frac{Z_{LBB}(p)}{p} \tag{2}$$

where b is the radius of the vacuum chamber. At low frequency, $Z_{LBB}(p)/p$ is inductive, therefore, $Z_{TBB}(p)$ is inductive.

The conventional space charge impedance is defined as proportional to the difference between the coherent and incoherent tune shifts [2,3],

$$Z_{TSC} = j\frac{RZ_0}{\beta^2\gamma^2}(\frac{1}{b^2} - \frac{1}{a^2}) \tag{3}$$

where Z_0 is the impedance of free space, a is the radius of the beam. Since $a < b$, this impedance is capacitive.

Let us define the coherent and incoherent space charge transverse impedances as follows,

$$Z_{TSCcoh} = j\frac{RZ_0}{\beta^2\gamma^2b^2} \tag{4}$$

and

$$Z_{TSCinc} = j\frac{RZ_0}{\beta^2\gamma^2a^2} \tag{5}$$

where both impedances are inductive.

In the following, we show that substituting the incoherent and coherent impedances (4,5) into the beam dynamic equation (1) yields the same incoherent and coherent tune shifts obtained by using the conventional Laslett approach.

For circular chamber, the incoherent and coherent tune shifts can be calculated by [4,5],

$$\Delta\nu_{coh} = \frac{-NRr_0}{\pi\nu_0B_f\beta^2\gamma^3}(\frac{1/2}{b^2}) \tag{6}$$

and

$$\Delta\nu_{inc} = \frac{-NRr_0}{\pi\nu_0B_f\beta^2\gamma^3}(\frac{1/2}{a^2}) \tag{7}$$

where N is the number of particles, r_0 is the classical radius of proton, and B_f is the bunching factor.

For space charge and broad band impedances, $Z_T(p)$ can be taken out of the summation, therefore for a Gaussian distribution we can use

$$\sum_{p=-\infty}^{\infty} (\Lambda_0^{(0)}(p))^2 \approx \frac{1}{\pi^{3/2} B_f} \qquad (8)$$

Substituting, say, Z_{TSCcoh} of (4) into (1), using (8) and the following relations,

$$I_0 = \frac{N e \omega_0}{2\pi} \qquad (9)$$

and

$$Z_0 = \frac{1}{\epsilon_0 c} \qquad (10)$$

where ϵ_0 is the permittivity in free space, and also

$$r_0 = \frac{1}{4\pi\epsilon_0} \frac{e^2}{m_0 c^2} \qquad (11)$$

we get the space charge coherent tune shift, which differs from the one obtained using (6) by a factor of 1.13. Similar result can be obtained for the incoherent tune shift. This shows that applying the space charge incoherent and coherent transverse impedances (4,5) into (1) generates consistent results with other approaches. Therefore, the transverse coherent frequency shifts and the Landau damping can be studied in a more unified formulation based on the equation (1).

Substituting Z_{TSC} of (3) into (1), and writing on the left side by following convention ,

$$\omega_\beta = \omega_{\beta 0} + \Delta\nu_{inc}\omega_0 \qquad (12)$$

and also replacing $\omega_\beta \approx \omega_{\beta 0}$ on the right side, we get

$$\omega - \omega_{\beta 0} - \Delta\nu_{inc}\omega_0 = \frac{j e I_0}{2 R m_0 \gamma \omega_{\beta 0}} \frac{1}{\pi^{3/2} B_f} (Z_{TSCcoh} - Z_{TSCinc}) = \Delta\nu_{coh}\omega_0 - \Delta\nu_{inc}\omega_0 \qquad (13)$$

where we observe that the incoherent tune shift is cancelled from both sides. This shows that ideally the incoherent tune shift plays no role in the transverse coherent motion. One reminds that in the longitudinal case, the space charge induced incoherent frequency shift directly affects the longitudinal focusing and therefore it has to be included in the longitudinal coherent motion. In addition, in the longitudinal case, the broad band impedance also gives rise to the incoherent frequency shift, while in the transverse case, the broad band impedance only contributes to the coherent tune shift. Therefore, the scenarios are quite different for the longitudinal and the transverse motions.

Note that the exact physical implication of applying the transverse impedance defined in (3) in a beam dynamic equation is not clear. A more straightforward approach is to use the impedances separately defined in (4,5). In this way, the space charge coherent impedance, together with the broad band impedances, is responsible for the beam coherent tune shift, and the space charge incoherent impedance, together with the chromatic and octupolar effects, etc, is responsible for the tune spread.

We note also that sometimes the transverse space charge impedance defined in (3) is used for the beam coherent tune shift, which is not only quantitatively incorrect, but also gives rise to a tune shift that is in the opposite polarity. For example, using the equation (6), or applying (4) into (1), we get the same space charge coherent tune, which is decreased. Applying (3) into (1), the obtained coherent tune is increased.

3 Transverse Landau Damping

Following the rule of thumb, i.e. the coherent tune shift must be smaller than the tune spread for the effective Landau damping, the transverse microwave instability can be determined by,

$$\Delta\nu > \frac{eI_{peak}}{4\pi R m_0 \gamma \omega_\beta \omega_0} |Z_T(p)| \tag{14}$$

where $\Delta\nu$ is the tune spread, and

$$I_{peak} = \frac{2\sqrt{2}}{\sqrt{\pi}B_f} I_0 \tag{15}$$

is the peak current for a Gaussian distribution.

Now for the coherent tune shift we use the space charge coherent impedance and the broad band impedance. For incoherent tune shift, we consider the space charge incoherent impedance, and also others, such as the chromatic and the octupolar tune spreads, and also the synchrotron oscillation frequency. The chromatic tune spread is not very effective for the weak instabilities with the growth rate comparable to the synchrotron oscillation period, it is effective, however, for the fast instabilities. The synchrotron tune is effective in stabilizing the weak instabilities.

Below transition at the lower energy, the space charge induced tune shift is large compared with contributions of the chromatic, the octupolar tune spread and the synchrotron frequency. The space charge incoherent tune spread therefore provides Landau damping. Since this tune spread is quite large, therefore the transverse microwave instability is not of serious concern at the low energy.

Above transition at the higher energy, both the incoherent and coherent space charge effects have been substantially reduced, the broad band impedance contributed coherent tune shift, therefore, becomes dominant. The stabilizing force has to come from other than the space charge incoherent tune spread. This implies that the situation above transition is much more serious than the one below transition.

Strictly speaking, only the chromatic and the octupolar tune spreads, and also the synchrotron oscillation frequency, which are not dependent on the beam current, can contribute to the microwave instability threshold. On the other hand, the space charge incoherent tune spread depends on the beam current, therefore it affects the growth rate. This situation is similar to the longitudinal case, where the incoherent frequency shifts affect the growth rate and the RF nonlinearity induced synchrotron oscillation frequency spread gives rise to the instability threshold.

The estimate of the microwave instability, however, is not exactly the same as the calculation of the instability threshold. In an estimate of the beam instabilities using the criterion (14), both the instability threshold and the growth rate have been accounted for. After all, the influence of the beam current on the both sides in (14) is already included in the calculation. This problem can also be looked in other ways. For example, given a beam current, the transverse beam size, which directly affects the space charge incoherent tune spread, can be seen as the variable. Thus, applying the equation (14) gives rise to an instability threshold in term of the beam size, rather than the current.

4 AGS as Proton Driver

We only present the longitudinal and the transverse microwave instabilities based on some incomplete upgrade of the AGS, using the usual Keil-Schnell criterion for the longitudinal case and the approach developed in this article for the transverse case. Under these upgrade, the repetition rate of the AGS becomes 2.5 Hz, and the RF peak voltage is increased to about 800 KV. The longitudinal broad band impedance is assumed to be a modest 30 Ω, and the longitudinal emittance is taken as 4 eVs. The 10^{14} protons are in 8 bunches, and the normalized transverse emittance including 95% particles is assumed to be 50 $\pi mm - mr$. For longitudinal microwave instabilities, the momentum spread and the required momentum spread are shown in Fig.1, where we used the 'momentum spread threshold' to indicate the required momentum spread. The valley of the momentum spread threshold at about 40 ms from the beginning of the cycle is due to the cancellation between the broad band and the space charge impedances. The peak of the momentum spread threshold indicates the transition. For transverse microwave instabilities, the required tune spread, and the space charge incoherent tune spread are shown in Fig.2, where also the chromatic tune spread is shown with $\xi = -0.2$.

The stability margins are barely met, yet the requirement for the beam for the 250 on 250 GeV muon collider proton driver is not satisfied. To meet the requirement, a bunch compression is needed. In order to do this, a new vacuum chamber, a more powerful RF system and other damping systems are required.

References

[1] S.Y. Zhang, '*Transverse Bunched Beam Instability*,' to be published.

[2] B. Zotter, '*Betatron Frequency Shift due to Image and Self Fields*,' CERN 85-19, p.253, 1985.

[3] A. Chao, *Physics of Collective Beam Instabilities in High Energy Accelerators*, Wiley, New York, 1994.

[4] P.J. Bryant, '*Betatron Frequency Shift due to Self and Image Fields*,' CERN 87-10, p.62, 1987.

[5] A. Hofmann, '*Tune Shifts from Self-fields and Images*,' CERN 94-01, p.329, 1994.

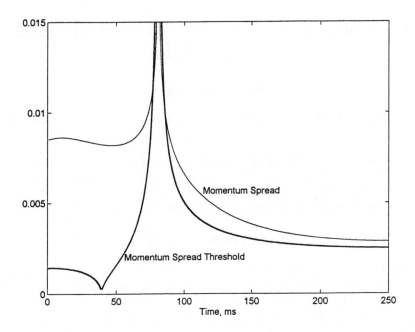

Fig.1. *Longitudinal Microwave Instabilities*

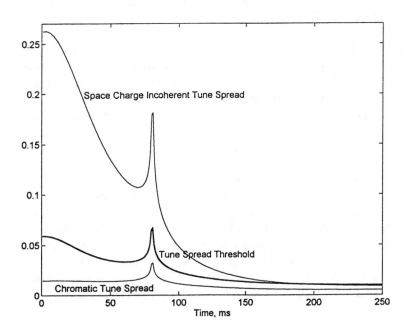

Fig.2. *Transverse Microwave Instabilities*

Beam Stability Issues in a Quasi-Isochronous Muon Collider

King-Yuen Ng

Fermi National Accelerator Laboratory, P.O. Box 500, Batavia, IL 60510*

Abstract

Beam instability issues are studied in a quasi-isochronous muon collider. The problems of beam breakup, strong head-tail, and longitudinal microwave instabilities are discussed. We find that in the present design, the muon beam may be susceptible to a microwave instability.

I. INTRODUCTION

To achieve the high luminosity required in a 2 TeV muon-muon collider and to overcome partially the hour-glass effect, the muon bunch has to be extremely short, within 3 mm (rms), or, equivalently, 10 ps. Therefore, the time slip per turn of a muon particle must be much less than 10 ps for the muons to remain bunched. This implies for the proposed 2-TeV muon-muon collider [1] — a ring with approximately 1 km radius and a full momentum spread of $\pm 0.5\%$ — that the phase-slip factor η must have an absolute value much less than 1×10^{-4}. For this reason, an isochronous collider ring is preferred. In this paper, first the problem of beam breakup in a perfectly isochronous ring is addressed in Sec. II. Then, in Sec. III, unpreventable deviations of a ring from isochronicity and the strong head-tail instability are discussed. Finally, in Sec. IV, it is shown that in the present design, the muon bunches may have surpassed the lower limit for excitation of longitudinal microwave instabilities.

II. BEAM BREAKUP

We first assume that the collider ring is isochronous. If so, the relative longitudinal positions of the bunch particles will not change. The tail of the bunch is continuously deflected by the wakefields emanating from the head increasing transverse emittances and eventually causing beam breakup. Intuitively this effect can be studied using a two-particle model [2]. Here, the N-particle bunch is modeled in Fig. 1 as two macro particles each with charge $eN/2$ separated by a distance $\hat{z} = 3$ mm apart. The transverse motion of the front or head particle, y_1, is just free betatron oscillation,

$$y_1''(s) + k_\beta^2 y_1(s) = 0 , \qquad (2.1)$$

---

*Operated by the Universities Research Association, Inc., under contract with the U.S. Department of Energy.

$$eN/2 \qquad \hat{z} \qquad eN/2$$

Figure 1: The two-particle model.

where $k_\beta = \nu_\beta/R$ is the betatron wave number, ν_β the betatron tune, R the ring radius, and s is the path length measured along the ideal closed orbit. The rear or tail particle sees the transverse wake $W_1(\hat{z})$ of the front particle and satisfies a driven equation,

$$y_2''(s) + k_\beta^2 y_2(s) = -e \left(\frac{Ne}{2}\right) \left(\frac{W_1(\hat{z})}{2\pi R}\right) \frac{1}{E} y_1(s) , \qquad (2.2)$$

where E is the energy of particles. The solution is

$$y_1(s) = \hat{y} \cos k_\beta s , \qquad (2.3)$$

$$y_2(s) = \hat{y} \left[\cos k_\beta s + \Upsilon \frac{s}{L} \sin k_\beta s\right] , \qquad (2.4)$$

where $\hat{y}$ is the initial betatron amplitude of the bunch, and

$$\Upsilon = -\frac{e^2 N W_1(\hat{z}) L}{8\pi R E k_\beta} \qquad (2.5)$$

denotes the growth in amplitude of the tail after traveling a length L.

A. Resistive Wall

The dipole wake due to the resistivity of the vacuum chamber wall is

$$W_1(\hat{z}) = -\frac{4c}{b^3} \sqrt{\frac{Z_0 \rho}{4\pi}} \frac{r}{|\hat{z}|^{1/2}} , \qquad (2.6)$$

where c is the velocity of light, b the beam pipe radius, $Z_0 \approx 377$ Ohms is the free-space impedance, and ρ the resistivity of the wall. Equation (2.6) is correct in the region

$$\frac{b}{\chi} \gg |\hat{z}| \gg b\chi^{1/3}, \qquad \chi = \frac{\rho}{bZ_0} . \qquad (2.7)$$

We can now compute the time $\tau = L/c$ for the amplitude of the tail particle to double by letting $\Upsilon = 1$. For a muon bunch containing $N = 2 \times 10^{12}$ particles in a stainless steel beam pipe with $\rho = 0.74$ $\mu\Omega$-m, the results are listed in

Table I: Tail amplitude doubling times for a stainless-steel beam pipe

b (cm)	χ	b/χ (m)	$\chi^{1/3}b$ (m)	τ (ms)	No. of Turns
1.0	1.96×10^{-7}	5.09×10^4	5.81×10^{-5}	0.12	6.1
2.5	7.85×10^{-8}	3.18×10^5	1.07×10^{-4}	1.90	95
5.0	3.93×10^{-8}	1.27×10^6	1.70×10^{-4}	15.2	760

Table II: Tail amplitude doubling times for an aluminum beam pipe

b (cm)	χ	b/χ (m)	$\chi^{1/3}b$ (m)	τ (ms)	No. of Turns
1.0	7.03×10^{-9}	1.42×10^6	1.91×10^{-5}	0.64	32
2.5	2.81×10^{-9}	8.89×10^6	3.52×10^{-5}	10.0	502
5.0	1.41×10^{-9}	3.55×10^7	5.60×10^{-5}	80.4	4018

Table I. For an aluminum beam pipe with $\rho = 0.0265$ $\mu\Omega$-m, the results are listed in Table II.

We see that Eq. (2.7) is satisfied in all the cases so that the wake potential described by Eq. (2.6) is valid. As is expected, the bunch is much more stable inside an aluminum beam pipe than in a stainless-steel beam pipe. Due to the short life-time of the muons, the bunch is designed to stay in the collider for only ~ 1000 turns. Even over this short length of time, as is clear from the table, the increase in the tail amplitude is serious for stainless-steel beam pipe, and is only marginal for aluminum beam pipe (and only then if the pipe radius is less than 2.5 cm). Note that the increase in tail amplitude is inversely proportional to the beam-pipe radius to the third power (b^3).

B. Broad-Band Impedance

For a ring with longitudinal broad-band impedance of peak value $Z_{\parallel}/n$, a figure of merit Q, and centered at angular frequency ω_r, the corresponding transverse broad-band impedance has a peak value given roughly by

$$Z_{\perp} = \frac{2R}{b^2} \frac{Z_{\parallel}}{n} .$$
(2.8)

Thus, the dipole wake at a short distance $\hat{z}$ is

$$W_1(\hat{z}) = -\frac{2\hat{z}R\omega_r^2}{Qb^2c} \frac{Z_{\parallel}}{n} .$$
(2.9)

If the broad band is centered at the cutoff frequency, or $\omega_r \approx c/b$, the doubling time τ for the tail amplitude can be obtained from the relation $\Upsilon = 1$:

$$\tau \approx \frac{2b^4 \nu_\beta (E/e)}{I_b R^2 \hat{z}} \frac{Q}{(Z_\parallel/n)} . \tag{2.10}$$

Putting in $N = 2 \times 10^{12}$, $R = 1$ km (or $I_b = 0.01529$ amp), and $\nu_\beta = 24$, one gets

$$\tau = \begin{cases} 3.3 \dfrac{Q}{Z_\parallel/n} \text{ turns} & b = 1.0 \text{ cm} \\[2ex] 130 \dfrac{Q}{Z_\parallel/n} \text{ turns} & b = 2.5 \text{ cm} \\[2ex] 2080 \dfrac{Q}{Z_\parallel/n} \text{ turns} & b = 5.0 \text{ cm} . \end{cases} \tag{2.11}$$

For $Z_\parallel/n = 1$ Ohm and $Q = 1$, we see that the emittance growth will be too large. Note, however, that this growth depends very sensitively on the beam-pipe radius (b^4).

C. BNS Damping

The rapid growth of the tail amplitude is a result of resonant driving of the tail by the head of the bunch. If the tail particle could be made to oscillate at a betatron frequency different from that of the head particle, this resonant interaction would be avoided and no emittance increase would occur. This idea is called BNS damping [3]. It is described by assuming that the betatron wave number is a function of length along the bunch, so that the equation of motion of the tail particle changes from Eq. (2.2) to

$$y_2''(s) + k_\beta^2(\hat{z}) y_2(s) = -e \left(\frac{Ne}{2}\right) \left(\frac{W_1(\hat{z})}{C}\right) \frac{1}{E} y_1(s) . \tag{2.12}$$

With $k_\beta(\hat{z}) = k_\beta + \Delta k_\beta$, the evolution of the tail amplitude becomes

$$y_2(s) = \hat{y} \cos(k_\beta + \Delta k_\beta)s + \left(\frac{\Upsilon}{L \Delta k_\beta}\right) \hat{y}[\cos(k_\beta + \Delta k_\beta)s - \cos k_\beta s] , \tag{2.13}$$

which is no longer linearly increasing. Furthermore, if we choose

$$\frac{\Upsilon}{L \Delta k_\beta} = -1 , \tag{2.14}$$

the trailing particle will follow the leading particle but without any amplitude growth.

To achieve BNS damping, the amount of tune spread required is given by

$$\frac{\Delta \nu_\beta}{\nu_\beta} = \frac{\Delta k_\beta}{k_\beta} = \frac{\Upsilon}{L k_\beta} . \tag{2.15}$$

In terms of the tail-amplitude doubling time τ ($\Upsilon = 1$), the required tune spread becomes

$$\frac{\Delta\nu_\beta}{\nu_\beta} = \frac{1}{\omega_0\nu_\beta\tau} \, . \tag{2.16}$$

If we substitute the growth time for a $N = 2 \times 10^{12}$ muon bunch in an aluminum beam pipe, we obtain

$$\frac{\Delta\nu_\beta}{\nu_\beta} = \begin{cases} 1.58 \times 10^{-4} & b = 1.0 \text{ cm} \\ 1.07 \times 10^{-5} & b = 2.5 \text{ cm} \\ 1.26 \times 10^{-6} & b = 5.0 \text{ cm} \end{cases}$$

These tune spreads are small and implementation of this idea should not be difficult. One method to provide the BNS focusing is to introduce a radio-frequency quadrupole whose strength changes as the head and the tail of the bunch pass through it. Another method is through chromaticity. The muons have a full momentum spread of $\pm 0.5\%$, so that only a small variation in energy from head to tail of the bunch would suffice to provide the required tune spread.

Instead of the two-particle model, when the whole bunch with a longitudinal distribution $\rho(z)$ is considered, the equation of motion becomes

$$y''(s,z) + [k_\beta + \Delta k_\beta(z)]^2 y(s,z) = -\frac{e^2 N}{2\pi RE} \int_z^\infty dz' \rho(z') W_1(z-z') y(s,z') \, , \tag{2.17}$$

where $y(s,z)$ is the transverse displacement of a point in the bunch z away from the bunch center as it passes the ring position s. The condition of BNS damping is that all particles in the bunch execute the same betatron motion, or

$$y(s,z) = \hat{y} \cos k_\beta s \, , \tag{2.18}$$

Substituting in Eq. (2.17), we arrive at the condition [4]

$$\frac{\Delta k_\beta}{k_\beta} = -\frac{e^2 N}{4\pi k_\beta^2 RE} \int_z^\infty dz' \rho(z') W_1(z-z') \, , \tag{2.19}$$

which is the equivalent of Eq. (2.15) for a bunch of particles.

III. STRONG HEAD-TAIL

In the Gallardo-Palmer design of the interaction region (IR) [5], the betatron function rises from the low-beta of $\beta^* = 3$ mm to a maximum of ~ 400 km in about 40 m along the ring. A particle executing a betatron oscillation will travel a much longer path length. In a perfectly isochronous ring, it is shown in the Appendix that a particle in a bunch with a rms normalized emittance of

228

$\epsilon_N = 50 \times 10^{-6}\pi$ m can lag as much as $\Delta \ell = 3.52 \times 10^{-5}$ m per turn, or 3.52 cm for a store of 1000 turns, an amount which exceeds the nominal bunch length. For this reason, the muon collider ring will only be quasi-isochronous. To avoid bunch lengthening, a synchrotron oscillation period of about $T_s = 150$ turns is suggested [6], which requires an rf voltage of $V_0 = 1.5$ GV and a momentum-compaction factor of $|\alpha| = 1.5 \times 10^{-5}$ for the collider ring.

With a synchrotron oscillation, the head and tail of the bunch interchange position and the resonant blowup of the tail can be avoided. The condition for stability can be inferred by requiring that the amplitude of the tail not double in a quarter of the synchrotron period. Putting in $L = cT_s/4$ into Eq. (2.5), we obtain the threshold

$$\Upsilon = -\frac{e^2 N W_1(\hat{z})c^2}{16 R E \omega_\beta \omega_s} < 1,\qquad (3.1)$$

where $\omega_\beta/2\pi$ and $\omega_s/2\pi$ are, respectively, the betatron and synchrotron frequencies. Physically, the two macro-particles oscillate coherently in the σ and π modes with distinct frequencies. At threshold, the two modes merge into one and become unstable. This instability has been referred to as a strong head-tail in the literature.

By putting in the wall impedance or the broad-band impedance given by Eqs. (2.6) and (2.9), Eq. (3.1) can be used as a check for instability. We can also infer the result from the calculations made in Sect. II. With the synchrotron period of $T_s = 150$ turns, stability against a strong head-tail interaction is equivalent to requiring that the tail amplitude double in a time longer than $T_s/4$, or 37.5 turns. From Table I and II as well as Eq. (2.11), we can see immediately that the beam will be stable if the beam pipe radius is not less than 2.5 cm.

IV. LONGITUDINAL MICROWAVE

The limit for longitudinal microwave instability is given by the Boussard-modified Keil-Schnell [7] or Krinsky-Wang [8] criterion:

$$\frac{|Z_\parallel|}{n} < \frac{2\pi|\eta|(E/e)}{I_p}\delta^2,\qquad (4.1)$$

where $I_p = eN/(\sqrt{2\pi}\sigma_\tau) = 1.28 \times 10^4$ Amp is the local peak current of the bunch having a rms length of $\sigma_\tau = 10$ ps or 3 mm. The rms momentum spread is $\delta \approx 0.0025$. With a phase-slip factor of $|\eta| = 1.5 \times 10^{-5}$, we find that for stability $|Z_\parallel|/n$ must be less than 0.092 Ohms. Because of the short bunch length, the muons see a frequency around $\omega_r/2\pi = 10$ GHz. Unfortunately, the impedance per harmonic does remain constant up to and over 10 GHz

229

provided that all the detailed contributions from the vacuum chamber are summed carefully [9]. Therefore, this microwave self-bunching may become a serious impediment to generating stable muon beams.

The Boussard modified Keil-Schnell criterion of Eq. (4.1) can be rewritten in the form

$$\omega_r \sqrt{\frac{|\eta||I_p||Z_{\parallel}|/n}{2\pi E}} < \omega_r |\eta| \delta. \tag{4.2}$$

The left side represents the growth rate without Landau damping and equals $\tau^{-1} = 7.76 \times 10^3 \text{ s}^{-1}$ if $|Z_{\parallel}|/n$ is taken as 1 Ohm. The right side of Eq. (4.2) denotes the spread in frequency of the 10 GHz self-bunching buckets, which leads to Landau damping. At this moment, this spread amounts to only $2.36 \times 10^3 \text{ s}^{-1}$, which is, of course, not enough to counteract resonant emittance growth.

To safeguard against microwave instability, we suggest that the vacuum chamber be made as smooth as possible so that the the broad-band impedance near 10 GHz can be kept well under, for example, $|Z_{\parallel}|/n \approx 0.1$ Ohm.

The Boussard-modified Keil-Schnell criterion as stated in Eq. (4.1) does not take into account whether the machine is operated below or above transition, or, in other words, the sign of the phase-slip factor η. In reality, the bell-shaped stability curve shows clearly that a machine should be more stable below transition. Recently, this fact has been demonstrated both analytically and numerically by Fang et al [10] and Ng [11]. As a result, the muon collider should be designed to have an imaginary transition gamma or a negative momentum-compaction factor [12].

V. CONCLUSION

Using current muon-collider specifications, the issue of beam breakup has been investigated under the assumption of a perfect isochronous ring. The only way to circumvent beam breakup may be to utilize the BNS focusing method. In this paper the path-length lag across the IR was computed and it was found that synchrotron-oscillation focusing must be implemented to avoid bunch lengthening. In other words, the ring must be quasi-isochronous. Also, although it was computed that the muon beam would be below the threshold of a strong head-tail instability, it may, however, be above the threshold for longitudinal microwave instability. To ensure a stable muon beam, it is suggested here that the coupling impedance be controlled accurately and that the collider ring be designed with a negative momentum-compaction factor.

APPENDIX

Let us start from a particle in a ring with uniform focusing channel. The transverse displacement is

$$y = a \sin k_\beta s \,, \tag{A.1}$$

where a is the maximum transverse displacement. For quarter of the betatron wavelength, $s = \pi/(2k_\beta)$, the path length along the sine curve is

$$\ell = \frac{\sqrt{1 + a^2 k_\beta^2}}{k_\beta} E \left(\frac{a k_\beta}{\sqrt{1 + a^2 k_\beta^2}} \right) \,, \tag{A.2}$$

where

$$E(\mu) = \frac{\pi}{2} \left(1 - \frac{\mu^2}{4} - \frac{3\mu^4}{64} + \cdots \right) \tag{A.3}$$

is the complete elliptic function of the second kind. The fractional increase in path length is therefore

$$\frac{\Delta \ell}{\ell_0} \approx 1 + \frac{1}{4} a^2 k_\beta^2 \,. \tag{A.4}$$

For a normalized emittance of $\epsilon_N = 50 \times 10^{-6} \pi$ m and a maximum betatron function of $\beta \sim 100$ m, a 2 TeV muon with one-sigma offset will have a maximum transverse displacement of $a \sim 0.5$ mm. Taking a tune of $\nu_\beta \sim 24$ and a ring radius of $R = 1$ km, we obtain

$$\frac{\Delta \ell}{\ell_0} \approx 0.75 \times 10^{-10} \,, \tag{A.5}$$

or for a turn, $\Delta \ell = 4.90 \times 10^{-7}$ m.

The actual muon collider contains two interaction regions (IR), each with high-beta point both upstream and downstream of the interaction point (IP). For the $\beta^* = 3$ mm IR [5], the betatron function increases from β^* to ~ 400 km in about $\ell_0 = 40$ m. The path-length difference in executing betatron oscillation in this region will contribute significantly. Let us assume that the betatron function increases to its highest value according to

$$\beta(s) = \beta^* + \frac{s^2}{\beta^*} \,. \tag{A.6}$$

In reality, in the situation when $\beta_x^* = \beta_y^*$, the first quadrupole tries to bend down one of the betatron functions while pushing the other betatron function to increase at a faster rate. Therefore, Eq. (A.6) is actually an underestimate.

Consider a particle at the edge of the unnormalized emittance ϵ passing through the IR. Its displacement can be written as

$$y = \sqrt{\frac{\beta \epsilon}{\pi}} \sin(\psi + \varphi) \,, \tag{A.7}$$

where ψ is the phase advance, which is set to zero right at the IP, and φ is a random phase. According to Eq. (A.6),

$$\psi(s) = \int_0^s \frac{ds}{\beta(s)} = \tan^{-1} \frac{s}{\beta^*} \, . \tag{A.8}$$

From this, the slope can be easily derived,

$$\frac{dy}{ds} = \left(\frac{\epsilon}{\beta^* \pi} \right)^{1/2} \cos \varphi \, , \tag{A.9}$$

which turns out to be s independent. For an advance of ℓ_0 along the ideal closed orbit from the IP to the high-beta point, the path length along the betatron-oscillation trajectory is

$$\ell = \left(1 + \frac{\epsilon}{\beta^* \pi} \cos^2 \varphi \right)^{1/2} \ell_0 \, . \tag{A.10}$$

Averaging over the random phase φ, we obtain

$$\ell = \frac{2}{\pi} \left(1 + \frac{\epsilon}{\beta^* \pi} \right)^{1/2} \ell_0 E \left(\sqrt{\frac{\epsilon/(\beta^* \pi)}{1 + \epsilon/(\beta^* \pi)}} \right) \, , \tag{A.11}$$

or a fractional increase of

$$\frac{\Delta \ell}{\ell_0} \approx \frac{\epsilon}{4 \beta^* \pi} \, . \tag{A.12}$$

There are two IR's in the ring. The path-length increase will becomes $\Delta \ell = 3.52 \times 10^{-5}$ m per turn for the 3 mm low-beta and $\ell_0 = 40$ m. Thus the contribution from the IR's dominates. To avoid bunch lengthening, synchrotron oscillation has to be introduced. The rule of thumb is that during half a synchrotron period, the amount lengthened must be less than the rms bunch length. Thus, for a rms bunch length of 3 mm, the synchrotron period has to be less than 170 turns. In another design [13] with $\beta^* = 1$ cm, $\ell_0 \approx 17$ m. Therefore, $\Delta \ell = 4.49 \times 10^{-6}$ m per turn and the synchrotron period can be made much longer.

References

[1] See for example, R.B. Palmer, "Beam Dynamics Problems in a Muon Collider," Beam Dynamics Newsletter No. 8, 1995, p. 27, ed. K. Hirata, S.Y. Lee, and F. Willeke.

[2] A.W. Chao, "Physics of Collective Beam Instabilities in High Energy Accelerators," Wiley and Sons, 1993.

[3] V. Balakin, A. Novokhatsky, and V. Smirnov, Proc. 12th Int. Conf. High Energy Accel., Fermilab, 1983, p. 119.

[4] V.E. Balakin, Proc. Workshop on Linear Colliders, SLAC, 1988, p. 55.

[5] J, Gallardo, private communication.

[6] R.B. Palmer, these proceedings.

[7] D. Boussard, CERN/LAB II/RF/75-2 (1975); E. Keil and W. Schnell, CERN/ISR/TH/RF/69-48 (1969).

[8] S. Krinsky and J.M. Wang, Part. Accel. **17**, 109 (1985).

[9] B. Zotter, private communication.

[10] S.X. Fang, K. Oide, Y. Yokoya, B. Chen, and J.Q. Wang, KEK Preprint 94-190, 1995.

[11] K.Y. Ng, "Potential-Well Distortion and Mode-Mixing Instability in Proton Machines," Proc. Int. Workshop on Collective Effects and Impedance for B-Factories, Tsukuba, Japan, 12-17 June, 1995, ed. Y. Chin, (to be published).

[12] S.Y. Lee, K.Y. Ng, and D. Trbojevic, Phys. Rev. **E48**, 3040 (1993).

[13] C. Johnstone, K.Y. Ng, and D. Trbojevic, "Interaction Regions with Increased Low-Betas for a 2-TeV Muon Collider," these proceedings.

Simulation of Backgrounds in Detectors and Energy Deposition in Superconducting Magnets at $\mu^+\mu^-$ Colliders

Nikolai V. Mokhov and Sergei I. Striganov[+]

*Fermi National Accelerator Laboratory**

P.O. Box 500, Batavia, Illinois 60510

+*Institute for High Energy Physics, Protvino, 142284, Russia*

Abstract. A calculational approach is described to study beam induced radiation effects in detector and storage ring components at high-energy high-luminosity $\mu^+\mu^-$ colliders. The details of the corresponding physics process simulations used in the MARS code are given. Contributions of electromagnetic showers, synchrotron radiation, hadrons and daughter muons to the background rates in a generic detector for a 2×2 TeV $\mu^+\mu^-$ collider are investigated. Four configurations of the inner triplet and a detector are examined for two sources: muon decays and beam halo interactions in the lattice elements. The beam induced power density in superconducting magnets is calculated and ways to reduce it are proposed.

INTRODUCTION

Recent studies on a high-energy high-luminosity $\mu^+\mu^-$ collider [1, 2] have shown the high physics potential and a feasibility of such a project. A candidate design for 2×2 TeV machine, based on the existing and near-term technology, with a luminosity as high as 10^{35} cm^{-2} s^{-1} is described in [3]. The two most serious beam-related problems envisioned on the way to the practical realization of a storage ring are enormous particle background levels in a detector and a high power density in the superconducting magnets [4, 5, 6] due to unavoidable muon decays and beam halo interactions. With 2×10^{12} muons in a bunch at 2 TeV one has 2×10^5 decays per meter in a single pass through an interaction region (IR), or 6×10^9 decays per meter per second.

Decay electrons with an energy of about 700 GeV and the enormous number of synchrotron photons emitted by these electrons in a strong magnetic field induce electromagnetic showers in the collider and detector components resulting in high

---

*Work supported by the U. S. Department of Energy under contract No. DE-AC02-76CH03000.

radiation and background rates. Another contribution comes from beam halo interactions at the limiting apertures.

A first-pass study [4] showed that the electromagnetic component of the backgrounds from $\mu \to e\nu\tilde{\nu}$ decays has the potential of killing the concept of the muon
collider without significant suppression via various shielding and collimators in the
detector vicinity. Beam induced energy deposition in the superconducting (SC) magnets may result in magnet quench and in high heat load to the cryogenic system
which also requires special protection measures and a serious design effort [5].

In this paper a calculational stream and the corresponding physics algorithms
used to study beam related effects in a muon collider lattice (Fig. 1) and a generic detector (Fig. 2) are described. Electron, positron and photon fluxes as well as photo-
hadron and photo-muon contributions are examined for four inner triplet and detector configurations. Detailed calculations and analysis are performed to study the
physics phenomena resulting in particle fluxes in detectors and energy deposition in
magnets. Results on beam induced heat load in the muon collider SC magnets are
presented. Efficacy of possible measures to reduce background and heat load levels
are analyzed quantitatively.

CALCULATIONAL STREAM

All the calculations are done with the MARS code [8]. Recent code modifications relevant to the studied problem include: a better description of muon cross-
sections, of muon decay and of the algorithm for electromagnetic fluctuations [9]; an
improved particle transport algorithm in a magnetic field; synchrotron radiation generation; generation of hadrons in photo-nuclear interactions and muons in the course
of electromagnetic shower development; modified geometry description; extended
histogramming and graphical possibilities.

Simulations with the MARS code are done for the realistic lattice extended
up to 300 meters from the interaction point (IP) including the detailed dipole and
quadrupole geometry and magnetic field maps [4, 5]. All the particle interaction processes in the lattice, the 1.45 m radius tunnel surrounded by rock, the experimental
hall 26 m long, 10 m radius) and the detector are taken into account.

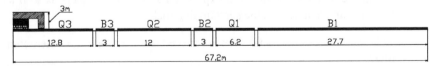

FIGURE 1. Schematic of muon collider inner triplet and detector [4] (Configuration 1) .
Dimensions are in meters.

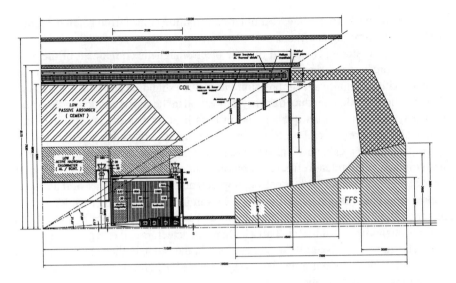

FIGURE 2. Schematic view of realistic detector [7] (Configurations 3 and 4).

A single MARS run includes:

- Forced $\mu \rightarrow e\nu\tilde{\nu}$ decays in the beam pipe (beam muon decay studies) or beam tail interactions with the limiting aperture beam pipe (beam halo studies).

- Tracking of created electrons in the beam pipe under the influence of the magnetic field with emission of synchrotron photons along the track.

- Simulation of electromagnetic showers in the collider and detector components induced by electrons and synchrotron photons hitting the beam pipe, with the appropriate hadron and prompt muon production (Bethe–Heitler pairs and direct positron annihilation).

- Simulation of muon interactions (bremsstrahlung, direct e^+e^- pair production, ionization, deep inelastic nuclear interactions and decays) along the tracks in the lattice, detector, tunnel and experimental hall components as well as with tunnel and hall air and the surrounding rock.

- Simulation of electromagnetic showers created in the above muon interaction vertices.

- Simulation of hadronic cascades generated in muon and photon interactions, with daughter electromagnetic showers, with muon production (π and K decays, prompt muons in hadronic and electromagnetic interactions), and with low–energy neutron transport.

- Histogramming and analysis of particle energy spectra, fluences and energy deposition in various detector and collider regions.

Energy thresholds are 1 MeV for muons and charged hadrons, 0.3 MeV for electrons and photons, and 0.00215 eV for neutrons. In this study we assume that a bunch of 2×10^{12} muons of 2 TeV energy enters the inner triplet moving toward the IP, creating showers responsible for backgrounds along its path. Only a single bunch is simulated to study the directionality which is masked in the case of two colliding beams. When studying the integrated effect, we assume 1000 turns as a beam life-time.

The physics model and corresponding calculational algorithms for particle interactions and transport and for geometry definition are described in detail elsewhere (see [8, 10] and bibliography there). The next section highlights muon interaction specifics.

SIMULATION OF MUON INTERACTIONS

Muons decay in flight producing energetic electrons. In the MARS code the synchrotron radiation spectrum for electrons in a magnetic field is taken from References [11] and [12] with the number of photons sampled from a Poisson distribution. These electrons and photons initiate electromagnetic showers in accelerator and detector components causing the severe background and radiation problems. Muons generated in electromagnetic showers and beam halo muons do penetrate through the bulk of material and contribute to the rates cumulatively from the extended regions of the collider. When high energy muons pass through matter all their interaction processes result in energy loss, production of photons, electrons and hadrons which accompany the muon track and are taken into account in the MARS code. For a 1 TeV muon in iron the mean energy loss rates due to ionization, bremsstrahlung, direct e^+e^- pair production and photo-nuclear interaction are 2.32, 2.92, 4.16 and 0.39 MeV/g/cm^2, respectively.

$\mu \rightarrow e\nu\tilde{\nu}$ Decays

Vector momentum of the emitted electron is sampled according to the differential decay probability of the *Vector–Axial* model of four-fermion interactions [12]. All measurements in direct muon decay $\mu \rightarrow e\nu\tilde{\nu}$ are successfully described by this model.

Ionization Energy Loss

Collisions of charged particles and atoms with energy transfer ε greater than some cutoff ε_c are considered in MARS as discrete events involving production of

δ-electrons, e^+e^--pair, bremsstrahlung. Energy losses with $\varepsilon < \varepsilon_c$ (so called restricted losses) are taken into account as continuous.

Several methods have been offered to simulate fluctuations of restricted losses for ionization [9, 13, 14]. All of them use Vavilov's function [15] for this purpose with redefined parameters

$$\xi = Bs$$

$$B = 0.1536 Z/A/\beta^2$$

$$\kappa_n = \xi/\varepsilon_G$$

$$\beta_n^2 = \beta^2 \varepsilon_G/\varepsilon_{max},$$

where Z and A are the absorber atomic and mass numbers, β is the particle velocity, s is the path length in g/cm^2, and ε_{max} is the maximum energy transferred in a single collision. With $\varepsilon_G < \varepsilon_c$ Vavilov's function becomes Gaussian asymptotically for $\kappa > 10$ [15]. Therefore for

$$\kappa_n > 10 \tag{1}$$

the restricted loss distribution becomes Gaussian with the mean

$$\bar{\Delta}_r = \alpha(\varepsilon < \varepsilon_G) \cdot s$$

and the variance

$$\sigma_r^2 = \frac{\xi^2}{\kappa_n}(1 - \frac{\beta_n^2}{2}),$$

where $\alpha(\varepsilon < \varepsilon_c)$ is the mean restricted energy loss per unit length.

For the simulation of δ-electrons with energy greater than ε_δ at any step, ε_G is calculated using (1) and the restricted energy loss with $\varepsilon_c = min(\varepsilon_G, \varepsilon_\delta)$ is sampled. Then, the number of δ-electrons is simulated using a Poisson distribution. The coordinates of δ-electron generation are obtained with the following recursive procedure

$$x_n = x_{n-1} + \left(1 - (1 - g_n)^{1/(k-n+1)}\right)(s - x_{n-1})$$

and

$$x_0 = 0,$$

where k is a number of δ-electrons, x_n are the coordinates of production points along the step ($1 \leq n \leq k$) and g_n are random numbers uniformly distributed on (0,1). The electron energies are sampled from the well-known Bhabha formulae. Total energy loss of a particle is the sum of the δ-electron energies and of the restricted energy loss.

Radiative Energy Loss

For high energy muons ($E \geq$ a few hundred GeV), the radiative mechanisms, bremsstrahlung and direct e^+e^- pair production, will dominate over the ionization losses. An exact but complicated expression has been given by Kel'ner [16]. One performs a two-fold numerical integration to calculate the differential cross section using this approach. Therefore approximations of Kel'ner's results are usually used in Monte Carlo calculations. However, the accuracy of these approximations is not very high. Even the mean energy loss differs by more than 10% from Kel'ner's calculations.

At the same time, the numerical integration shows that only the first two moments of the pair production cross section give a sizable contribution to the moments of the total electromagnetic cross section. Relative contributions for higher moments do not exceed a few percent. Therefore, in MARS a simple function [17] is used to approximate the pair production cross section:

$$\frac{d\sigma_p}{d\nu} = b_p \frac{a(1+a)}{\nu(\nu+a)^2}, \tag{2}$$

$$b_p = (\frac{1}{E} \frac{dE}{dx})_p,$$

where $(dE/dx)_p$ is the mean energy loss per unit length. The following expression approximates b_p with a few percent accuracy:

$$b_p = 1.689 \cdot 10^{-5} \frac{m_e}{M} \frac{Z(Z+1)}{A} \left[b_1 ln \left(\frac{b_3 Z^{-1/3}}{1 + 4 b_3 Z^{-1/3} M/E} \right) - b_2 \right], \ [\frac{cm^2}{g}]. \tag{3}$$

Here M and m_e are the incident particle and electron masses, respectively. The parameter a is determined from the second moment of the cross section

$$a = \begin{cases} a_1 \cdot 10^{-3} m_\mu/M \,, & E \leq a_3; \\ (a_1 + a_2 ln(E/a_3)) \cdot 10^{-3} m_\mu/M, & E > a_3, \end{cases}$$

where m_μ is the muon mass. The parameters a_i, b_i for some charged particles are given in Table 1. To sample the angles of e^+, e^- and muons after the interaction, the method proposed in [18] is used.

There is a number of different approaches for the calculation of the muon bremsstrahlung cross section (see [19] for recent discussion). These methods differ mainly in the treatment of screening corrections. In MARS a general expression proposed in [19] is used, which allows arbitrary nuclear and atomic formfactors to be applied.

Table 1: Parameters of the approximation for pair production cross section.

	Muon $E > 20$ GeV	Pion $E > 20$ GeV	Kaon $E > 50$ GeV	Proton $E > 90$ GeV
a_1	5.2	5.5	5.45	5.4
a_2	.13	.18	.35	.43
a_3 (GeV)	200	200	100	100
b_1	.787	.791	.819	.833
b_2	1.1	1.09	1.14	1.17
b_3	2986	3017	2773	2532

For a small energy transfer $\varepsilon < \varepsilon_\gamma = 10^{-3} E$ at a muon energy $E > 10$ GeV, the bremsstrahlung differential cross section reaches the complete screening limit

$$\Sigma_\gamma(E, \varepsilon) = \frac{d_\gamma}{\varepsilon},$$

where $d_\gamma = 4b_\gamma/3$ and

$$b_\gamma = (\frac{1}{E} \frac{dE}{dx})_\gamma.$$

With this, the restricted energy loss distribution can be approximated by

$$f_c^\gamma(\Delta_c, E, s) \simeq \frac{d_\gamma s}{\varepsilon_c^{d_\gamma s}} \frac{1}{\Delta_c^{1-d_\gamma s}}. \tag{4}$$

The continuous energy loss Δ_c at a step s is sampled from (4) providing $\Delta_c < \varepsilon_c < \varepsilon_\gamma$. The production of bremsstrahlung photons with energies greater then ε_c is considered as a discrete process. Fig. 3 shows muon differential cross-sections for bremsstrahlung calculated by four different programs.

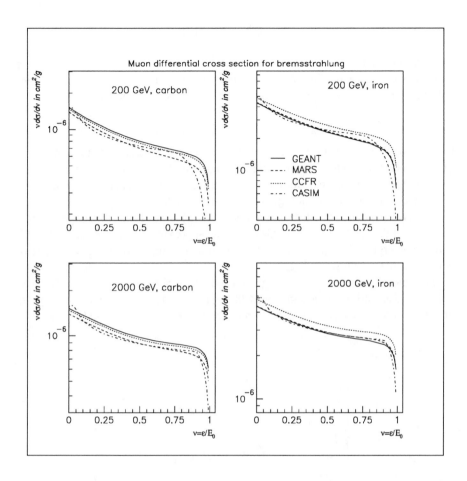

FIGURE 3. Muon differential cross-section for bremsstrahlung in carbon and iron at E_0 = 200 GeV and 2000 GeV *vs* energy transferred to photons, as used in GEANT [20], MARS [19], in CCFR collaboration [22], and in CASIM [18] (approximation of Tsai formula).

Deep Inelastic Interactions

Different models of deep inelastic muon-nucleus scattering are consistent at the 30% accuracy level. At the same time, the relative mean energy loss for this process is $\leq 10\%$ of the total, even at very high energies. The corresponding mean free path exceeds ~ 100 meters of iron for muons in the TeV energy region. So, high precision in a deep inelastic interaction description is not of primary importance. We choose the formula evaluated in [21], which is in good agreement with recent experimental data [22].

$\gamma A \to hX$ **Reactions**

Hadroproduction in photon-nucleus interactions at $E_\gamma \geq 0.14\,\mathrm{GeV}$ is simulated in an approximate way. The photon is replaced with a real pion of random charge with the same kinetic energy. The total cross section is calculated as

$$\sigma_{\gamma A} = R_A \cdot (Z\sigma_{\gamma p} + (A - Z)\sigma_{\gamma n}). \tag{5}$$

For the total γp cross section experimental data are used from [23] at momentum below 4.215 GeV/c and the fit from Review of Particle Properties [24] at higher energies. For the total γn cross section experimental data from [25] at momentum below 4 GeV/c and a fit from [26] at higher energies are used.

The A-dependence of the cross section R_A in (5) is extracted from experimental data: recent data [27] at $E \leq 1.15\,\mathrm{GeV}$, and the approximation $R_A = 1.047A^{-0.085}$ at higher energies. The quality of this description in comparison with available experimental data is shown in Fig. 4.

Photoneutron production in the giant resonant energy region $6 \leq E_\gamma \leq 60\,\mathrm{MeV}$ is described according to the algorithm [28, 29] extended for light nuclei $4 \leq A \leq 56$ on the basis of the latest data. An interpolation is used between 60 and 140 MeV.

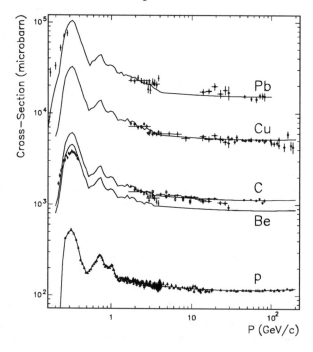

FIGURE 4. Calculated photon-nucleus cross-section in comparison with experimental data *vs* photon momentum for lead, copper, carbon, beryllium and hydrogen.

DAUGHTER MUON PRODUCTION

Processes responsible for muon generation at the above stages in the induced hadronic and electromagnetic cascades are included in the MARS code.

Hadronic Interactions

Simulation algorithms for $\pi^\pm \to \mu^\pm \nu(\tilde{\nu})$ and $K^\pm \to \mu^\pm \nu(\tilde{\nu})$ decays and for prompt muon production (single muons in charmed meson decays, $\mu^+\mu^-$ pairs in vector muon decays, and the dimuon continuum) with forced generation of weighted muons are described in [8, 10].

Electromagnetic Showers

Prompt muons produced by electrons and photons are handled by the current MARS version [8, 30]. Bethe-Heitler pairs $\gamma Z \to Z\mu^+\mu^-$ are produced at $E_\gamma \geq 0.25\,\text{GeV}$ at a rate of $(m_e/m_\mu)^2$ times that for e^+e^- with the appropriate statistical weights and a complete simulation of the electromagnetic showers, similar to the MUSIM code [31]. It was shown [30] that this approach gives results in remarkable agreement with those based on the numerical integration in the Tsai formalism [32].

At $E \geq 45\,\text{GeV}$ direct positron annihilation $e^+e^- \to \mu^+\mu^-$ is simulated according to [12] with the cross section $86.8/s$ nb, where s is in GeV^2, and the $(1 + cos^2\theta)$ angular distribution.

LATTICE AND DETECTOR CONFIGURATIONS

Beam muon decays and beam halo interactions in the inner triplet and adjacent regions have been studied in four lattice-detector configurations for a 6 km circumference 2×2 TeV storage ring.

Configuration 1

The first configuration is examined for a distance $L \leq 70\,\text{m}$ from the IP for the inner triplet with $\beta^* = 3$ mm and $\beta_{peak} = 400$ km [4]. Superconducting dipole magnets B1, B2 and B3 have a central field 8 T (Fig. 1). Combined function superconducting quadrupoles Q1 and Q2 are have a 2 T dipole component and a gradient of 50 T/m. All of the SC components have an 8 cm radius aperture. The Q3 quadrupole is resistive with a 0.5 T dipole field, except the first 1.3 m near the IP where the dipole component is equal to zero. Its aperture is reduced toward the IP from R=4.5 cm at L=12.8 m to R=0.45 cm at L=1.2 m, with the gradient increasing from 33.3 T/m to

333.3 T/m, appropriately. Geometries and materials of beam pipes, collars, yokes and cryostats for the SC dipoles and quadrupoles as well as the 2-D POISSON calculated magnetic fields in these components are embedded in the calculational package. A copper bucking coil is placed on the outside of the Q3 to neutralize the effect of the solenoidal field in the quadrupole.

A rather simple model detector is used (Fig. 1) [4]: a two-region silicon tracker with volume averaged density $\rho = 0.15$ g/cm^3, a central calorimeter (CH, $\rho = 1.03$ g/cm^3) and a solenoid magnet with 2 Tesla magnetic field.

Configuration 2

This configuration is based on the first complete lattice of the 2×2 TeV $\mu^+ \mu^-$ storage ring. It was found [5] that a suppression of synchrotron radiation generation is possible by using single function quadrupoles in the triplet and by keeping the high field dipoles as far from the IP as possible. In a prototype lattice [5] with β^*=3 mm, β_{peak}=200 km, the low-β quadrupoles have a gradient of 262 T/m and the nearest 8 T SC dipole starts at L=130 m from the IP. The detector configuration is as in the first case. Calculations have been carried out for the region L $\leq$ 300 m from the IP with and without some collimators there.

Configuration 3

The previous lattice with a much more realistic detector (Fig. 2) is used in this configuration. Detector design is based on GEM and ATLAS detector ideas adopted to the muon collider [7]. The detector consists of an inner tracker, Pb-LAr/Cu-LAr electromagnetic calorimeter followed by Al-Sci hadronic calorimeter, low-Z passive absorber and a superconducting solenoid surrounded with muon chambers. In the forward direction it has an endcap calorimeter and forward muon spectrometer. Studies are performed for L $\leq$ 300 m from the IP.

Configuration 4

A more practical lattice of lower luminosity designed by K. Y. Ng is adopted. The final focus is a doublet with β^*=3 cm, β_{peak}=31.4 km. The first 6.3 m long low-β quadrupole, still inside the detector, starts at 6.5 m, which is much father compared to the previous cases. A second 6 m quadrupole, just 0.13 m away, is followed by about 20 m of drift. Both quadrupoles have a tapered SC coil aperture of 2.5 cm radius at L=6.5 m and 6.1 cm radius at L=18.93 m. The pole tip field is 9.5 T. The coil aperture radius of the rest of the machine is assumed to be 7 cm. A tungsten liner (flared in the doublet) occupies the allowable space in the aperture. The realistic detector description of Fig. 2 is used. To study in detail the importance of various parts of the final focus, the contribution to the backgrounds is explored for a short region L $\leq$ 30 m.

BACKGROUNDS IN DETECTORS

Muon Decays

Typical particle spectra integrated over the whole system are shown in Fig. 5. The results are normalized to a single pass, so the relative importance of different components is clearly seen. The peak sitting around 700 GeV in the e^+e^- spectrum represents the $\mu \rightarrow e\nu\bar{\nu}$ decay spectrum with a tail at lower energies enriched by electrons and positrons of electromagnetic showers induced in the beam pipe and superconducting coils. Photons emitted due to synchrotron radiation along e^+e^- tracks in a strong magnetic field have a peak around 1 GeV. Neutron spectra (not shown) have pronounced peaks at ≈ 80 MeV and ≈ 0.8 MeV, and a 1/E slope down to the thermal peak. Neutrons, along with photons, dominate in total flux, especially at large radii.

95/12/10 15.42

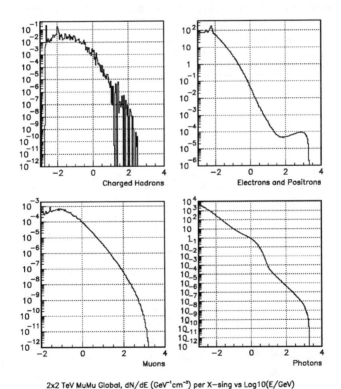

2x2 TeV MuMu Global, dN/dE (GeV^{-1}cm^{-2}) per X–sing vs Log10(E/GeV)

FIGURE 5. Particle spectra averaged over the interaction region and detector with 2 TeV muon decays as a source (Configuration 4).

Due to the very high energy of electrons and photons in the large aperture, the whole triplet is a source of backgrounds in the detector. As calculated in [4], e^+e^- and photon fluxes and energy deposition density in detector components are well beyond current technological capabilities if one applies no measures to bring these levels down. The most effective collimation includes a limiting aperture about one meter from the IP, with an interior conical surface which opens outward as it approaches the IP (Fig. 6). These collimators have the aspect of two nozzles spraying electromagnetic fire at each other, with the charged component of the showers being confined radially by the solenoidal magnetic field and the photons from one nozzle being trapped (to whatever degree possible) by the conical opening in the opposing nozzle.

A few collimator configurations occupying the cone θ < 150 mrad in the 15<L<120 cm region on the either side of the IP are studied (Fig. 6). Collimators made of tungsten as well as of a combination of various materials (aluminum, copper and tungsten) with different hole shapes have been considered. It turns out that for the main source of the backgrounds, direct muon decays, the best choice is a tungsten nozzle with an aperture radius R=0.45 cm at L=120 cm and R=1 cm at L=15 cm. The background reduction is amazing. The e^+e^- fluence calculated in [4] for Configuration 1 is given in Table 2. There is a significant difference in the background levels in the left (outward) and in the right (inward) parts of the tracker and calorimeter.

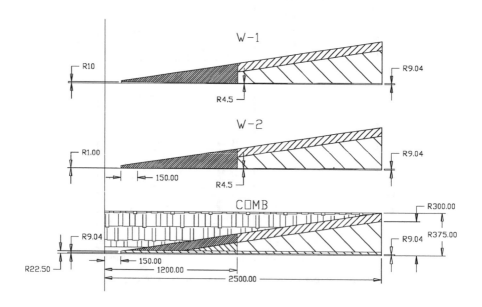

FIGURE 6. Configurations of collimating nozzles examined. Dimensions are in millimeters.

246

Table 2: e^+e^- fluence (cm^{-2} per crossing of 10^{12} muons per bunch) in the outward (L) and inward(R) parts of the inner (T1) and outer (T2) trackers and the central calorimeter (CH) for the cases with and without tungsten nozzle and with and without copper collimator in Configuration 1. For 2×10^{12} muons per bunch multiply by 2.

Nozzle	Collim	T1 (L)	T1 (R)	T2 (L)	T2 (R)	CH (L)
No	No	8.54×10^4	2.04×10^4	1.57×10^3	6.21×10^2	2.86×10^1
No	Yes	1.68×10^4	6.74×10^3	3.46×10^2	2.38×10^2	4.00×10^0
Yes	No	1.40×10^3	1.10×10^3	4.11×10^1	3.02×10^1	3.93×10^0
Yes	Yes	1.74×10^2	4.81×10^2	3.15×10^0	2.31×10^0	1.73×10^0
	$K_{max} =$	491	43	498	269	17

An additional way to suppress further the background levels can be a collimator between Q3 and B3 (see Fig. 1) with the smallest possible aperture. A copper collimator 50 cm long with 2.5σ radius aperture in these calculations provided up to a factor of 10 additional background reduction. The total maximum reduction K_{max} defined as a ratio of background levels in the given detector region without protective measures to that with a tungsten nozzle and copper collimator between Q3 and B3 magnets is given in the last row of the Table 2. It is as high as a factor of 500 at best, but for the larger radii in calorimeter it is "only" a factor of 20 to 40. Particle spectra in the detector are very soft in such a configuration: for all the regions considered background photons and electrons have energies below 100 MeV, being on the average just a few MeV. With a tungsten nozzle and copper collimator between Q3 and B3, the peak charged particle flux in the silicon tracker is of the order of 1000 cm^{-2} per crossing of two single bunches through the IR in opposite directions. The fluxes fall off very rapidly with radius.

Transverse distributions of e^+e^- flux in the central tracker are shown in Fig. 7 for Configurations 1 and 2. Backgrounds in the part of the tracker toward the ring center are significantly lower in Configuration 2 with a strong dipole field starting at 130 m from the IP [5]. On the outside the reduction is about a factor of 2 to 5. Fig. 8 shows the contribution to the energy deposition in the central tracker ($6 \le r \le 100$ cm) from muon decays along the IR in Configuration 2 [5]. Collimation and spoiling in the second focus region and in the long drift between the detector and the first dipole magnet can provide additional background reduction, but one needs to be very careful with beam halo handling in this case.

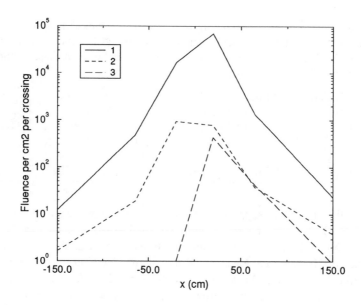

FIGURE 7. Distribution of e^+e^- flux in the central tracker horizontal plane. 1 - Configuration 1, no collimation, 2 - Configuration 2, with tungsten nozzle, 3 - Configuration 2, with tungsten nozzle.

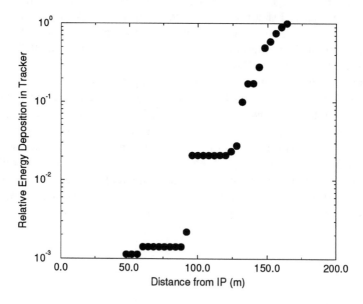

FIGURE 8. Cumulative energy deposition in the central tracker *vs* shower origin coordinate in Configuration 2.

Table 3: Particle fluence (cm^{-2} per crossing) in the central tracker (r =20 cm) and at the outer side of the endcap calorimeter (r =150 cm) due to 2 TeV muon decays (2×10^{12} muons per bunch) in Configuration 3.

Detector	n >0.00215 eV	γ >0.3 MeV	$e^{\pm}$ >0.3 MeV	$h^{\pm}$ >1 MeV	$\mu^{\pm}$ >1 MeV
Tracker	5×10^4	1.5×10^4	1000	80	3
Calorimeter	800	3×10^4	7000	0.3	0.2

The above results are confirmed for Configurations 3 and 4 with a much more realistic detector. In addition to e^+e^- and γ, hadrons and muons generated in the course of the electromagnetic shower development are taken into account. Fig. 9 shows a tagged decay distribution of energy deposited in the tracker as a function of the decay electron creation coordinate. Even with the optimal tungsten nozzle, the contribution from the $3 \leq L \leq 30$ m region to the background rates is very high. Particle fluxes in the central tracker and at the downstream end of the endcap calorimeter are presented in Table 3 for Configuration 3. In the endcap calorimeter the radial distributions are rather flat.

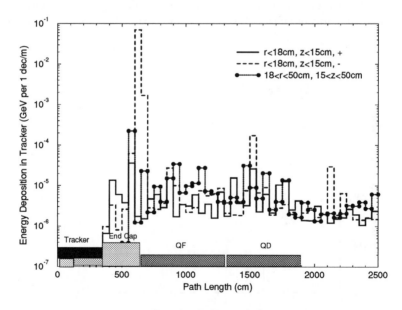

FIGURE 9. Tagged energy deposition in different regions of the central tracker due to $\mu \rightarrow e\nu\bar{\nu}$ decays happened at given path length from the IP in Configuration 4.

Beam Halo

Beam halo backgrounds arise from muons which are lost some distance upstream of the detectors. These muons induce electromagnetic or hadronic showers either upstream or inside of the detector and can cause more serious problems. Beam particles injected with large momentum errors or betatron amplitudes will be lost within the first few turns. After this, an equilibrium level of losses will be attained as particles are promoted to larger betatron amplitudes via beam disruption from the collision point, beam-gas scattering, etc.

In simulations the beam is assumed to enter the IR with a non-truncated Gaussian profile. Muons outside $\pm 3\sigma$ will then interact and be scraped by the final arc magnets, low-beta quads, collimators, and detector components. The energy spectrum of muons averaged over the IR due to beam loss is presented in Fig. 10. The distribution of the muon interaction vertices in the vertical plane in the vicinity of the IP is rather symmetric, but in the horizontal plane there is a strong asymmetry related to the magnetic field. With an energy cut-off equal to 50 MeV for particles produced in muon interaction vertices, more than 95% of the vertices are direct e^+e^- pair production. Other processes, such as muon bremsstrahlung, deep inelastic nuclear interactions, muon decays and energetic knock-on electron production, are not so numerous, but secondary particles are more energetic (Fig. 11). Therefore the total energy going to these channels can be comparable to that of the pair production.

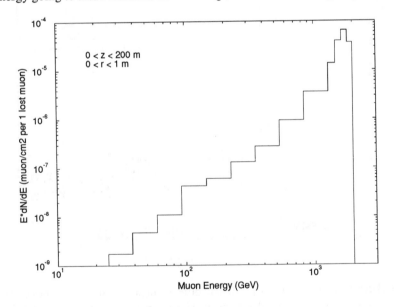

FIGURE 10. Muon energy spectrum due to 2 TeV muon beam loss at a limiting aperture at L = 200 m averaged over the IR region (Configuration 3).

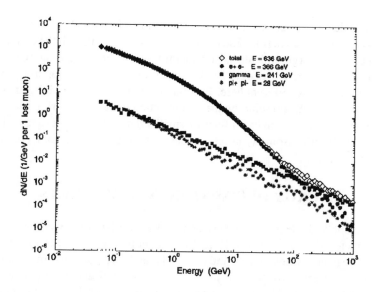

FIGURE 11. e^+e^-, γ and hadron energy spectra in muon interaction vertices for 2 TeV muon beam loss at a limiting aperture at L = 200 m (Configuration 3). Energies carried out by different particles are indicated.

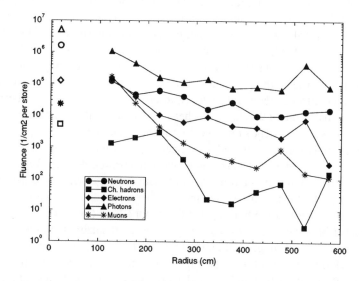

FIGURE 12. Particle fluences in the detector for 2 TeV beam halo loss (1% per store) at a limiting aperture at L = 200 m (Configuration 3): central tracker (open symbols) and radial distribution at the endcap calorimeter (filled symbols).

Particle fluences in the detector (Fig. 2) are shown in Fig. 12 for a 1% per store beam loss at a limiting aperture 200 m from the IP. One sees that all the components contribute to the background rates which for the few first turns (crossings) can significantly exceed those due to decays for the model considered. It is important that the protection strategy be different here. Collimators in the detector vicinity are drastically less effective. Beam scraping prior to injection to the storage ring and a dedicated beam cleaning system (with aperture smaller than any in the IR) a few kilometers from the IR are absolutely necessary elements of the collider to overcome the beam halo problem.

ENERGY DEPOSITION IN SC MAGNETS

Due to unavoidable $\mu \to e\nu\tilde{\nu}$ decays, about 300 to 900 W of power are deposited in every meter of the ring with a 10 to 30 Hz repetition rate. This results in a heat load to cryogenics which significantly exceeds the levels tolerated in existing SC magnets. This energy is deposited via electromagnetic showers induced in the beam pipe and in SC coils by high energy synchrotron photons and by decay electrons.

The calculations of energy deposition distributions in the storage ring components are performed for muon beam decays in Configurations 1 and 2. Even with a longitudinally uniform source, there is an increased rate at the β_{peak} location. With an 8 cm radius aperture, the photon flux at the 7.5 cm radius beam pipe and beam side of the SC coil is $\approx 10^9\,\mathrm{cm}^{-2}$ and e^+e^- flux is $\approx 5 \times 10^6\,\mathrm{cm}^{-2}$ per store. There is a significant azimuthal dependence of energy deposition density due to the effect of the strong magnetic field. The peak energy deposition $\approx 3\,\mathrm{mJ/g}$ exceeds the quench limits for the magnet of the assumed type by almost an order of magnitude.

The way to mitigate the problem would be to intercept most of the shower energy, say at the nitrogen temperature level, inserting a liner between the beam pipe and the SC coils. A rather thin layer of a heavy material would do a good job both to reduce the peak energy deposition density (quench) and the total energy deposited at the liquid helium level (heat load to cryogenics). A 5-mm tungsten liner provides a factor of 8 reduction of the maximum energy deposition density. Fig. 13 shows the energy deposition azimuthal distribution with such a liner. Being azimuthally averaged, the effect is two times smaller. The lateral gradient of energy deposited in the SC coil is very strong both with and without a liner (Fig. 14).

A thicker liner has a bigger effect. Fig. 15 shows the azimuthally averaged energy deposition versus tungsten liner thickness in the 7.5 cm to 9.0 cm radial region in the first IR dipole. Even for the averaged deposition the reduction can be as high as a factor of 18 with a 10-mm liner and exceeding a factor of 30 reduction for the peak energy density. For the fixed inner radius (=7.5 cm) the effect of a thicker liner (>10 mm) is weaker. For a particular lattice, the liner thickness might be non-uniform: thicker in a horizontal plane (and in a vertical plane in the quadrupoles), consisting of rod-like insertions, and thinner in the rest of the aperture (see Fig. 13).

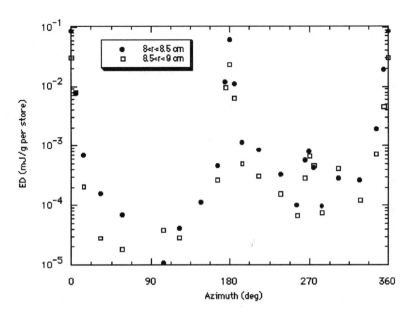

FIGURE 13. Azimuthal distribution of energy deposition density in the first SC cable shell of the first IR dipole with 5-mm tungsten liner.

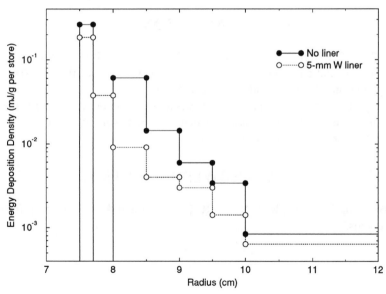

FIGURE 14. Azimuthally averaged radial distribution of energy deposition density in the first IR dipole with and without 5-mm tungsten liner.

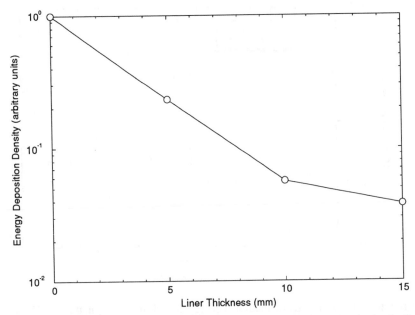

FIGURE 15. Attenuation of azimuthally averaged energy deposition density in the first SC cable shell as a function of the tungsten liner thickness.

CONCLUSIONS

A 2×2 TeV high-luminosity $\mu^+\mu^-$ collider offers exciting physics opportunities. The calculational tools developed allow reliable and detailed analyses of the background and radiation fields formed in machine and detector components. In the studies performed, the beam induced effects look severe, but can be mitigated with the proposed measures. There is a hope that with more work the design goals of this new generation project can be achieved.

ACKNOWLEDGMENTS

We thank K. Y. Ng, R. J. Noble, R. B. Palmer and A. V. Tollestrup for useful discussions.

References

[1] "Physics Potential and Development of $\mu^+\mu^-$ colliders", Sausalito–94, ed. by D. Cline, *AIP Conference Proceedings* **352** (1996).

[2] Palmer, R. B. et al., "Beam Dynamics Problems in a Muon Collider", in *ICFA Beam Dynamics Newsletter*, No. 8, pp. 27–33, August 1995; also BNL–61580 (1995).

[3] Palmer, R. B. et al., "High-Energy High-Luminosity $\mu^+\mu^-$ Collider Design", BNL–62041 (1995).

[4] Foster, G. W. and Mokhov, N. V., "Backgrounds and Detector Performance at a 2×2 TeV $\mu^+\mu^-$ Collider", in [1] pp. 178–190; also Fermilab–Conf–95/037 (1995).

[5] Gelfand, N. M. and Mokhov, N. V., "2×2 TeV $\mu^+\mu^-$ Collider: Lattice and Accelerator-Detector Interface Study", in *Proc. of 1995 Particle Accelerator Conference*, Dallas, May 1995; also Fermilab–Conf–95/100 (1995).

[6] Mokhov, N. V., "The Radiation Environment at Muon Colliders", in *Proc. of the 2nd Workshop on Simulating Accelerator Radiation Environments (SARE2)*, CERN, Geneva, October 1995.

[7] Polychronakos, V., presented at *2×2 TeV $\mu^+\mu^-$ Collider Collaboration Meeting*, Brookhaven, February 1995.

[8] Mokhov, N. V., "The MARS Code System User's Guide, Version 13(95)", Fermilab–FN–628 (1995).

[9] Striganov, S., *Nucl. Instruments and Methods* **A322**, 225-230 (1992); "Fast Precise Algorithm for Simulation of Ionization Energy Losses", *IHEP 92-80*, Protvino (1992).

[10] Kalinovskii, A. N., Mokhov, N. V., and Nikitin, Yu. P., *Passage of High-Energy Particles through Matter*, AIP, New York, 1989.

[11] Jackson, J. D., *Classical Electrodynamics*, 2nd Ed., J. Wiley, New York, 1975.

[12] "Review of Particle Properties", *Phys. Rev.* **D50**, No. 3 (1994).

[13] Bukin, A.D. and Grozina, N.A., *Comp. Phys. Communications* **78**, 287–290 (1994).

[14] Van Ginneken, A., "Fluctuations of Ionization Energy Loss and Simulation of Ionization Cooling", Fermilab–Pub–94/384 (1995).

[15] Vavilov, P. V., *JETP* **32**, 920 (1957).

[16] Kel'ner, S. R., *Yad. Fiz* **5**, 1092 (1967).

[17] Kabayakawa, K., *Nuovo Cimento* **B47**, 156–184 (1967).

[18] Van Ginneken, A., *Nucl. Instr. Meth.* **A251**, 21–39 (1986).

[19] Andreev, Yu. M., Bezrukov, L. B., and Bugaev, E. V., *Phys. of Atomic Nuclei* **57**, 2066–2074 (1994).

[20] Petrukhin, A. V. and Shestakov, V. V., *Canad. J. Phys.* **46**, 337 (1968).

[21] Bezrukov, L. B. and Bugaev, E. V., *Sov. J. Nucl. Phys* **33**, 635 (1981).

[22] Sakumoto, W. K. et al., *Phys. Rev.* **D45**, 3042–3050 (1992).

[23] Bloom, E. D. et al., SLAC–PUB–653 (1969); Armstrong, T. A. et.al., *Phys. Rev.* **D5**, 1640 (1972).

[24] Montanet, L. et al., *Phys. Rev.* **D50**, 1335 (1994).

[25] Armstrong, T. A. et al., *Nucl. Phys.* **B41**, 445 (1972).

[26] Caldwell, D. O. et al., *Phys. Rev.* **D7**, 1362 (1973).

[27] Bianchi, N. et al., *Phys. Lett.* **B325**, 333 (1994).

[28] Baishev, I. S., Kurochkin, I. A., and Mokhov, N. V., IHEP–91–118, Protvino (1991).

[29] Berman, B. L., and Fultz, S. C., *Rev. Mod. Phys.* **47**, 713 (1975).

[30] Mokhov, N. V., "Muons at LHC. Part 1: Beam Dumps", CERN/TIS–RP/TM/95–27 (1995).

[31] Van Ginneken, A., "MUSIM: Program to Simulate Production and Transport of Muons in Bulk Matter", Fermilab–FN–594 (1992). ERRATUM: Fermilab–FN–610 (1993).

[32] Tsai, Y., *Rev. Mod. Phys.* **46**, No. 4, 815–851 (1974).

The Effect of Muon Decay on the Design of Dipoles and Quadrupoles for a Muon Collider

Michael A. Green

E. O. Lawrence Berkeley National Laboratory
University of California
Berkeley, CA. 94720

Abstract. The decay of muons to neutrinos and electrons can cause heating in the superconducting dipoles and quadrupoles in the muon collider acceleration rings and the colliding beam ring. The problem is particularly acute in the colliding beam ring where heating in the magnets can be as high as 2.4 kW per meter in the bending magnets of muon collider ring with 2 TeV mu plus and mu minus beams with 2.22×10^{12} particles per bunch at a repetition rate of 30 Hz. The energy deposited within the helium temperature region must be reduced at least three orders of magnitude in order for the refrigeration system to begin to keep up with the heat load. Beam heating from muon decay will require changes in dipole design from the traditional cosine theta (or intersecting ellipse) design used in the SSC magnets. Some dipole and quadrupole design options are presented in this report for both the accelerator rings and the colliding beam rings.

BACKGROUND

Muons decay after they are produced. The decay time constant for a muon at rest is 2.197×10^{-6} seconds. Muons traveling near the speed of light have a decay time constant equal to the rest decay time constant multiplied by the gamma factor. All through the processes of cooling, acceleration and storage the muons are decaying to two neutrinos and an electron or positron (depending on the charge of the muon that is decaying). The energy of the muon is split between the three particles. In our calculations, 40 percent of the muon energy is assumed to end up in electrons and positrons. The energy that ends up in the neutrinos is not of concern, but the energy that is in the electrons and positrons is of serious concern because it can end up in the superconducting elements of the machine.

The electrons and positrons will emit synchrotron radiation as they are bent by the magnetic field. The synchrotron radiation power will be deposited on the outside of the magnet on the beam orbital plane. The power that remains in the electrons will be deposited on the inside of the magnet on the beam orbital plane. The energy from the synchrotron radiation and the electrons and positrons can have negative consequences on the superconducting magnets that are used to bend and focus the charged muons as they are accelerated and stored in the collider rings. As a result, the superconducting magnets will probably have a different design from those that are used in the Tevatron, HERA or the LHC.

MUON DECAY IN THE ACCELERATION SECTIONS AND THE COLLIDER RING

Once the muons are produced, bunched and cooled they must be accelerated quickly to the final energy of the collider. This report assumes that the final beam energy where the muons collide is 2 TeV. It is further assumed that negative and positive muons will be carried in the same accelerator structure. The muons are assumed to leave the muon cooling system at an energy 0.2 GeV. Two bunches (one with positive muons, the other with negative muons) containing 3×10^{12} muons leave the cooling section at a repetition rate of 30 Hz. Thus the assumed muon flux entering the first acceleration section is 1.8×10^{14} muons per second. In order to achieve the design luminosity for the colliding beam ring, less than one third of the muons will decay during the acceleration process. (It is not known if one muon bunch is stacked on top of the previous bunch when the design luminosity was calculated.) The remainder of the muons will decay in the colliding beam ring with a time constant of 41.6 ms (for muons at 2 TeV). About 41 percent of the muons from the previous cycle will be left when the new bunches are injected into the collider 33.3 ms after the previous bunch was put into the ring.

The system for accelerating the muons from the cooler at an energy of 0.2 GeV to a final energy of 2 TeV is assumed to contain the following components: 1) a linac to accelerate the muons from 0.2 to 2 GeV, 2) A nine turn recirculation ring with two superconducting linacs in the straight sections to accelerate the muons from 2 GeV to 20 GeV, 3) an eighteen turn recirculation ring with two superconducting linacs in the straight sections to accelerate the muons from 20 GeV to 200 GeV, and 4) an eighteen turn recirculation ring with two superconducting linacs in the straight sections to accelerate the muons from 200 GeV to their final energy of 2 TeV. The magnets for the recirculation rings will be superconducting and all of the magnet bores share a common cold iron flux return system. The collider will be a single separated function ring of superconducting magnets that carries both the negative and the positive muons. The beta star at the collision point must be about 3 mm in order for the desired design luminosity of 10^{35} cm^{-2} s^{-1} to be achieved.

The number of muons that will decay in a given length L can be estimated using the following expression:

$$N_d = \frac{N L E_o}{\tau_o (E_T + E_o) c}) \tag{1}$$

258

where N is the number of muons transported through a structure per second: N_d is the number of muon that decay in the structure per second; L is the length of the structure: E_T is the muon energy; E_0 is the muon rest mass ($E_0 = 105.7$ MeV); c is the velocity of light ($c = 2.998 \times 10^8$ m s^{-1}): and τ_0 is the muon decay time constant at rest ($\tau_0 = 2.197 \times 10^{-6}$ s). Equation 1 is applicable when the transit time for the muon through the structure of length L is much less than the decay time constant of the muon at its energy E_T.

The power available for deposition into the structure from the muon decay can be estimated using the following expression:

$$P = {\sim}0.4 \, N_d \, E_{ave} \, e \tag{2}$$

where N_d is number of muon that decay per second (See Equation 1); E_{ave} is the average energy of the muon in the structure; e is the unit charge for the electron (muon have the same charge as an electron. $e = 1.602 \times 10^{-19}$ C s^{-1}). The factor 0.4 at the start of Equation 2 represents the assumed portion of the muon energy that ends up in the decay electrons or positrons. The remainder of the muon energy is transported to the universe by the decay neutrinos.

Table 1 presents calculations for muon decay in each of the accelerator components and the collider ring. Included in the table 1 is the number of turns through the component and the total transit length L_T through the structure. Table 1 gives an estimate of the decayed muon power that is transferred to electrons and positrons. This is the portion of the decayed muon power that can end up in the superconducting magnet system. The beam flux of μ^+ and μ^- that enters the accelerator section is assumed to be 1.8×10^{14} muons per second (one bunch of each type with 3×10^{12} muons per bunch at a repetition rate of 30 Hz). The peak bending induction in all of the rings is assumed to be 7 T.

Table 1 Muon Decay Parameters for Various Parts of a Muon Collider

Component	Energy (GeV)	Turns	L_T (km)	Decay Rate (μ s^{-1})	Power* (kW)
Linac	2	-NA-	0.12	1.938×10^{13}	1.37
First Ring	20	9	1.37	0.653×10^{13}	2.49
Second Ring	200	18	20.13	1.088×10^{13}	31.9
Third Ring	2000	18	201.3	1.011×10^{13}	324
Collider Ring	2000	1356**	9.2	1.331×10^{14}	17100

* Power from e$^+$ and e$^-$, assumed to be 40 percent of the muon beam power
** during one decay time constant of 41.6 ms

THE EFFECT OF DECAYED MUONS ON THE COLLIDER MAGNETS

Table 1 shows the estimated beam power that ends up in the form of decay electrons and positrons in various components of the muon collider acceleration system and the storage ring. Depending on electron positron beam power per unit length and the energy of the decay electron and positrons, the effect on various components can range from not serious to the factor that determines the design of that component. It appears that the superconducting magnets are most severely affected by the decay of the muons in the machine. The superconducting RF structure is less likely to be affected by the products of muon decay. The detectors around the collision point will be greatly affected by the muon decay products, but this report does not deal with that issue.

The Linac to Accelerate from 0.2 to 2.0 GeV

About 1.4 kW from electron and positron decay products will be deposited in the linac channel and the components downstream from the linac. The average energy of the electrons produced will be about 400 MeV. The synchrotron radiation critical energy that comes from the electrons in a string of 1.5 T bending magnet downstream from the linac will be about 160 eV. The decay electrons and positrons will probably be accelerated along with the muons in the linac. Thus, it is likely that little of the electron energy will end up in the linac RF structure. The energy from electrons will, for the most part be deposited in the string of conventional bending magnets and quadrupoles between the linac and the first acceleration ring. It is expected that the electrons, positrons and photons will be absorbed by the vacuum chamber wall.

The First Ring to Accelerate from 2 to 20 GeV

About 2.5 kW of power from electron and positron decay products will be deposited in the bending magnets and quadrupoles of the first accelerator ring. The average energy of the electrons and positrons produced by muon decay will be about 4.4 GeV. The synchrotron radiation critical energy that comes from the electrons and positrons in the 7 T bending magnets will be about 89 KeV. It is probable that the decay electrons and positrons will be accelerated along with the muons within the linacs. Therefore it is unlikely that much of the decay energy will end up in the RF structure of the linacs. The power per unit length that would be deposited within 0.65 km of magnet bore from the decay products would be about 4.6 watts per meter. From an overall refrigeration standpoint, this is not serious, but to keep the superconductor from overheating, it is desirable to absorb the decay product energy in a cooled bore tube at a temperature above 4 K. An 80 K bore tube can be designed to absorb eighty to ninety percent of the energy from the electrons, positrons and the synchrotron radiation.

The Second Ring to Accelerate from 20 to 200 GeV

About 32 kW of power from the muon decay products will be deposited in the bending magnets and quadrupoles of the second accelerator ring. The average energy of the electrons and positrons produced by muon decay will be in the range of 40 GeV. The synchrotron radiation critical energy that comes from the electrons and positrons in the 7 T bending magnets will be about 7.4 MeV. The decay electrons and positrons will be accelerated along with the muons within the linacs. Therefore, it is expected that little of the decay energy will end up in the RF structure. The power per unit length that would be deposited within 12.9 km of magnet bore from the decay products is estimated to be about 2.5 watts per meter. From an overall refrigeration standpoint, this begins to become a problem. To keep the dipole and quadrupole superconductor from overheating, it is desirable to absorb eighty to ninety the decay product energy in a cooled bore tube at a temperature of 80 K or higher. If eighty to ninety percent of the decay energy is deposited at a higher temperature, the overall sizes of the helium refrigeration plants become reasonable.

The Third Ring to Accelerate from 0.2 to 2.0 TeV

About 324 kW of power from the muon decay products will be deposited in the bending magnets and quadrupoles of the third accelerator ring. The average energy of the electrons and positrons produced by muon decay will be about 400 GeV. The synchrotron radiation critical energy that comes from the electrons and positrons in the 7 T bending magnets will be about 740 MeV. The decay electrons and positrons will be accelerated along with the muons in the linacs. It is expected that little of the decay energy will end up in the RF structure. The power per unit length that would be deposited within 130 km of magnet bore from the decay products is estimated to be about 2.5 watts per meter. From an overall refrigeration standpoint, 324 kW at 4.4 K is a serious problem (by an order of magnitude given the size of the ring). In addition, one would like to reduce the beam losses into the superconductor to something around 0.3 watts per meter. A simple cooled vacuum chamber is probably not adequate for absorbing the energy from the muon decay products. One is forced to look at magnet designs where the coils are split on the mid plane so that a heavy muon decay product absorption system at 80 K or above can be installed. The quadrupoles will also have to be split on the mid plane so muon decay product energy can be absorbed.

The Muon Collider Ring

It is estimated that about 17100 kW of power from the muon decay products will be deposited in the collider ring. Most of this energy will be deposited within the dipole and quadrupoles magnet strings. The average energy of the electrons and positrons produced by muon decay will be about 800 GeV. The synchrotron radiation critical energy that comes from the electrons and positrons in the 7 T bending magnets will be about 4.3 GeV. The power per unit length that would be deposited within 7 km of magnet structure from the decay products is estimated to be about 2400 watts per meter. This is unacceptable from the total helium refrigeration required and the energy deposition per unit volume in the superconductor of the magnets. From both standpoints, the energy deposited into the 4 K region of the magnets should be reduced by more than three orders of magnitude. One is forced to look at dipole and quadrupole magnet designs where the coils are split on the mid plane so that the electrons, positrons, and synchrotron gamma rays can be moved away from the cold regions of the magnet into an energy absorption system at 300K that is capable of removing 17.1 MW of thermal energy. The long straight sections in the collider ring will also see some of the muon decay products. The muon decay products are a serious problem for the detectors.

SUPERCONDUCTING DIPOLE AND QUADRUPOLE DESIGNS FOR THE MUON COLLIDER RINGS

The size of the region where the decay electrons, positrons and synchrotron radiation strike the wall of the vacuum chamber is determined by the vertical emittance of the beam, and the beam dispersion caused by inhomogeneties of the magnetic field. The stay clear region for the decay products as far as the magnet coils are concerned is determined by the size of the region where most of the decay energy comes from (two or three sigma beam size from the mid plane in the vertical direction), the physical size of the magnet coil supports at the mid plane (This depends on the structure, but the minimum size is about 1 mm.), and the minimum thickness of multilayer insulation between the coil mid plane support structure and the warm parts of the vacuum chamber that absorbs the energy from the muon decay products (3 mm is a minimum sort of number). The field quality has a larger effect than beam emittance on the size of the fan of muon decay products from the beam. If the field within the dipole is good to 1 part in 1000, the decay product fan angle from the mid plane is about 1 milliradian. The minimum separation of the magnet, coils at the mid plane, surrounding a warm decay energy absorber will be about 12 mm for a magnet that has an inner coil radius of 30 mm.

The design of the dipoles and quadrupoles is dependent on the percentage of the muon decay product energy that can be deposited within the 4 K mass of the dipole. For low energy decay products, such as those from the first acceleration ring, all but ten percent of the muon decay product energy will be deposited in the beam vacuum chamber. As the energy of the muon decay product increases, the depth where the energy is deposited increases. Thus muon decay product energy absorption has to extend further along the mid plane of the dipole before there is only ten percent of the decay product energy left to be deposited at 4 K. The extra mass needed to absorb decay product energy at some point will impact the superconducting coils at the magnet mid plane. The magnets of the collider ring, where only one tenth of one percent of the decay product energy can end up in the 4 K region, must have an open structure to allow the decay product energy to be absorbed in a water cooled structure that is at or near room temperature

At least three design approaches for can be considered for the superconducting dipole magnets in the various rings for the muon collider:

1) For first acceleration ring (perhaps the second ring as well), a conventional cosine theta dipole design can be employed with a warm vacuum chamber that has a cooling system to remove the energy from the muon decay products. The temperature of the vacuum chamber can be as low as 80 K. This type of magnet does not require coils that are split on the mid plane because at least eighty percent of the decay product energy can be removed within the vacuum chamber. Figure 1 illustrates this type of magnet concept with a cooled vacuum chamber in the bore.

2) For the second and third acceleration rings a design with the coils split on the mid plane should be pursued. In this instance, most the energy from the decay products can not be absorbed in the vacuum chamber unless it extends into the space between the coils and perhaps beyond into the iron return yoke. The iron around the superconducting coils could still be cold, but alternative methods for supporting the mid plane forces between the upper and lower coils must be found. Since it is allowable for ten to fifteen percent of the energy from the muon decay products to be absorbed in a 4 K structure, a cold bridge structure between the coils to carry force across the mid plane can be considered. The remainder of the muon decay product energy must be removed at 80 K or above. Figure 2 illustrates a dipole magnet of this type.

3) The collider ring must have coils that are completely separated on the mid plane. The iron return yoke, in all probability, must be at room temperature. The forces pulling the coil together across the mid-plane must be carried by 300 to 4 K supports. The coils must be separated so that less than 0.1 percent of the energy from the muon decay products ends up in the superconducting coils or its surrounding support structure that is at 4 K. The rest of the muon decay product energy ends up in the water cooled vacuum chamber and water cooled iron at the mid plane. Figure 3 illustrates the design of a dipole with maximum separation between the coils. The dipole shown in Figure 3 has reasonably good field quality (about 4 parts in 10000 at a 10 mm radius).

The quadrupoles mirror the dipoles on the mid plane because the decay products from the upstream dipole will end up in the quadrupoles and most of the decay products produced within the quadrupoles will travel along the mid plane of the magnet. The quadrupoles may have some energy deposition from the decay products along the vertical line of symmetry. In general, the energy of the decay products deposited in this region will represent less than one percent of the energy deposited along the mid plane. Mid plane coil separation has a worse affect on field quality in a quadrupole or sextupole than it does in a dipole.

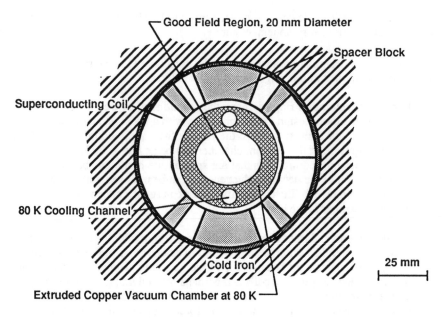

Figure 1 The Dipole Coils, Cold Iron and 80 K Vacuum Chamber for the
Low Energy Accelerator Sections of a Muon Collider

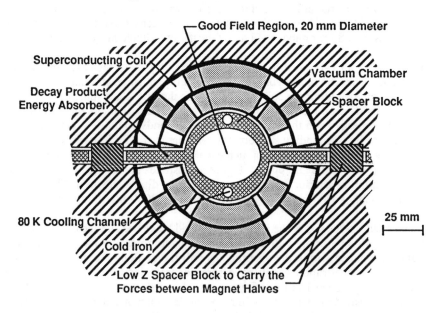

Figure 2 The Dipole Coils, Cold Iron, Vacuum Chamber and Decay Energy Absorbers
for the high Energy Accelerator Sections of a Muon Collider

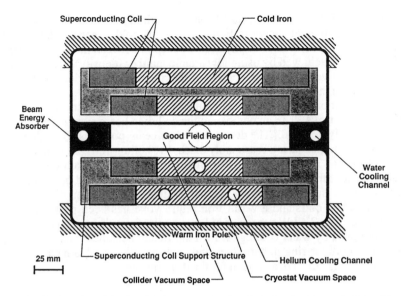

Superconducting Coil — — Cold Iron

Beam
Energy
Absorber

Good Field Region

Water
Cooling
Channel

25 mm

Superconducting Coil Support Structure

Warm Iron Pole

Helium Cooling Channel

Collider Vacuum Space

Cryostat Vacuum Space

Figure 3 A warm Iron Dipole with Superconducting Coils for the Muon Collider Ring
(Less than 0.1 percent of the decay energy ends up in the 4 K region.)

CONCLUDING COMMENTS

The decay of the muons will deposit energy in the form of electrons, positrons, and photons in all of the superconducting magnets in the collider. Within the collider dipoles and quadrupoles, the decay energy will be deposited along the acceleration plane (the mid plane of the magnets). The region where the muon decay energy is deposited will be only a few millimeters wide. How one deals with the energy from the muon decay products depends on the muon energy and the muon residence time in the magnet structure.

The accelerator sections of the muon collider have a muon residence time that is from 5 to 8 percent of the muon decay time constant. It is estimated that 26 percent of the muons will decay between the muon cooling system and the entry of the muons into the collider ring. The amount of energy in the decay products is about ten times the energy that is allowed to be deposited in the superconducting magnets of the three acceleration sections. At low energies the penetration distance for the decay products is relatively short The dipole and quadrupole superconducting coils can be on the mid plane if less than ten percent of the muon decay product energy ends up in that region. As the energy of the muon decay products increases, the penetration distance will also increase. The extra mass needed to attenuate the muon decay product energy begins to impact on the superconducting magnet coils. The dipole and quadrupole coils must be moved off of the mid plane. The coil geometry changes if one is to maintain the field quality required in the bending and focusing elements. The magnet shown in Figure 2, has the cold upper and lower magnet halves completely separated. A physical connection between the upper lower parts of the magnet is achieved by using low atomic number (density) supports between the two cold halves. Up to ten percent of the muon decay product energy can be absorbed within these interconnect members.

The amount of energy deposited by the muon decay products is highest in the colliding beam rings. The muons stay in the colliding beam storage ring until they have completely decayed. The decayed product beam power is about three orders of magnitude higher than a reasonably sized 4 K helium refrigeration system can remove from the magnet string. The muon decay product beam power per unit volume is about three orders of magnitude higher than it should be for reasonable operation of a superconducting magnet system. A collider dipole that has superconducting coils and a warm iron return yoke has been postulated for the collider ring. The 2.4 kW per meter of beam power from the muon decay products can be removed by water cooling through the vacuum chamber and the iron return yoke. The collider ring quadrupoles and sextupoles must also have warm iron. Superconducting coils are an option for the these magnets.

ACKNOWLEDGMENTS

The author acknowledges the discussions with R. B. Palmer and G. H. Morgan of the Brookhaven National Laboratory, and with D. B. Cline of the UCLA Physics Department concerning magnets and the possible introduction of heat into the superconductor from muon decay. This work was performed with the support of the Office of High Energy and Nuclear Physics, United States Department of Energy under contract number DE-AC03-76SF00098.

REFERENCES

1. Transparencies presented at the *Second Workshop on the Physics Potential and Development of $\mu^+ \mu^-$ Colliders*, Alta Mira Hotel, Sausalito California, 17-19 November 1994

2. Transparencies by R. B. Palmer from the *First High Luminosity 2+2 TeV $\mu^+ \mu^-$ Collider Collaboration Meeting,*, Brookhaven National Laboratory, 6-8 February 1995

3. Transparencies presented at the *A Second High Luminosity 2+2 TeV $\mu^+ \mu^-$ Collider Collaboration Meeting*, Fermi National Laboratory 11-13 July 1995, compiled by Robert Noble of FNL

4. Transparencies presented at the *Ninth Advanced ICFA Beam Dynamics Workshop: Beam Dynamics and Technology Issues for $\mu^+ \mu^-$ Colliders*, Montauk, New York, 15-20 October 1995, compiled by Juan C. Gallardo of BNL

MUON DOSES IN SOIL

A. Van Ginneken
Fermi National Accelerator Laboratory
P. O. Box 500, Batavia, Illinois 60510

Abstract

Results from the program MUSIM for the transport of a
2 TeV muon beam through homogeneous soil.

Fig.1 shows a set of isodose contours obtained with the program MUSIM[1] for
a pencil beam of 2 TeV muons incident on soil with a density of $2.24\,\mathrm{g/cm^3}$.
For reasons of clarity contours above 10^{-10} rem/muon are not shown. Likewise
contours below 10^{-19} are omitted for lack of statistics. Some crowding of the
low dose contours is evident in Fig.1 and the physics of muon energy loss
suggests increased crowding for contours of 10^{-20} and below. The picture that
emerges is that of a narrow beam of muons which travels some 2.7 km before
coming to rest while spreading out no more than about 13 m radially. Fig.1
must be interpreted with some care since heavier materials and magnetic fields
can cause considerable distortion to the contours. Both may be present when
beam is dumped whether intentional or accidental. Attention must be given
to the topography of the terrain surrounding the collider (on a scale of 2-3 km)
to ensure that beam loss does not result in a muon beam emerging anywhere
above the surface.

Acknowlegement

This research was supported by the U.S. Department of Energy under Contract
No. DE-AC02-76CHO3000.

References

1. A. Van Ginneken, Fermilab Report FN-594 (1992); ibid., FN-610 (1993).

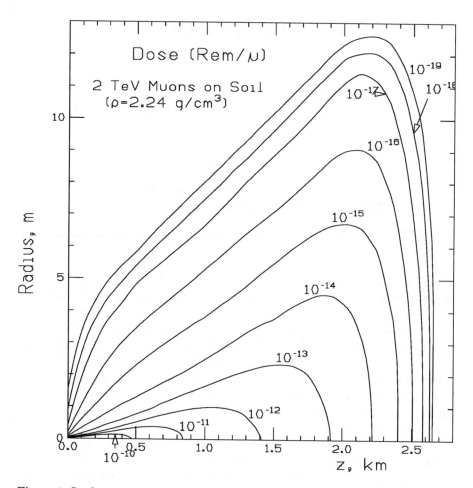

Figure 1: Isodose contours due to a beam of 2 TeV muon on homogeneous soil (density of 2.24 g/cm^3).

New Physics Potential of Muon-Muon Colliders

V. Barger
Department of Physics, University of Wisconsin, Madison, WI 53706

Abstract

The development of muon-muon colliders would open up new physics opportunities at both the low energy ($E_{CM} = 500$ GeV) and high energy ($E_{CM} = 4$ TeV) frontiers. An unique possibility is s-channel Higgs boson production and measurement of its width. Heavy supersymmetric particles may be discovered and their properties determined. Alternatively, strong electroweak symmetry breaking can be revealed.

The possibility that muon colliders can be constructed[1, 2, 3] opens the important question of their potential to discover or study new physics[4, 5]. In particular, what can be accomplished at a $\mu^+\mu^-$ collider compared to an e^+e^- linear collider or the Large Hadron Collider? A general observation is that a $\mu^+\mu^-$ collider can at a minimum explore the same physics as an e^+e^- collider of the equal CM energy and luminosity provided that the larger backgrounds in the detector can be managed. But are there unique possibilities of a muon collider that would make it complementary to an e^+e^- linear collider? In fact, the finer beam energy resolution, smaller bremsstrahlung and beamstrahlung, possibly higher energy reach than an e^+e^- collider, and Higgs coupling proportional to m_μ offer distinct and possibly crucial advantages that will be discussed in this report, which is largely drawn from the studies in Refs [4, 5].

First, we briefly outline the advantages of a $\mu^+\mu^-$ collider for fundamental physics and then later discuss the prospects in more detail. The major areas of interest are as follows:

- Higgs bosons
  The fact that $m_\mu \gg m_e$ allows s-channel Higgs production to occur at an interesting level in $\mu^+\mu^-$ collisions, whereas the corresponding rate at e^+e^- colliders is insignificant. Higgs width measurements are possible[5], even for a light Higgs with $m_h \approx M_Z$, provided that excellent resolution

is achieved. Fortunately, the muon beam energy resolution is not limited by beamstrahlung. Also, s-channel discovery of the heavy neutral bosons H, A of supersymmetry is possible at a muon collider.

- Supersymmetry
  High CM energy ($\gtrsim 2$ TeV) may be needed for new particle searches, especially if the mass scale of supersymmetric particles is of order 1 TeV. Synchrotron radiation does not limit the circular acceleration of muons and multi-TeV energies can be realized through a linear acceleration array. In comparison, the parameters of an e^+e^- collider with CM energy > 2 TeV may be difficult to accommodate.

- Strong WW scattering
  If supersymmetry and Higgs bosons with mass $m_h \lesssim 0.8$ TeV do not exist, then the scattering of longitudinally polarized W-bosons must become strong[6] at CM energies ~ 1 TeV. A multi-TeV muon collider would provide improved signals of strong WW scattering, since the cross section grows with energy. Also, the virtual photon backgrounds will be less than at an e^+e^- machine of comparable energy.

- Precision threshold measurements
  The rms deviation R of the Gaussian energy spectrum of each beam can be as small as $R = 0.01\%$ at a $\mu^+\mu^-$ collider, as compared with $R > 1\%$ at an e^+e^- collider. Thus the threshold curves in $\mu^+\mu^-$ collisions suffer less smearing and the top and W masses can be measured with greater precision[7].

- CP violation and flavor-changing neutral currents in the Higgs sector
  New CP violating phases or flavor-changing neutral current couplings may be present in the Higgs sector which could be studied via s-channel Higgs production at a $\mu^+\mu^-$ collider[8].

In the following we address the first three of these physics issues in the framework of the

First Muon Collider (FMC)
(250 GeV) $\times$ (250 GeV) $\mathcal{L} = 2 \times 10^{33}$ cm^{-2} s^{-1}

Next Muon Collider (NMC)
(2 TeV) $\times$ (2 TeV) $\mathcal{L} = 10^{35}$ cm^{-2} s^{-1}

which are the two design targets of current accelerator development efforts[2, 3].

Design Luminosity Requirements

The figure of merit for luminosity considerations is the QED cross section for $\mu^+\mu^- \to \gamma^* \to e^+e^-$,

$$\sigma_{\text{QED}} \simeq \frac{100 \text{ fb}}{s(\text{TeV}^2)}.$$

As a rule of thumb, to find and study new physics the luminosity needed is typically

$$\left(\int L dt\right) \sigma_{\text{QED}} \gtrsim 1000 \text{ events}.$$

At the CM energies under consideration, this corresponds to

	necessary L	design L
$\sqrt{s} = 500$ GeV		
$\int L dt \gtrsim 1$ fb^{-1},	$L \gtrsim 10^{32}$ cm^{-2} s^{-1}	2×10^{33} cm^{-2} s^{-1}
$\sqrt{s} = 4$ TeV		
$\int L dt \gtrsim 100$ fb^{-1},	$L \gtrsim 10^{34}$ cm^{-2} s^{-1}	10^{35} cm^{-2} s^{-1}

The proposed machine designs have luminosity targets that exceed the minimum requirements.

Linear e^+e^- collider capabilities

It is appropriate to consider the physics potential of a $\mu^+\mu^-$ collider in the context of what can be accomplished at a future linear e^+e^- collider[9]. First, the SM model Higgs boson would be discovered at a linear e^+e^- machine through the $e^+e^- \to Z^* \to Zh$ process for $m_h < 0.7\sqrt{s}$, where $\sqrt{s}$ is the center-of-mass energy; the discovery limit for the lightest Higgs boson of the minimal supersymmetric standard model (MSSM) is somewhat lower than the SM Higgs, because its coupling to the Z is less. There is an ironclad result that the mass of the lightest MSSM Higgs boson in minimal SUSY grand unified theory models (GUT) is bounded by[10, 11]

$$m_h < 120 \text{ GeV}.$$

The next linear e^+e^- collider (NLC) with $\sqrt{s} = 300$ to 500 GeV can find or exclude the MSSM Higgs boson in this mass range and thereby confirm or exclude such SUSY GUT theories.

The other Higgs bosons in the MSSM are predicted to be almost mass degenerate and heavy

$$m_{H^0} \approx m_{A^0} \approx m_{H^\pm} \geq 300 \text{ GeV}.$$

At an e^+e^- collider these scalars can be discovered via $e^+e^- \to Z^* \to H + A$ and $H^+ + H^-$, which are kinematically accessible only for $m_H \lesssim \sqrt{s}/2$.

Other sparticles with mass $< \sqrt{s}/2$ could be discovered as well at an e^+e^- collider, although event rates for scalar particles are p-wave suppressed near threshold. The energy reach of the NLC is probably adequate for $\chi_1^+ \chi_1^-$ pair production of the lightest chargino and the for production $\chi_1^0 + \chi_2^0$ of the lightest neutralinos. The NLC energy may also be sufficient for pair production of the sleptons $(\tilde{e}, \tilde{\mu}, \tilde{\tau})$ and the lightest $(\tilde{t}_1)$. However the NLC energy could well be inadequate for production of the heavier chargino and neutralinos and squarks other than stop.

Light Higgs strategy at a muon collider

First, the light Higgs discovery will likely be made at LEP 2, the LHC or the NLC before a muon collider is constructed. At the LHC the detection can be made via the $h \to \gamma\gamma$ decay mode, with production by $W^* \to Wh$ and $t\bar{t}h$. The signal is deemed viable for a SM Higgs in the mass range $80 \lesssim m_{h_{\mathrm{SM}}} \lesssim 150$ GeV, with a 1% Higgs mass resolution[12]. At an e^+e^- or $\mu^+\mu^-$ collider discovery will be made via the $Z^* \to Zh$ process, which is self-scanning in energy. The Higgs mass resolution with an e^+e^- detector is[13]

$$\delta m_h = \pm 4 \text{ GeV}/\sqrt{N} \qquad (\text{for } m_h < 2M_W) ,$$

where N is the number of events

$$N = \epsilon L \sigma(Zh) \, \mathrm{BF}(\text{effective}) .$$

Here ϵ is the efficiency and BF the effective branching fraction for the joint Z and h decays. An e^+e^- detector can determine the SM Higgs mass to an accuracy

$$\delta m_{h_{\mathrm{SM}}} \simeq \pm 0.4 \text{ GeV} \left(\frac{10 \text{ fb}^{-1}}{L} \right)^{1/2}$$

giving $\delta m_{h_{\mathrm{SM}}} = \pm 200$ MeV for $L = 50$ fb^{-1}.

Once the Higgs boson is discovered the next paramount issue is to precisely measure its mass and to determine its width. The s-channel process $\mu^+\mu^- \to h \to b\bar{b}$ is uniquely suited to this task. The light quark background can be rejected by b-tagging (hereafter we assume a 50% b-tagging efficiency). The s-channel resonance cross section is

$$\sigma_h = \frac{4\pi \Gamma(h \to \mu\bar{\mu}) \, \Gamma(h \to b\bar{b})}{(s - m_h^2)^2 + m_h^2 \Gamma_h^2}$$

with $\Gamma(h \to \mu\mu) \propto m_\mu^2$. Since the storage ring is a modest fraction of the overall muon collider cost, Palmer has suggested that a special purpose ring could be designed to optimize the luminosity at $\sqrt{s} = m_h$, with m_h known from the $Z^* \to Zh$ process.

A Gaussian shape is expected to be a good approximation for the energy spectrum of each muon beam[14]. The primary effect of bremsstrahlung is to reduce the peak luminosity by about 40%, roughly preserving the Gaussian shape, while giving a lower energy tail[5]. The root mean square deviation R of the Gaussian is expected to be naturally in the range $R = 0.04\%$ to 0.08% with $R = 0.01\%$ to 1% possible[14]. The corresponding rms error σ in $\sqrt{s}$ is

$$\sigma = (6 \text{ MeV}) \left(\frac{R}{0.01\%}\right) \left(\frac{\sqrt{s}}{100 \text{ GeV}}\right) .$$

Thus for $m_h = 100$ GeV, $\sigma = 50$ MeV for $R = 0.06\%$ ($\sigma = 7.7$ MeV for $R = 0.01\%$). The effective cross section $\bar{\sigma}$ is given by the convolution of the Breit-Wigner with the Gaussian CM energy spread. Figure 1(a) compares the SM Higgs signal vs. m_h with the backgrounds. Figure 1(b) shows the luminosity required at $\sqrt{s} = m_h$ for a 5 standard deviation SM Higgs detection.

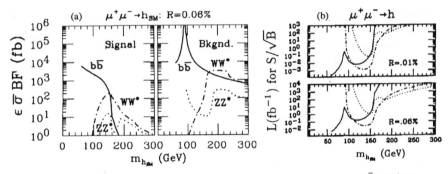

Figure 1. (a) The SM Higgs signal and background in the $b\bar{b}$, W^+W^- and ZZ^* channels for an efficiency $\epsilon = 0.5$; (b) luminosity required for $S/\sqrt{B} = 5$, with $R = 0.01\%$ and $R = 0.06\%$ [Ref. 5].

Scanning for Higgs

Starting from the m_h determination from the $Z^* \to Zh$ process, a scan could be made over the $\sqrt{s} = m_h \pm \delta m_h$ band with steps of size $\sim \sigma$. The backgrounds vary slowly over the energy resolution. A fit to the resulting line shape will determine m_h. For sufficiently small $\sigma \sim \Gamma_h$, the line shape will be that of a Breit-Wigner and Γ_h can be determined from the fit. Figure 2 shows SM and MSSM Higgs profiles for $m_h = 110$ GeV[15]. The width of the MSSM Higgs exceeds the SM Higgs width and grows as $(\tan \beta)^2$; the SM places the most exacting requirements on the resolution that could be needed to determine Γ_h.

273

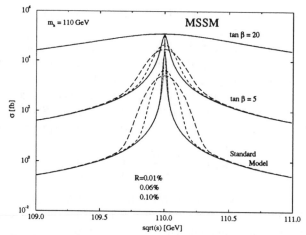

Figure 2. Profile of the Higgs resonance shape for several values of the resolution.

Figure 3(a) shows a sample SM Higgs resonance scan[15], for $m_h = 110$ GeV and $\Gamma_h = 3.2$ GeV. The 'experimental' values were generated by Monte Carlo. For a coarse scan, with resolution $R = 0.06\%$ ($\sigma = 50$ MeV) and integrated luminosity 5 fb^{-1} distributed over 11 bins of 30 MeV, the fit to the Breit-Wigner yields

$$\delta m_h = \pm 9 \text{ MeV}.$$

Figure 3(b) illustrates a fine scan ($R = 0.01\%$) over 5 bins of 2 MeV[15]. In this case a fit yields

$$\delta m_h = \pm 0.3 \text{ MeV}$$
$$\delta \Gamma_h = \pm 0.6 \text{ MeV}$$

The measurement of Γ_h can distinguish between the SM the MSSM Higgs.

High longitudinal polarization of both beams would be beneficial for s-channel Higgs physics, if 85% polarization could be reached with $\gtrsim 1/10$ of the luminosity in the absence of polarization[16].

Heavy MSSM Higgs search

The s-channel search techniques also applies to the H^0, A^0 MSSM Higgs bosons. These are relatively narrow resonance at high mass with widths[5]

$$\Gamma_A \approx \Gamma_H \approx 0.1 \text{ to } 10 \text{ GeV}$$

that are comparable or broader than the expected resolution for $R = 0.06\%$. Thus measurement of their widths in s-channel production should be relatively

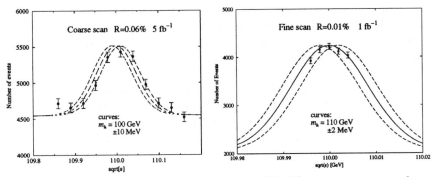

Figure 3. Illustration of scans across a SM Higgs resonance peak; (a) coarse scan, (b) fine scan [Ref. 15].

straightforward with a scan over several $\sqrt{s}$ settings once the states are found. The decays of H^0 and A^0 into $b\bar{b}$ are dominant for $\tan\beta > 5$. Figure 4 gives the cross section and significance $S/\sqrt{B}$ of the $\mu^+\mu^- \to A \to b\bar{b}$ signal versus m_A. An integrated luminosity of $L = 20$ fb^{-1} would allow a scan over 200 GeV at intervals of 1 GeV with $L = 0.1$ fb^{-1} per point. For $\tan\beta \gtrsim 5$ the significance of the signal is $\gtrsim 5$ standard deviations for all m_A.

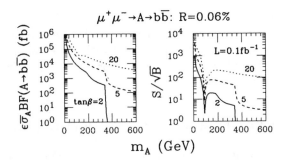

Figure 4. (a) Effective s-channel cross section of the MSSM A^0 boson for the $b\bar{b}$ final state; (b) corresponding statistical significance $S/\sqrt{B}$ for $L = 0.1$ fb^{-1} delivered at $\sqrt{s} = m_h$ [Ref. 5].

Next Muon Collider as a SUSY factory

We turn our attention now to the physics at a $\sqrt{s} = 4$ TeV $\mu^+\mu^-$ collider with luminosity $L = 10^{35}$ cm^{-2}s^{-1}. The LHC will provide sufficient CM energy to produce heavy squarks and gluinos, but disentangling the particle spectrum and measuring the masses will be a real challenge at a hadron collider due to complex cascade decays and QCD backgrounds. If $M_{SUSY} \sim 1$ TeV many of the sparticles could be beyond the range of linear e^+e^- colliders, where designs up to $\sqrt{s} = 2$ TeV are under consideration. The p-wave suppression of spin-0 pair production in e^+e^- or $\mu^+\mu^-$ collisions means that energies well above threshold will be required: see Fig. 5. At the NMC, with integrated annual luminosity 1000 fb^{-1}, sparticles of mass 1 TeV would be produced with the following cross sections and event rates:

$$\sigma_{\tilde{u}_{L,R}} = 4\beta^3 \text{ fb} \rightarrow 2500 \text{ events (one flavor)}$$
$$\sigma_{\tilde{d}_{L,R}} = 1\beta^3 \text{ fb} \rightarrow 600 \text{ events (one flavor)}$$
$$\sigma_{\chi^\pm} = 6\beta \text{ fb} \rightarrow 5000 \text{ events}$$
$$\sigma_t = 8 \text{ fb} \rightarrow 8000 \text{ events}$$

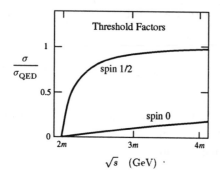

Figure 5. The kinematic suppression of heavy particle threshold cross sections in e^+e^- or $\mu^+\mu^-$ collisions.

The production of heavy SUSY particles leads to spherical events near threshold characterized by multijets, missing energy (associated with the escaping lightest supersymmetric particle) and leptons. The separation of these signals from backgrounds should not be a problem.

Study of a strongly interacting electroweak sector (SEWS)

In the event that there is no low-energy supersymmetry and no Higgs boson of mass $m_H \lesssim 0.8$ TeV, then the scattering of longitudinally polarized W-bosons must become strong at the TeV energy range. At a $\mu^+\mu^-$ collider SEWS can be studied through the process in Fig. 6. We can estimate the size of the SEWS

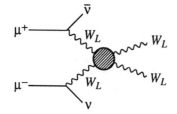

Figure 6. Diagram for strong $W_L W_L$ scattering in $\mu^+\mu^-$ collisions.

276

effect from a model with a 1 TeV Higgs boson,

$$\Delta\sigma = \sigma(m_H = 1 \text{ TeV}) - \sigma(m_H = 0).$$

The results for a 4 TeV $\mu^+\mu^-$ collider and a 1.5 TeV e^+e^- collider are compared below.

	$\sqrt{s}$	$\Delta\sigma(W_L W_L)$	$\Delta\sigma(Z_L Z_L)$	L
$\mu^+\mu^-$	4 TeV	80 fb	50 fb	10^3 fb
e^+e^-	1.5 TeV	8 fb	6 fb	200 fb

The SEWS contribution grows with energy so the energy reach and high luminosity are especially critical here.

Conclusion

In summary, muon collider offer exciting opportunities to explore new physics at both the low ($\sqrt{s} = 500$ GeV) and high ($\sqrt{s} = 4$ TeV) energy frontiers. The most exciting prospects include the s-channel production of Higgs bosons, allowing precision measurements of Higgs masses and widths, the discovery and study of heavy SUSY particles, or in the absence of a weakly interacting Higgs sector, the study of strong scattering of W-bosons at high energies.

Acknowledgments

I thank M. Berger, J. Gunion and T. Han for collaboration on the results reported here. This work was supported in part by the U.S. Department of Energy under Grant No. DE-FG02-95ER40869.

References

[1] *Proceedings of the First Workshop on the Physics Potential and Development of $\mu^+\mu^-$ Colliders*, Napa, California, 1992, Nucl. Instr. and Meth. **A350**, 24 (1994).

[2] *Proceedings of the Second Workshop on the Physics Potential and Development of $\mu^+\mu^-$ Colliders*, Sausalito, California, 1994, ed. by D. Cline (to be published).

[3] Reports in these proceedings.

[4] V. Barger et al., Physics Goals Working Group Report in Ref. [2].

[5] V. Barger, M.S. Berger, J.F. Gunion and T. Han, Phys. Rev. Lett. **75**, 1462 (1995) and report in preparation.

[6] M.S. Chanowitz and M.K. Gaillard, Nucl. Phys. **B261** 379 (1985); J. Bagger et al., Phys. Rev. **D49**, 1246 (1994).

[7] M.S. Berger, in *Proceedings of the International Symposium on Particle Theory and Phenomenology (XVIII Kazimierz Meeting on Particle Physics and 1995 Madison Phenomenology Symposium)*, Iowa State University, Ames, May 1995; S. Dawson, these proceedings.

[8] D. Atwood and A. Soni, hep-ph/9505233 (SLAC-PUB-95-6877); D. Atwood, L. Reina and A. Soni, hep-ph/9507416 (SLAC-PUB-95-6962).

[9] For references to in-depth studies at future e^+e^- colliders, see e.g. *Proceedings of the Fermilab Linear Collider Workshop: Physics with High Energy e^+e^- Colliders*, Fermilab, November 1995; *Proceedings of the International Workshop on Physics and Experiments at Linear e^+e^- Colliders*, Iwate, Japan, September 1995; *Proceedings of the Workshop on Physics and Experiments with Linear e^+e^- Colliders*, Waikoloa, Hawaii, ed. by F. Harris et al. (World Scientific, 1993); JLC Group, KEK Report 91-16 (1992); *Proceedings of the Workshop on Physics and Experiments with Linear Colliders*, Saariselkä, Finland, September 1991, ed. by R. Orava et al. (World Scientific, 1992).

[10] M. Drees, Int. J. Mod. Phys **A4**, 3635 (1989); J. Ellis et al., Phys. Rev. **D39**, 844 (1989); L. Durand and J.L. Lopez, Phys. Lett. **B217**, 463 (1989); P. Binétry and C.A. Savoy, Phys. Lett. **B277**, 435 (1992); T. Morori and Y. Okada, Phys. Lett. **B295**, 73 (1992); G. Kane et al., Phys. Rev. Lett. **70**, 2686 (1993); J.R. Espinosa and M. Quirós, Phys. Lett. **B302**, 271 (1993).

[11] V. Barger et al., Phys. Lett. **B314**, 351 (1993); P. Langacker and N. Polonsky, Phys. Rev. **D50**, 2199 (1994).

[12] ATLAS and CMS Collaborations, Technical Design Reports CERN/LHCC/94-13 and CERN/LHCC/94-38.

[13] T. Barlow and D. Burke, private communication.

[14] G.P. Jackson and D. Neuffer, private communication.

[15] V. Barger and M.S. Berger (unpublished).

[16] Z. Parsa (unpublished)

PHYSICS POTENTIAL OF A POLARIZED $\mu^+\mu^-$ COLLIDER– A SCALAR COLLIDER

D. Cline

Center for Advanced Accelerators, University of California Los Angeles
Box 951547, Los Angeles, CA 90095-1547 USA

Abstract

We discuss the possible physics with a fully polarized 250 sf x 250-GeV $\mu^+\mu^-$ collider. Provided adequate luminosity can be obtained, this will be a very powerful machine to study the scalar-Higgs sector for both standard and SUSY models. In principle, it will yield a Higgs/Top Factory. In addition, tests of CP and FCNC processes could be carried out. We outline two schemes for obtaining high polarization in such a collider.

1 PHYSICS POTENTIAL OF THE 250 × 250-GeV $\mu^+\mu^-$ COLLIDER

Over the past three years, several workshops have been organized to study the physics potential (1–3) and machine design of a $\mu^+\mu^-$ collider. Two energy ranges have been identified, 2 × 2 TeV (largely by R. Palmer and others) and 250 × 250 GeV (studied first by the author) (3,4). In this paper, we update the physics potential for the low-energy machine with recent LEP and other recent results and discuss the physics potential if polarized $\mu^+\mu^-$ collisions can be achieved.

At the Napa workshop, the possibility of developing a $\mu^+\mu^-$ collider in the range of luminosity of 10^{31} cm^{-2} s^{-1} was considered and appears feasible. However, higher luminosity will be required for the Higgs sweep. It is less certain that the high-energy resolution required for the Higgs sweep can be obtained. We summarize in Table 1 some of the key issues in the Higgs search.

With a high-mass t quark, precision LEP/SLD data and the theorists' dreams of a SUSY world, the scalar (pseudoscalar sector) is possibly very complex and may require several types of colliders. Consider:

- If the low-mass Higgs has $m > 130$ GeV, mssm is not allowed.

- If $m > 200$ GeV, there are constraints from the requirement that perturbation theory be useful up to very high energy and from the stability of the vacuum.

- If $m < 130$ GeV, mssm is possibly ok, but we may expect other particles (H,A) and the width of the low mass Higgs may change.

- The scalar sector may be extremely complex, requiring pp (LHC) and $\mu^+\mu^-$ colliders (and possibly NLC and $\gamma\gamma$ colliders).

- In high energy collisions, vector states are allowed unless a special method is used. Consider $\mu^+\mu^-$ colliders with polarized $\mu^\pm$

$$\mu^+\mu^- \begin{cases} (100\text{–}500) \text{ GeV - scalars (h,A, ...)} \\ \geq 2 + \text{TeV} \qquad\qquad\qquad\qquad W^+W^- \end{cases}$$

$\qquad\qquad\qquad\qquad\qquad\qquad\qquad\qquad\qquad Z^0 Z^0$ production in scalars

This cannot be done for pp or e^+e^- colliders.

- A $\mu^+\mu^-$ collider is complimentary to the LHC/CMS detector.

Table 1: The Scalar Sector

At the Sausalito (5) and at this workshop it seems that luminosity of perhaps 10^{33} cm^{-2} s^{-1} is feasible, if we combine polarized $\mu^\pm$ beam and, thus, we will have a most powerful Higgs factory and scalar collider.

A SCALAR COLLIDER

The most interesting question in particle physics now is associated with the origin of mass. It is generally assumed that the exchange of fundamental scalar particles, called the "scalar sector" is somehow responsible for this. For supersymmetry modes, this scalar sector is even more complex and interesting (see Table 1).

In this section, we highlight one of the most interesting goals of a $\mu^+\mu^-$ collider: the discovery of a Higgs Boson in the mass range beyond that to be covered by LEP I&II ($\sim 80\text{–}90$ GeV) and the natural range of the supercolliders

($\gtrsim 2 \ M_Z$) (4). In this mass range, as far as we know, the dominant decay mode of the h⁰ will be

$$H^0 \rightarrow b\bar{b} \ , \tag{1}$$

whereas the Higgs will be produced by the direct channel

$$\mu^+ \mu^- \rightarrow h^0 \ , \tag{2}$$

which has a cross section enhanced by

$$\left(\frac{M_\mu}{M_e}\right)^2 \sim (200)^2 = 4 \times 10^4 \tag{3}$$

larger than the corresponding direct product at an $e^+ e^-$ collider. However, we will see that the narrow width of the Higgs partially reduces this enhancement.

There is growing evidence that the Higgs should exist in this low-mass range from:

1. The original paper of Cabibbo, *et al.*, which shows that, when $M_t > M_Z$ and assuming a Grand Unification of Forces, $M_h < 2 \ M_Z$.

2. Fits to LEP data imply that a low mass h^0 could be consistent with $M_t > 150$ GeV.

3. The extrapolation to the GUT scale that is consistent with SUSY also implies that one of the Higgs should have a low mass, perhaps below 130–150 GeV.

This evidence implies the exciting possibility that the Higgs mass is just beyond the reach of LEP II and in a range that is very difficult for the LHC to detect.

We expect the supercollider LHC to extract the signal from background (*i.e.*, seeing either $h^0 \rightarrow \gamma\gamma$ or the very rare $h^0 \rightarrow \mu\mu\mu\mu$ in this mass range, since $h \rightarrow b\bar{b}$ is swamped by hadronic background. However, detectors for the LHC are designed to extract this signal. Figure 1 gives a picture of the various physics thresholds that may be of interest for a $\mu^+ \mu^-$ collider.

In this low mass region, the Higgs is also expected to be a fairly narrow resonance and, thus, the signal should stand out clearly from the background from

$$\mu^+ \mu^- \rightarrow \gamma \rightarrow b\bar{b} \rightarrow Z_{\text{tail}} \rightarrow b\bar{b} \ . \tag{4}$$

For masses above 180 GeV, the dominant Higgs decay is

$$H^0 \rightarrow W^+ W^- \quad \text{or} \quad Z^0 Z^0 \ , \tag{5}$$

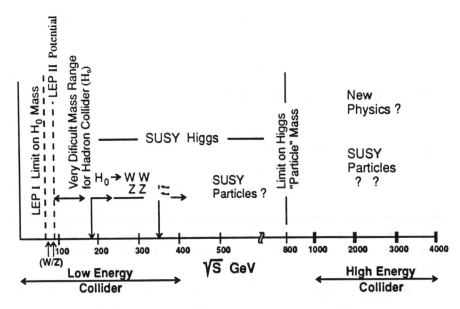

Figure 1. Physics threshhold for a $\mu^+\mu^-$ collider.

giving a clear signal and a larger width; the machine energy resolution requirements could be relaxed somewhat!

Another possibility for the intermediate Higgs mass range is to search for

$$\mu^+\mu^- \rightarrow Z^0 H^0 \tag{6}$$

using a broad energy sweep. The corresponding cross section is small. Once
an approximate mass is determined, a strategy for the energy sweep through
the resonance can be devised. The study of the t quark through $t\bar{t}$ production
would also be interesting.

Finally, another possibility is to use the polarization of the $\mu^+\mu^-$ particles
oriented so that only scalar interactions are possible. However, there would be
a trade-off with luminosity and, thus, a strategy would have to be devised to
maximize the possibility of success in the energy sweep through the resonance.

2 EXTRACTION OF A SCALAR SIGNAL IN $t\bar{t}$ PRODUCTION

In Fig. 2, we show tha Feynman diagrams for the production of $t\bar{t}$ for both
e^+e^- and $\mu^+\mu^-$. Because of the larger mass of the μ compared to the e, the

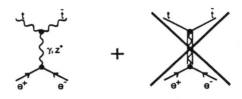

1) e^+e^- COLLISIONS

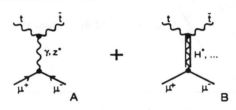

2) $\mu^+\mu^-$ COLLISIONS

A   B

— IF PROCESS B CAN BE ISOLATED
IT COULD GIVE A POWERFUL METHOD
TO STUDY THE SCALAR SECTOR —

Figure 2. $t\bar{t}$ production at $\mu^+\mu^-$ colliders.

diagram with a scalar intermediate state can be important. If we fully polarize the $\mu^+\mu^-$ system to give a net zero scalar state, we believe the scalar sector will be enhanced to the point that a measurable asymmetry will be generated. Thus, one could search for evidence of a scalar particles far from the central mass. This is a unique feature of polarized $\mu^+\mu^-$ colliders.

3 HIGGS FACTORY AND HIGGS WIDTH

There are several ways to determine the approximate mass of the Higgs Boson in the future (4). Suppose it is expected to be at a mass of 135 ± 2 GeV, the energy spread of a $\mu^+\mu^-$ collider can be matched to the expected width (see Fig. 3). An energy scan could yield a strong signal to background especially with polarized $\mu^+\mu^-$ in the scalar configuration. Once the Higgs is found, the following could be carried out:

1. Measurement of width, to separate standard model Higgs from SUSY or other Higgs models,

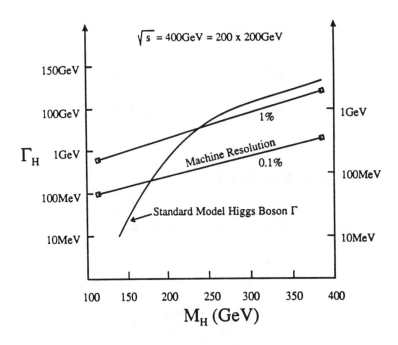

Figure 3. Higgs search at a $\mu^+\mu^-$ collider (required machine resolution and the expected Higgs width).

2. Measurement of the Branching fractions, the rare decay will involve loop effects that can sample very high energies.

We believe this is another unique feature of a polarized $\mu^+\mu^-$ collider.

4 REFERENCES

1. Early references for $\mu\mu$ colliders are: E. A. Perevedentsev and A. N. Skrinsky, *Proc., 12th Int. Conf. on High Energy Accelerators*, R. T. Cole and R. Donaldson, eds., (1993), p. 481; D. Neuffer, *Part. Accel.* 14 (1984) 75; D. Neuffer, in *Adv. Accel. Concepts*, AIP Conf. Proc. 156 (1987) p. 201.
2. D. V. Neuffer, "$\mu^+\mu^-$ Colliders: Possibilities and Challenges," *Nucl. Instrum./Meth.* A 350 (1994) 27.
3. D. Cline, *Nucl. Instrum./Meth.* A 350 (1994) 24, and the following 4 papers constitute a mini-conf. proceeding of the Napa meeting.
4. D. Cline, "Physics Potential and Development of $\mu^+\mu^-$ Colliders, UCLA preprint CAA-115-12/94(1994).
5. *Proceedings of the 2nd $\mu^+\mu^-$ Workshop*, D. Cline, ed., AIP Press (1995).

PRECISION MEASUREMENTS AT A MUON COLLIDER

S. Dawson
Physics Department
Brookhaven National Laboratory
Upton, New York 11973

Abstract

We discuss the potential for making precision measurements of M_W and M_T at a muon collider and the motivations for each measurement. A comparison is made with the precision measurements expected at other facilities. The measurement of the top quark decay width is also discussed.

1 INTRODUCTION

A $\mu^+\mu^-$ collider with high luminosity and narrow beam spread offers the possibility of performing high precision measurements of fundamental masses and decay widths occurring in the Standard Model and in some extensions of the Standard Model. We discuss precision measurements of the W mass and the top quark mass and width. Measurements of the Higgs boson mass and width, both in the Standard Model and in SUSY models are discussed in Ref. [1]. We pay particular attention to the motivation for making each precision measurement and the experimental precision which is necessary in order to test the theoretical consistency of the Standard Model or to verify the existence of new physics.

Each of these precision measurements depends on knowing the relevant energy and building a storage ring to maximize the luminosity at that energy. The mass measurements of the W and top quark are made by scanning the threshold energy dependences of the cross sections. The threshold energy dependences will be smeared by radiation from the initial state particles, limiting the precision of the measurements. Because the muon is much heavier than the electron, there will be less initial state radiation and the beam energy resolution may be better in a muon collider than in an electron collider, leading to the possibility of more precise measurements.

2 MEASUREMENT OF THE W MASS

A precision measurement of the W mass is of fundamental importance to our understanding of the Standard Model. Combined with a precision measurement of the top quark mass, the consistency of the Standard Model can be checked since the W mass is predicted as a function of the top quark mass.

The current world average on the W mass is obtained by combining data from UA2, CDF, and D0: [2]

$$M_W = 80.23 \pm .18 \ GeV. \tag{1}$$

With more data from CDF and D0, both the systematic and statistical errors will decrease and it has been estimated [2] that with 100 pb^{-1} it will be possible to obtain:

$$\Delta M_W^{\text{Tevatron}} \sim 110 \pm 20 \ MeV, \tag{2}$$

while 1000 pb^{-1} will give

$$\Delta M_W^{\text{Tevatron}} \sim 50 \pm 20 \ MeV, \tag{3}$$

where the first error is statistical and the second is systematic.

At LEP-II, the error on M_W can be reduced still further. There are two general strategies for obtaining a mass measurement. The first is to reconstruct the decay products of the W, while the second method is to measure the excitation curve of the W pair production cross section as the energy is varied. Both methods give approximately the same precision. By reconstructing the W decay products with 500 pb^{-1} (3 years running) at $\sqrt{s} = 190 \ MeV$[2],

$$\Delta M_W^{\text{LEP-II}} \sim 40 \ MeV. \tag{4}$$

The precision is ultimately limited by the knowledge of the beam energy, $\Delta E^{\text{ beam}} \sim 20 \ MeV$.

It is possible that a muon collider could obtain a more precise measurement of M_W than is possible at LEP-II. We will discuss the design restrictions on a muon collider in order to make this the case. The beam spread at a lepton collider can be roughly assumed to have a Gaussian energy resolution with a rms deviation,[1]

$$\delta \sim 60 \ MeV \left(\frac{R}{.06\%} \right) \left(\frac{\sqrt{s}}{2M_W} \right), \tag{5}$$

leading to an energy resolution smaller that the W decay width. A typical parameter for a high energy e^+e^- collider is $R = 1$, while a lower value is envisioned for a muon collider, (say $R \sim .06$ %), due to the fact that a muon collider will have less initial state radiation (ISR) than an e^+e^- collider. Here we investigate the requirements on R in order for a measurement of M_W to be made at a muon collider which will improve on the precision expected at LEP-II.

The procedure is to study the shape of the $l^+l^- \rightarrow W^+W^-$ cross section as a function of the center-of-mass energy, $\sqrt{s}$, and to fit a theoretical expectation to the cross section. The many theoretical effects which must be included are discussed by Stirling [3] and we follow his discussion closely.

We compute the cross section for off-shell W pair production including Coulomb effects as,[3]

$$\sigma_a(s)(l^+l^- \rightarrow W^+W^-) = \left[1 + \delta_C(s)\right] \int ds_1 \int ds_2 \rho(s_1)\rho(s_2)\sigma_0(s,s1,s2), \quad (6)$$

where

$$\rho(s) = \frac{\Gamma_W}{\pi M_W} \left[\frac{s}{(s - M_W^2)^2 + s^2\Gamma_W^2/M_W^2}\right] \quad (7)$$

and $\sigma_0(s, s_1, s_2)$ is the Born cross section for producing a W^+W^- pair with $W^\pm$ energies $\sqrt{s_1}$ and $\sqrt{s_2}$. (Electroweak radiative corrections are negligible in the threshold region.) This procedure defines what we mean by the W mass

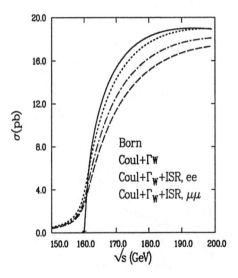

Figure 1: Contributions to the $l^+l^- \rightarrow W^+W^-$ cross section. The solid curve includes only the tree level cross section, while the dotted curve includes the Coulomb and finite W width effects. The long-dashed and dot-dashed curves include the effects of ISR for e^+e^- and $\mu^+\mu^-$ colliders, respectively.

287

and is in fact the same definition as used in LEP measurements. The Coulomb corrections are included in the factor $\delta_C(s)$ and arise from the fact the W^+W^- cross section diverges as $1/v$ at threshold. The analytic expression for $\delta_C(s)$ can be found in Ref. [4].

Finally, the corrections due to initial state radiation, which are sensitive to the lepton mass, m_l, must be included. The ISR is the first correction which differentiates between an electron and a muon collider. These corrections are included with a radiator function $F(x, s)$ (given in Ref. [5]) to obtain our final result for the W^+W^- pair production cross section:

$$\sigma(s)(l^+l^- \to W^+W^-) = \frac{1}{s}\int ds' F(x,s)\sigma_a(s') \tag{8}$$

where $x \equiv 1 - \frac{s'}{s}$ and

$$F(x,s) \sim tx^{t-1} + \left(\frac{x}{2} - 1\right)t + ... \tag{9}$$

with $t = \frac{2\alpha}{\pi}\left[\log(\frac{s}{m_l^2}) - 1\right]$. We see that the ISR is potentially much larger at an e^+e^- machine than at a muon collider. The various contributions to $l^+l^- \to W^+W^-$ are shown in Fig. 1. Near threshold, $\sqrt{s} \sim 2M_W$, the cross section rises rapidly with $\sqrt{s}$. As expected, there is less reduction of the cross section due to ISR at a muon collider than at an electron collider. In Fig.

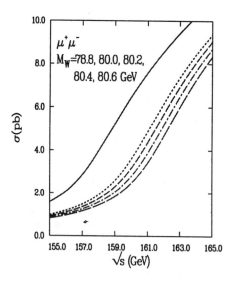

Figure 2: Cross section for $\mu^+\mu^- \to W^+W^-$. The solid curve has $M_W = 78.8\ GeV$.

288

2, we see that the cross section is very sensitive to the precise value of M_W assumed. It is straightforward to find the statistical error for a given efficiency, ϵ, and luminosity, $\int \mathcal{L}$,[3]

$$\Delta M_W^{stat} = \frac{1}{|\frac{d\sigma}{dM}|}\sqrt{\frac{\sigma}{\epsilon \int \mathcal{L}}}. \tag{10}$$

From Fig. 3, we find that the minimum statistical error occurs at $\sqrt{s} \sim 2M_W$ (where the cross section has the steepest dependance on energy). The statistical error is obviously not much different at a muon collider than at an electron collider, so our results correspond to those of Ref. [3]. That this must be the case can be seen from Fig. 1; at threshold, the effects of initial state radiation are small. At the minimum:

$$\Delta M_W^{stat} \sim 90 \; MeV \left[\frac{\epsilon \int \mathcal{L}}{100 \; pb^{-1}}\right]^{-1/2}. \tag{11}$$

If it were possible to have $1 \; fb^{-1}$ concentrated at $\sqrt{s} \sim 2M_W$ with an efficiency $\epsilon = .5$, then a muon collider could find $\Delta M_W^{stat} \sim 40 \; MeV$! Unfortunately, the luminosity of a muon collider decreases rapidly away from the design energy, so $1 \; fb^{-1}$ at $\sqrt{s} \sim 160 \; GeV$ is probably an unrealistic goal.

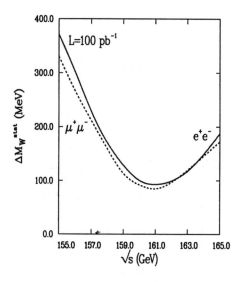

Figure 3: Statistical error on M_W from $l^+l^- \rightarrow W^+W^-$ from an absolute measurement of the rate with an integrated luminosity, $\mathcal{L} = 100 \; pb^{-1}$.

289

It is also necessary to consider the systematic error, which is primarily due to the uncertainty in the beam energy. From Fig. 2, changing the beam energy is equivalent to a shift in M_W, $\Delta M_W^{sys} \sim \Delta E^{beam}$. Therefore, to obtain a measurement with the same precision as LEP-II, a muon collider must be designed with $\Delta E^{beam} \sim 20\ MeV$, which corresponds to $R \sim .02\%$.

The bottom line is that a $\mu^+\mu^-$ collider must have on the order of $1\ fb^{-1}$ at $\sqrt{s} \sim 2M_W$ and an extremely narrow beam spread, $R \sim .02\ \%$, in order to be competitive with LEP-II for a measurement of ΔM_W.

3 PRECISION MEASUREMENT OF THE TOP QUARK MASS AND WIDTH

A muon collider will also be able to obtain a very precise measurement of the top quark mass, as has been discussed in detail by Berger.[6] Here, we concentrate on the physics motivations for making a precision measurement of the top quark mass.

Since at the moment we have no firm prediction for the top quark mass, a precision measurement of M_T is not particularly interesting in itself. However, when combined with a precision measurement of M_W, it tests the consistency of the Standard Model. This is because the prediction for M_W in the Standard

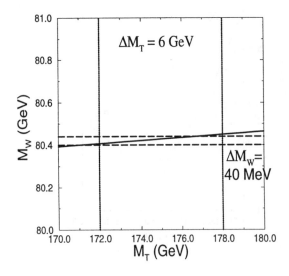

Figure 4: Dependence of the predicted W mass on the top quark mass in the Standard Model.

Model depends on M_T,[7]

$$M_W^2 = M_Z^2 \left[1 - \frac{\pi \alpha}{\sqrt{2} G_\mu M_w^2 (1 - \Delta r)} \right]^{\frac{1}{2}} \qquad (12)$$

with $\Delta r \sim \frac{M_T^2}{M_W^2}$, ($\Delta r$ also depends logarithmically on the Higgs boson mass). In Figure 4, we show the relationship between M_T and M_W in the Standard Model (where we have assumed $M_H = 100 \ GeV$ and included only contributions to Δr which depend quadratically on the top quark mass). For $M_T = 175 \ GeV$, a measurement of M_W to $\Delta M_W = 40 \ MeV$ requires a measurement of M_T to $\Delta M_T = 6 \ GeV$ in order to check the consistency of the Standard Model, while $\Delta M_W = 20 \ MeV$ requires $\Delta M_T = 3 \ GeV$. Given the expected precision on ΔM_W at LEP-II, it is clear that there is no motivation for a more precise measurement of M_T than several GeV.

A precise measurement of M_T and M_W also gives some information on the Higgs mass. For example, if $\Delta M_W = 40 \ MeV$, $\Delta M_T = 4 \ GeV$ and the true value of M_H were 100 GeV, then one could deduce from the electroweak measurements that at the 1 σ level, $50 < M_H < 200 \ GeV$.[2]

From the Tevatron, we will have $\Delta M_T \sim 8 \ GeV$ with 100 pb^{-1} and $\Delta M_T \sim \pm 4 \ GeV$ with 1000 pb^{-1}.[2] The LHC experiments are designed such that with 1 year of running, $\int \mathcal{L} = 10 \ fb^{-1}$, a value on the order of, $\Delta M_T \sim 3 \ GeV$ will be obtained.[2] In contrast, the values which would be obtained from a muon collider [6]

$$\Delta M_T^{\mu^+ \mu^-} \sim 300 \ MeV \qquad (13)$$

and from an electron collider [8]

$$\Delta M_T^{e^+ e^-} \sim 520 \ MeV \qquad (14)$$

are considerably more precise.

A precision measurement of the top quark width conveys significantly more information than a precision measurement of the mass. This is because the width is sensitive through loop effects to new particles contained in extensions of the Standard Model. Single top production at the Tevatron will measure the top quark width to roughly,[9]

$$\frac{\Delta \Gamma_T^{\text{Tevatron}}}{\Gamma_T} \sim .3, \qquad (15)$$

while a 500 GeV $e^+ e^-$ collider might obtain[8]

$$\frac{\Delta \Gamma_T^{e^+ e^-}}{\Gamma_T} \sim .2 \quad . \qquad (16)$$

Presumably, a $\mu^+ \mu^-$ collider will do even better. Such measurements will be capable of limiting the low mass particle spectrum of supersymmetric models and it would be interesting to have a systematic comparison of the capabilities of an $e^+ e^-$ and $\mu^+ \mu^-$ collider and the corresponding limits on SUSY particles.

4 REFERENCES

1. See, for example, V. Barger, M. Berger, J. Gunion, and T. Han, Phys. Rev. Lett. **75** (1995) 1462, hep-ph/9504330 ; V. Barger, contribution to this workshop.
2. F. Merritt, H. Montgomery, A. Sirlin, and M. Swartz, *Precision Tests of Electroweak Physics*, report of the DPF Committee on Long Term Planning, 1994.
3. W. Stirling, DTP-95-24 (1995), hep-ph/9503320.
4. V. Fadin, V. Khoze, A. Martin, and W. Stirling, Phys. Lett. **B363** (1995) 112, hep-ph/9507422.
5. F. Berends, in *Z Physics at LEP I*, CERN Yellow Report No. 89-08, Geneva, 1989, Vol. 1, edited by G. Altarelli, R. Kleiss, and C. Verzegnassi.
6. M. Berger, presented at *International Symposium on Particle Theory and Phenomenology*, Ames, Iowa (1995), hep-ph/9508209.
7. W. Marciano, Ann. Rev. Nucl. Part. Sci **41** (1991) 469.
8. P. Igo-Komines, in *Proceedings of the Workshop on Physics and Linear e^+e^- Colliders*, Waikoloa, Hawaii, 1993, edited by F. Harris, S. Olsen, S. Pakvasa, and X. Tata (World Scientific, Singapore, 1993); K. Fujii, T. Matsui, and Y. Sumino, Phys. Rev. **D50** (1994) 4341.
9. D. Carlson and C.P. Yuan, presented at *International Symposium on Particle Theory and Phenomenology*, Ames, Iowa (1995), hep-ph/9509208.

SIGNALS FROM FLAVOR CHANGING SCALAR NEUTRAL CURRENTS AT $\mu^+\mu^-$ COLLIDERS

L. Reina*

Brookhaven National Laboratory

P. O. Box 5000, Upton, New York 11973-5000

Abstract

We illustrate the possibility of observing signals from Flavor Changing Neutral Currents, originating from the scalar sector of a Two Higgs Doublet Model. In particular, we focus on the tree level process $\mu^+\mu^- \to \bar{t}c + \bar{c}t$, via scalar exchange in the s-channel, as a distinctive process for $\mu^+\mu^-$ colliders.

1 INTRODUCTION

A high energy lepton collider will be a good environment to look for *new physics* beyond the Standard Model (SM), taking advantage of the large enough energy, which will become available at low background rates. This is the reason why many people look at LEP II, NLC and $\mu^+\mu^-$ colliders with the specific aim of testing some crucial points of the proposed extensions of the SM. Supersymmetric Theories are among the favorite candidates, as we heard in one of the plenary talks at this meeting[1].

We want to present here a different analysis, focusing on minimal extensions of the SM, which might produce extremely distinctive signals at the next generation of lepton colliders. Our attitude being phenomenological, we will look for those extensions of the SM, whose consequences turn out to be particularly challenging both from the theoretical and the experimental point of view. In this context, our aim will be to point out where a $\mu^+\mu^-$ collider would be able to get a better performance with respect to an e^+e^- collider.

The possibility of extending the scalar sector of the SM has very often been considered in the literature, since this sector of the Model is still rather poorly known at the moment. The natural extension of adding one more

---

*Work done in collaboration with D. Atwood (CEBAF) and A. Soni (BNL).

scalar SU(2) doublet, resulting in the so called Two Higgs Doublet Models
(2HDM's), has the considerable distinction of introducing Flavor-Changing
Neutral Currents (FCNC) at the tree level in the scalar sector. Since FCNC
are forbidden in the SM context, this would be an extremely promising field to
look for unambiguous signals from *new physics*. However, severe constraints
are imposed by the low energy physics of the K- and B-mesons, such that
FCNC have practically to be avoided in this sector of the theory. This is
naturally accomplished by the SM itself, as we said, and has to be imposed
ad hoc in any 2HDM, by introducing a discrete symmetry[2], which limits the
possible Yukawa couplings between fermions and scalars.

Apart from the experimental constraints coming from K- and B-physics,
there is no *a priori* theoretical reason not to have FCNC. Therefore, the assumption of this *ad hoc* discrete symmetry may be dropped in favor of a more
natural one, which takes any Flavor Changing (FC) coupling to a scalar field
to be proportional to the mass of the coupled quarks. The basic idea is that
a natural hierarcy is provided by the observed fermion masses and this may
be transfered to the couplings between fermions and scalar fields, even when
they are not the ones directly involved in the mass generation mechanism[3].
In this way, FCNC are naturally suppressed in the light sector of the theory,
while dramatic effects may be seen in processes which involve the heavy quark
fields of the third generation.

We will illustrate these ideas at work in the following sections, first presenting the theoretical model we propose and then focusing on some FC signals,
namely $(\bar{t}c + \bar{c}t)$-production, which, if possible at an e^+e^--collider, will be even
more enhanced and characteristic at a $\mu^+\mu^-$-collider.

2 GENERAL FRAMEWORK

2.1 The Model

Let us recall that the 2HDM's with no FCNC at the tree level are normally
distinguished, in the literature, as Model I and Model II, depending on the
way the fermion fields couple to the two scalar SU(2) doublets. Model I refers
to the case in which both the Up-type and the Down-type components of a
fermion doublet couple to the same scalar field, while in Model II the Up-type
components couple to one scalar doublet and the Down-type components to
the other. In both cases, the particular pattern of fermion-scalar couplings is
obtained imposing that *ad hoc* discrete symmetry we were talking about in
the Introduction.

We want to consider here the case in which all the natural couplings introduced by a 2HDM are kept[3] and see if, under any assumption on the new FC
couplings, the model stays compatible with the experiments. We will refer to

the present model as Model III. The Yukawa Lagrangian of Model III, limited to the quark fiels, can be written as [4, 5]:

$$\mathcal{L}_Y^{III} = \eta_{ij}^U \bar{Q}_i \tilde{\phi}_1 U_j + \eta_{ij}^D \bar{Q}_i \phi_1 D_j + \xi_{ij}^U \bar{Q}_i \tilde{\phi}_2 U_j + \xi_{ij}^D \bar{Q}_i \phi_2 D_j + h.c. \quad (1)$$

where ϕ_i for $i = 1, 2$ are the two scalar doublets of a 2HDM, $\tilde{\phi}_i = i\sigma_2\phi_i$, while $\eta_{ij}^{U,D}$ and $\xi_{ij}^{U,D}$ are the non diagonal coupling matrices. Since the two scalar doublet are completely independent , by a suitable rotation of the quark fields, we can chose the two scalar doublets in such a way that only the $\eta_{ij}^{U,D}$ couplings generate the fermion masses, i.e. such that:

$$< \phi_1 > = \begin{pmatrix} 0 \\ v/\sqrt{2} \end{pmatrix}, \quad < \phi_2 > = 0 \quad (2)$$

The physical mass spectrum consists of two charged $\phi^\pm$ and three neutral spin 0 bosons, two scalars (H^0, h^0) and a pseudoscalar (A_0):

$$
\begin{aligned}
H^0 &= \sqrt{2}[(\text{Re }\phi_1^0 - v)\cos\alpha + \text{Re }\phi_2^0 \sin\alpha] \\
h^0 &= \sqrt{2}[-(\text{Re }\phi_1^0 - v)\sin\alpha + \text{Re }\phi_2^0 \cos\alpha] \\
A^0 &= \sqrt{2}(-\text{Im }\phi_2^0)
\end{aligned}
\quad (3)
$$

where α is a mixing phase (for $\alpha = 0$, H^0 corresponds exactly to the SM Higgs field, and $\phi^\pm$, h^0 and A^0 generate the new FC couplings).

In principle the $\xi_{ij}^{U,D}$ FC couplings are arbitrary, and we may have different attitudes in choosing their form. Some major proposals exist in the literature[3], which suggests to take them proportional to the mass of the quarks involved in the coupling, i.e. roughly as follows

$$\xi_{ij} = \lambda_{ij} \frac{\sqrt{m_i m_j}}{v} \quad (4)$$

where for the sake of simplicity we take the λ_{ij} to be real (for more details see[7]). Or, in view of the outstanding role played by the top quark within the SM framework, we might even adopt a more phenomenological point of view and assume that the effect of the FCNC on the first generation of quarks is negligible, focusing only on the t- and b-couplings and constraining them from experiments.

In both cases, the most interesting signals of these non-standard couplings will come from the physics of the top quark, both production and decays. Therefore, we would like to single out the right processes and the right environment in which we could already have the possibility of testing the proposed model. Of particular interest will be those quantities which either receive only tiny contributions in the SM framework or show some kind of disagreement between theory and experiments.

2.2 The process: $\mu^+\mu^- \to \bar{t}c + \bar{c}t$

In this section we will focus on the production of top-charm pairs at lepton colliders, i.e. e^+e^-, $\mu^+\mu^- \to t\bar{c} + c\bar{t}$, whose branching ratio turns out to be extremely suppressed not only in the SM ($Br(t \to cZ) \sim 10^{-13}$), but also in the 2HDM without tree level FCNC ($10^{-14} - 10^{-9}$)[7]. The final state for this process has a unique kinematics, with a very massive jet against an almost massless one. This quite peculiar signature may allow to work even with relatively low statistics, as can be the case for a lepton collider. The much better statistics one could get at an hadron collider, would come at a cost of a much higher background (mostly, tree level SM background for a one-loop process). We think the two effects somehow compensate, but the kind of analysis requested for the hadronic case would be much more complicated. Hence, it is worthwhile to consider a cleaner environment as an e^+e^- or a $\mu^+\mu^-$ machine.

With this respect, the case of a $\mu^+\mu^-$ with respect to an e^+e^- collider singles out, both because it will offer the possibility of exploring much higher energy regimes and because for the first time the possibility of an s-channel production will be available. In fact, for an e^+e^- collider, the tree level s-channel top-charm production is strongly suppressed and the top-charm production process arises as a one-loop correction to the Ztc-vertex. On the other hand, a $\mu^+\mu^-$ collider ($m_\mu \sim 200\, m_e$) may be able to produce scalar bosons in the s-channel in sufficient quantity to study their properties directly, i.e. to study directly the effect of a tree-level FC Higgs-$t\,c$ vertex. The possibility of direct scalar boson production in the s-channel constitutes, indeed, one of the challenging and interesting kinds of physics available in the future of these machines (for more details see [1, 6, 8, 9]).

The crucial point about a $\mu^+\mu^-$ collider is also that, if it is run on the Higgs resonance, $\sqrt{s} = m_\mathcal{H}$, Higgs bosons may be produced at an appreciable rate [11, 8, 6].

At $\sqrt{s} = m_\mathcal{H}$, the cross section for producing $\mathcal{H}$, $\sigma_\mathcal{H}$, normalized to $\sigma_0 = \sigma(\mu^+\mu^- \to \gamma \to e^+e^-)$, is given by

$$R(\mathcal{H}) = \frac{\sigma_\mathcal{H}}{\sigma_0} = \frac{3}{\alpha_e^2} B_\mu^\mathcal{H} \tag{5}$$

where $B_\mu^\mathcal{H}$ is the branching ratio of $\mathcal{H} \to \mu^+\mu^-$ and α_e is the electromagnetic coupling.

If the Higgs is very narrow, the exact tuning to the resonance implied in equation (5) may not in general be possible. Let us suppose then that the energy of the beam has a finite spread described by δ

$$m_\mathcal{H}^2(1 - \delta) < s < m_\mathcal{H}^2(1 + \delta) \tag{6}$$

where we assume that s is uniform about this range. The effective rate of Higgs production will thus be given by

$$\tilde{R}(\mathcal{H}) = \left[\frac{\Gamma_\mathcal{H}}{m_\mathcal{H}\delta} \arctan \frac{m_\mathcal{H}\delta}{\Gamma_\mathcal{H}} \right] R(\mathcal{H}) \qquad (7)$$

Let us now consider that a Higgs $\mathcal{H}$ of mass $m_\mathcal{H}$ is under study at a $\mu^+\mu^-$collider. For illustrative purposes we take $\mathcal{H} = h^0$ in the model discussed in sec. 2.1. In our model (see (1) and (4)), the coupling of h^0 to $f\bar{f}$ can be written as

$$C_{h^0 ff} = -\frac{g}{2}\frac{m_f}{m_W}\sin\alpha + \frac{\mathrm{Re}\xi_{ff} + i\gamma_5\mathrm{Im}\xi_{ff}}{\sqrt{2}}\cos\alpha \equiv \frac{gm_f}{2m_W}\chi_f e^{i\gamma_5\lambda_f} \qquad (8)$$

while the coupling to ZZ and WW is given by

$$C_{h^0 ZZ} = \frac{g\sin\alpha}{\cos\theta_W} m_Z g^{\mu\nu} \qquad C_{hWW} = g\sin\alpha m_W g^{\mu\nu} \qquad (9)$$

Finally the flavor changing Higgs $-t\bar{c}$ coupling is given by

$$C_{h^0 tc} = \frac{1}{\sqrt{2}}\left[\xi_{tc}P_R + \xi_{ct}^\dagger P_L\right]\cos\alpha \equiv \frac{g\sqrt{m_t m_c}}{2m_W}(\chi_R P_R + \chi_L P_L) \qquad (10)$$

where χ_L and χ_R are in general complex numbers and of order unity if (4) applies.

The decay rates to these modes given the above couplings can be readily calculated at tree level by using the results that exist in the literature [10], while the decay rate to $\bar{t}c$ is given by [6]

$$\Gamma(\mathcal{H} \to t\bar{c}) = \frac{3g^2 m_t m_c m_\mathcal{H}}{32\pi m_W^2}\left(\frac{(m_\mathcal{H}^2 - m_t^2)^2}{m_\mathcal{H}^4}\right)\left(\frac{|\chi_R|^2 + |\chi_L|^2}{2}\right) \qquad (11)$$

and, $\Gamma(\mathcal{H} \to t\bar{c}) = \Gamma(\mathcal{H} \to c\bar{t})$ at the tree level that we are considering for now. The decay rate to $\mu^+\mu^-$ which we require in equation (5) is

$$\Gamma(\mathcal{H} \to \mu^+\mu^-) = \frac{g^2 m_\mu^2 m_\mathcal{H}}{32\pi m_W^2}\chi_\mu^2; \qquad B_\mu^\mathcal{H} = \Gamma(\mathcal{H} \to \mu^+\mu^-)/\Gamma_\mathcal{H} \qquad (12)$$

This are the basic ingredients we need in order to proceed to study the phenomenology of top-charm production at a $\mu^+\mu^-$-collider as will be elaborate in the next section.

297

3 PHENOMENOLOGICAL SCENARIO: THE RESULTS

In our analysis, we let $M_{\mathcal{H}}$ vary between 100 GeV and 800 GeV and consider two possible values for the mixing phase α: $\alpha = 0$ (Case 1) and $\alpha = \pi/4$ (Case 2), i.e. α equal or different from zero. The only difference between Case 1 and Case 2, as we can read in (9), is that the scalar neutral boson h^0 does or does not couple to the gauge bosons, this implying a difference at the level of $\Gamma_{\mathcal{H}}$, i.e. of $B_\mu^{\mathcal{H}}$.

We summarize our results in Fig. 1, where the following sets of parameters for Case 1 and Case 2 are used

- Case 1: $\alpha = \lambda_c = \lambda_t = 0$, $\chi_\mu = \chi_b = \chi_t = 1$ and $\chi_L = \chi_R = 1$

- Case 2: $\alpha = \pi/4$, $\lambda_c = \lambda_t = 0$, $\chi_\mu = \chi_b = \chi_t = 1$ and $\chi_L = \chi_R = 1$

We plot $\tilde{R}(\mathcal{H})$ with $\delta = 0$, 10^{-3} and 10^{-2} in the two cases as well as

$$\tilde{R}_{tc} = \tilde{R}(\mathcal{H})\,(B_{t\bar{c}}^{\mathcal{H}} + B_{c\bar{t}}^{\mathcal{H}}) \tag{13}$$

Note that in Case 1 if $m_{\mathcal{H}}$ is below the $t\bar{t}$ threshold $\tilde{R}_{tc}$ is about $.01 - 1$ and in fact tc makes up a large branching ratio. Above the $t\bar{t}$ threshold $\tilde{R}_{tc}$ drops. For Case 2 the branching ratio is smaller due to the WW and ZZ threshold at about the same mass as the tc threshold and so $\tilde{R}_{tc}$ is around 10^{-3}. For a specific example if $m_{\mathcal{H}} = 300 GeV$, then $\sigma_0 \approx 1pb$. For a luminosity of $10^{34} cm^{-2} s^{-1}$, a year of $10^7 s$ (1/3 efficiency) and for $\delta = 10^{-2}$, Case 1 will produce about $5 \times 10^3 (t\bar{c} + \bar{t}c)$ events and Case 2 will produce about 150 events. Given the distinctive nature of the final state and the lack of a Standard Model background, sufficient luminosity should allow the observation of such events.

If such events are observed, we will also have the possibility to extract the values of χ_L and χ_R, provided we determine the helicity of the produced top quark, expressed in termes of χ_L and χ_R as

$$\mathbf{H}_t = -\mathbf{H}_{\bar{t}} = \frac{|\chi_R|^2 - |\chi_L|^2}{|\chi_R|^2 + |\chi_L|^2} \tag{14}$$

The helicity of the t quark cannot be determined directly, but has to be obtained from the decay distributions of the top[12], the number of events required to observe it with a significance of $3\,\sigma$ being

$$N_{3\sigma} = \frac{36}{\mathcal{E}_t^2 \mathbf{H}_t^2} \approx \frac{107}{\mathbf{H}_t^2} \tag{15}$$

Thus at least 10^2 events are required to begin to measure the helicity of the top and hence the relative strengths of χ_L and χ_R. In the above numerical examples it is clear that for some combinations of parameters, particularly if the luminosity is $10^{34}cm^{-2}s^{-1}$, sufficient events to measure the helicity may be present.

4 Acknowlegement

This research was supported by the U.S. Department of Energy under Contract DE-ACO2-76CH0016(BNL) and DE-AC03-765F00515 (SLAC).

References

[1] V. Barger, talk presented at this Meeting.

[2] S.L. Glashow and S. Weinberg, *Phys. Rev. Lett.* **69** (1977) 1958.

[3] T.P. Cheng and M. Sher, *Phys. Rev.* **D35** (1987) 3484;M. Sher and Y. Yuan, *Phys. Rev.* **D44** (1991) 1461;A. Antaramian, L.J. Hall, and A. Rasin, *Phys. Rev. Lett.* **69** (1992) 1871;L.J. Hall and S. Weinberg, *Phys. Rev.* **D48** (1993) R979.

[4] W.S. Hou, *Phys. Lett.* **B296** (1992) 179.

[5] M. Luke and M.J. Savage, *Phys. Lett.* **B307** (1993) 387;M.J. Savage, *Phys. Lett.* **B266** (1991) 135.

[6] D. Atwood, L. Reina and A. Soni, *Phys. Rev. Lett.* **75** (1995) 3800.

[7] D. Atwood, L. Reina and A. Soni, preprint SLAC-PUB-95-6927, 1995 (to appear in *Phys. Rev. D*).

[8] V. Barger, M. Berger, J. Gunion, and T. Han, *Phys. Rev. Lett.* **75** (1995) 1462;V. Barger et al., *Proceedings of the 2nd Workshop on Physics Potential and Development of $\mu^+\mu^-$ Colliders*, Sausalito, California, Nov. 1994, [hep-ph/9503258].

[9] J.F. Gunion, Proceedings of the *International Europhysics Conference on High Energy Physics*, Brussels, July 1995, UCD-95-35, [hep-ph/9510226].

[10] J. Gunion, H. Haber, G. Kane and S. Dawson, "The Higgs Hunters Guide," published by Addison Wesley.

[11] D. Cline, *Nucl. Instr. Meth.* **350** (1994) 24.

[12] D. Atwood and A. Soni, SLAC-PUB-95-6877, to appear in Phys. Rev. D.

Figure 1

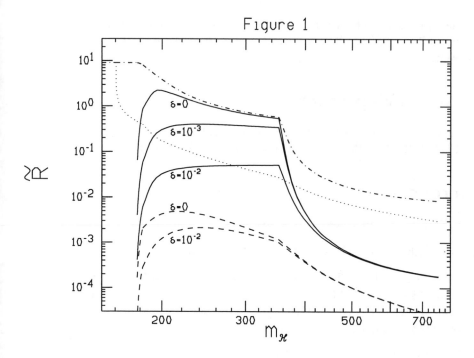

Figure 1: The value of $\tilde{R}(\mathcal{H})$ is shown as a function of $m_\mathcal{H}$ for scenario 1 (dash-dot) and for scenario 2 (dots). The value of $\tilde{R}_{tc}$ is shown in case 1 for $\delta = 0$ (upper solid curve); $\delta = 10^{-3}$ (middle solid curve) and $\delta = 10^{-2}$ (lower solid curve). The value of $\tilde{R}_{tc}$ is shown in case 2 for $\delta = 0$ (upper dashed curve) and $\delta = 10^{-2}$ (lower dashed curve).

AN INTENSE LOW ENERGY MUON SOURCE FOR THE MUON COLLIDER

D. Taqqu

Paul Scherrer Institut, Villigen, CH

Abstract

A scheme for obtaining an intense source of low energy muons is described. It is based on the production of pions in a high field magnetic bottle trap. By ensuring efficient slowing down and extraction of the decay muons an intense intermediate energy muon beam is obtained. For the specific case of negative muons a novel technique called frictional accumulation provides efficient conversion into a 10 keV μ^- beam whose emittance is then reduced in a configuration providing extended frictional cooling. The result is a beam of very small transverse and longitudinal emittance that can be used together with an equivalent μ^+ beam as a compact intense muon source for the $\mu^+\mu^-$ collider. A final luminosity around 10^{34} cm^{-2} s^{-1} is expected to be obtained at 2 TeV.

1. Introduction

The muon source for the $\mu^+\mu^-$ collider is still in its conceptual stage and the door has not yet been fully closed to competing ideas. It is therefore timely to throw into the race what may become a real alternative to the standard high energy muon production concept, namely the pathway of low energy muon production and cooling. Although this pathway has been considered and discussed in the initial meetings on the muon collider, it suffered from the absence of a scheme providing simultaneous high intensity and low emittance beams. The present contribution aims to improve the situation by presenting basic schemes for efficient muon production, slowing down and cooling. It has quite a preliminary aspect because no full Monte-Carlo simulation of the various stages considered have yet been undertaken so that a reliable final value for the muon intensity is not available. Nevertheless it should provide sufficient motivation for a more detailed investigation and lead to optimized most efficient designs.

Because of the radically different interaction of low energy positive and negative muons with matter the required final low emittance beams will be obtained via quite different pathways. For the μ^+ the existing scheme of muonium formation and ionization [1] can in principle provide a phase space compression (PSC) factor of 10^9 in one step. No such effective cooling operation is available for the μ^- branch. We therefore concentrate here on the apparently handicapped μ^- branch leaving the μ^+ sector to a short discussion near the end of the paper.

2. Basic aspects of the μ^- source

One most useful technique to obtain an efficient and compact muon source is to force pion decay to take place within a trapping configuration. With a magnetic bottle trapping both pions and muons the intrinsic high longitudinal phase space density of a trap [2] can be made used of to provide efficient slowing down of the high energy muons [3]. This is done by letting the muons lose their energy in an suitable moderator material. The associated loss in transverse energy provides an efficient transverse PSC. With B_0 being the magnetic field in the central trap region and B_m the field maximum at both ends, the mirror action of the field increase results in trapping all muons whose angle θ to the axis at the center of the trap is greater than a cut-off angle

$$\theta_c = \text{arc sin } \sqrt{B_o / B_m} \, . \tag{1}$$

In order to induce muon escape during slowing down an electric potential difference ΔV is applied at one side of the trap, modifying thereby the escape condition to [4].

$$\theta > \theta'_c = \text{arc sin } \sqrt{B_o/B_m(1 + e\Delta V / E)} \tag{2}$$

where E is the energy of the muon as it comes out from a moderator assumed to be placed at $B=B_o$ [4]. This leads to a selective extraction of the lower energy muons enabling their escape before stopping in the moderator.

For efficient μ^- production a most intense deuterium beam at energy around 2 GeV will be shot on a thin carbon rod at the center of the trap, and by operation at high magnetic field and a relatively high B_m/B_o (of the order of 2), it will be possible to trap a large fraction of the produced π^-. By using a relatively short trap, a large number of round trips will be made possible for the decay muon so that a quite high initial muon energy can be slowed down to the escape before muon decay. A schematic configuration for this muon production stage will be presented in § 3.

The function of the next stage is best understood if we jump first to the last stage and explain how a minimal final beam emittance is obtained. In a recent experiment at PSI [5] it has been demonstrated that the quality of a low energy μ^- beam can be improved by letting it pass a stack of thin foils with an accelerating electric field between them. Both longitudinal and transverse cooling have been observed at energies below 10 keV where the stopping power falls with energy. This cooling effect (called ionization cooling in this book or frictional cooling by its promoter [6]), which required the use of a large magnetic field to prevent beam size blow-up, has here to be implemented in a much more efficient configuration in order to provide the very large PSC factors needed to convert the wide μ^- beam exiting the magnetic trap into a beam fulfilling the collider requirements. The way to do this is to extract the muons from their magnetic field entanglement (a nontrivial task by itself) and form muon beams that can be focused by lenses from one foil to the next. By making the operation at optimal beam energy and divergence and allowing for adequate demagnification between one foil and the next, the beam size can be reduced step by step and after a large number of cooling stages the beam emittance can be brought to its final small value.

Furthermore, by converting the few µs long muon pulse into a large number of microbunches the time spread of each microbunch can be reduced between one foil and the next. The associated increase in energy spread is then cooled down in the next foil crossing. Repeated bunching between two successive foils leads to the full conversion of the energy cooling capacity of the foils into longitudinal emittance decrease.

This extended cooling concept will be shown in §6 to result in a final 6-dimensional emittance that is in principle by far better than what is required for the muon collider. On the other hand the beam intensity budget (§8) comes out somewhat poorer than what the standard high energy muon beam concept obtains. Because of the well known tune shift problematic in colliders, the low intensity results in strict luminosity limitations that cannot be compensated by a small emittance are used. It is therefore essential to minimize the muon (and pion) losses.

In this context the intermediate stage between the initial muon production bottle and the final cooling operation plays an essential role. It is made necessary because efficient final cooling requires muons of energy around 10 keV and below while efficient muon production in the considered trap requires the extraction to take place at energies greater than the MeV. How can the MeV to keV energy conversion be efficiently achieved? Here steps in a recently proposed technique called frictional accumulation [7]. It makes use of a stack consisting of a large number (50 - 100) of thin foils to which a monotonically increasing voltage is applied. High energy muons that are made to cross this stack (one or many times) along a magnetic field perpendicular to it, slowly lose their energy until they reach such a low energy that they are on the verge of stopping in one of the foils. However, because of the reduced stopping power at the lowest energies and also because the muon may have a dominantly transverse energy before stopping, the electric field between the foils can re-accelerate the muon and give it a sufficient longitudinal energy kick that it escapes stopping in a foil. With its transverse energy rapidly falling down and its longitudinal energy increasing it can reach an energy of 10 keV or less at low divergence and under these conditions, for optimally selected potential difference between the foils, have almost equal energy loss in the foils and energy gain between them. It therefore hops from one foil to next without much change in energy. This results in an accumulation of muons in the low energy region. These muons exit as a low energy muon beam at one side of the stack.

The efficiency of the scheme can be obtained from Monte-Carlo simulations. Preliminary results indicate that for muons entering a stack of 40 foils on its decelerating side, an initial energy distribution extending to 150 keV (with a Lambertian angular distribution) can be converted in a muon beam of energy less than 10 keV with an efficiency of 30%. This corresponds to a PSC factor of 1000.

By increasing the number of foils and letting the muons cross the stack many times (as provided in a trapping configuration) the energy acceptance can be increased to many MeV. This leads to the scheme described in §4 where the accumulation stack is introduced in a second trap.

The general discussion of the full chain of operations is herewith completed. We return now to a more detailed description of each of the successive stages.

3. The production bottle

The deuteron beam is obtained from an accelerator of the kind recently proposed for European Spallation Source [8]. It should be able to provide a 5mA average current at a repetition rate of 30/sec. The relatively short deuteron pulses could be significantly lengthened at the advantage of a smaller final transverse beam emittance. This would allow the use of a thinner carbon target rod and reduce the pion and muon losses in the magnetic bottle. With a 30 cm long carbon target more than the half of the deuterons undergo nuclear interaction. In order to insure efficient trapping of the higher energy pion in a compact trap, relatively high magnetic fields should be used. These can be obtained by producing an almost constant high field via a wide shielded superconducting solenoid and using two high field normal conducting coils of small size to provide the mirror action. The resulting configuration with the field distribution is shown in fig. 1.

The carbon target is to be replaced after each pulse. Trapped pions have cyclotron trajectories than return to the axis periodically within about 1 ns. Many of them lose significant energy in the carbon rod itself. Lower energy pions lose a relatively higher energy in the degrader disks so that they rapidly undergo a slight increase of the closest distance to the axis reducing the chance of a stop in the target rod.

Most decays of the trapped pions lead to trapped muons. Due to the momentum kick in the decay a large part of the trapped muons follow trajectories that escape crossing the target. These muons lose their energy in the degrader disks whose thickness is optimized in relation to the extraction voltage in such a way that the probability of escaping before stopping is maximized. Because of multiple scattering, the angle to the axis θ increases with slowing down so that already before extraction time most muons have reached the region right to the foil F1. This allows the extraction field to be limited to the region between F1 and the extraction electrode. The electrode is pulsed to a high voltage during the extraction process. For a 1 MV voltage most muons entering the region right of F1 with an energy less than a few MeV will be extracted.

The transverse size of the outcoming beam is controlled by the diameter of the main slowing down disks. Small disk sizes reduce the outcoming beam diameter at the cost of intensity.

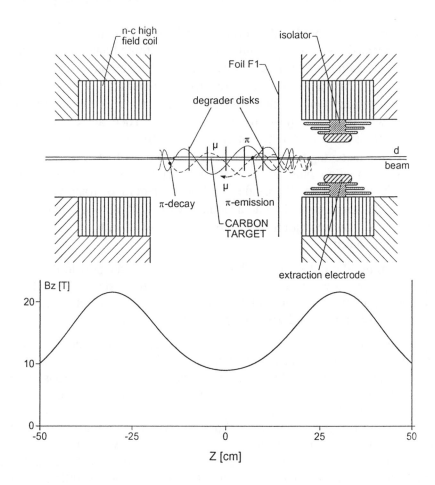

Fig.1: a) A schematic view of the magnetic mirror trap for efficient muon production together with the trajectory of a 120 MeV/c π followed by the decay muon. The shielded large diameter superconducting coils are not shown. b) Axial magnetic field distribution

4. The accumulation trap

A high field curved solenoidal channel transports the muon beam between the production and the accumulation trap. An efficient injection of an external beam in a trap requires the use of some special trapping feature. If, as in the present case, the beam has wide energy, angular and time spreads, the injection operation will necessary lead to a large blow up of the beam size. Living with this fact, we consider off-axis injection followed by beam rotation via the curvature drift. In an axially symmetric magnetic field configuration, the muon beam enters a first high field coil at a distance r from the axis. It then crosses a low field region where the magnetic field lines curve further away from the axis and then return toward the axis to flow into an intermediate field solenoid. Under adiabatic conditions the guiding center of the trajectory is subjected to the curvature drift which gives rise to a rotation around the axis by an angle θ . With the accumulation stack placed in the

305

intermediate field solenoid and a terminating high field coil acting as a magnetic mirror, a reflected higher energy muon (that lose a limited amount of energy in the stack) crosses again the low field region, rotates further by an angle close to θ and reaches the entrance high field coil displaced azimutally by about 2rθ. If this quantity is sufficiently greater than the beam diameter D it is possible to ensure that the muon will remain trapped by introducing an adequate change in electric potential. This is done by having an entrance electrode at a high positive voltage (that sucks the incoming μ⁻ into the trap) and a fast fall to zero of the potential in the azimutal direction outside the electrode. Optimization of the magnetic and electric field distribution will provide a highly efficient muon trapping. It is possible to improve the design by adding an adequate $\vec{E} \times \vec{B}$ drift along the muon trajectory. With both $\vec{E} \times \vec{B}$ and curvature drift the variation of θ with muon energy can be significantly reduced.

The allowed number of round trips (~π/θ) can be adapted to the muon beam energy and the slowing down in the accumulation stack. The downstream outcoming beam of energy around 10 keV (or higher if accelerated) has the shape of an annulus of radius r and width D (at maximum field).

A technical difficulty has to be solved before the accumulation process at relatively high muon current can be implemented. Its origin is the very efficient secondary-electron emission by the foils of the stack. These electrons, accelerated by the potential between the foils, will cross them and re-emit more secondary electrons. The resulting significant electron multiplication (observed at PSI in a stack of 10 foils [6]) is too high to be allowed. One way to eliminate this effect is to introduce between any two foil planes a high transparency grid (made of ultra thin wires) at a relatively low potential (< 100V) that forces the emitted secondary electrons (of ~ 10eV energy) to be reflected back to their emitting foil without affecting the transport of the higher energy muons.

Taking in account the various loss processes the efficiency of the accumulation stage will be between 10% and 20% depending of the muon beam energy and the optimization effort. This corresponds to a significant intensity loss but remains the price to pay to allow the mandatory muon cooling to proceed.

5. The extraction from the magnetic field

The long-standing problem of extracting a low energy beam from a high magnetic field has recently found two different solutions [9, 10]. We consider here the use of one of them whose practicability has been recently demonstrated with an electron beam [11].

In a magnetic field the true momentum of a particle is $\vec{P} = \vec{p} - e\vec{A}$ with $\vec{p}$ being the usual mechanical momentum and $\vec{A}$ the vector potential which in a solenoid with field B follows transverse circular lines centered on the axis. On these circles $|\vec{A}| = Br/2$ (r = circle radius). At high B and large r, $|\vec{A}|$ will be much greater than $p_\perp$ (component of $\vec{p}$ perpendicular to $\vec{B}$) and angular momentum conservation implies that a plain extraction form the solenoidal field is necessarily associated with an almost uncorrectable transverse momentum kick of the order of the value of $e\vec{A}(r)$ in the solenoid.

By breaking the field symmetry and inducing $|\vec{A}| = 0$ regions near almost each point of the extraction plane, the amount of transverse kick can be reduced at will. This is done by first adiabatically reducing the magnetic field to a very low value and then terminating the field onto a grid consisting of thin parallel bars made from a high permeability material. Between the bars separated by a distance d the maximum excursion of the transverse kick is reduced to $Bd/2$.

For a muon beam the lowest field that can be reached adiabatically (at an increased radius and a decreased transverse momentum) within a few 100 ns is about 0.05 T. This is low enough to achieve high transparency by using as bars thin longitudinal foils, and for a distance of 5 mm between the foils the transverse kick is distributed between 0 and 0.04 MeV/c which is less than the average adiabatically decreased transverse beam momentum.

At the exit the muon beam has a large area and small divergence. It has to be divided in a large number of separate beams of smaller size so that it can be focused on foils of diameter less than a few cm diameter starting thereby the extended cooling operation.

6. The cooling stage

The optimal cooling action is obtained at energies (~5 keV) where straggling and scattering induce relatively large $\Delta E/E$ and large divergence. These are not adequate conditions for good muon transport. The concept proposed here is to induce a strong acceleration of the muons at their immediate exit from a foil so that a higher energy beam with small $\Delta E/E$ and divergences is obtained. This allows high quality focusing (electric or magnetic) to be used. Also the accelerating field applied on the conductive foil acts as a high quality cathode lens (or "immersion" lens) that has minimum aberration effects [12]. As the muons are refocused on the next foil a decelerating electric field will reduce the muon energy to the optimally required foil entrance muon energy. This "inverted" cathode lens is also of high quality so the muons are imaged from one foil to the next under optimal conditions. Aberration effects can practically be kept small by using sufficiently strong lenses. Under these conditions the cooling remains inaffected by the beam transport. The immense improvement relative to the standard configuration is however that at each step the cooling can be converted in a decrease of beam radius (for transverse cooling) or a decrease in pulse length (for a bunched beam with a cavity inserted in the beam line) so that the cooling action can be repeated at optimal efficiency from one stage to the next. This leads to introduction of the notion of "PSC factor of a

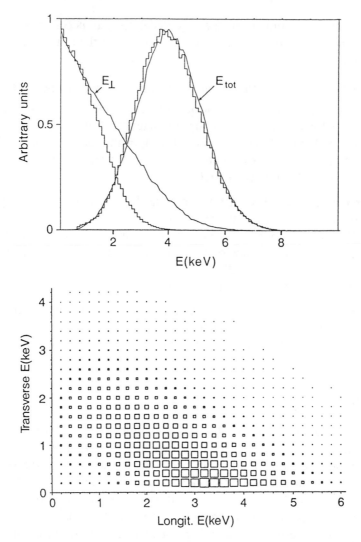

Fig.2: a) Simulation of the conditions for pure transverse cooling with a 4 μg/cm2 carbon foil. Upper graph: distributions of transverse ($E_\perp$) and total energies (E_{tot}) before (point to point connected curve) and after (histogram kind curve) foil crossing. Equalization of average E_{tot} is obtained at a 2.4 kV potential difference. The intensity loss is about 2 %. Lower graph: Transverse versus longitudinal muon energy at the exit of the foil.

single foil" which can be investigated and optimized. It comes out that longitudinal cooling is significantly less efficient than transverse cooling. A simulation of the foil action is presented in Fig. 2 for a beam entering the foil with a large divergence. Under these conditions extensive cooling can be achieved without any beam bunching requirement.

The PSC factor of a single 4 μg/cm^2 carbon foil comes out to be near 0.66 (or a 0.81 decrease in emittance) at the lowest energy for which straggling losses can be neglected. This gives the relation between the number of stages n and the PSC factor:

$$f_{PSC} = (0.66)^n \qquad (3)$$

The muon beam cooling will first consist a large number of cooling stages acting in parallel on each of the separate beamlets previously extracted. As the size of the beams becomes sufficiently small (a few mm on the foil) they are merged together transversally (in a few successive steps interspaced by further cooling) until a final single intense beam is obtained. Once transverse cooling has approached its limit, the 2 μs long muon pulse enters a highly efficient buncher which generates a large succession of microbunches of larger energy spread.

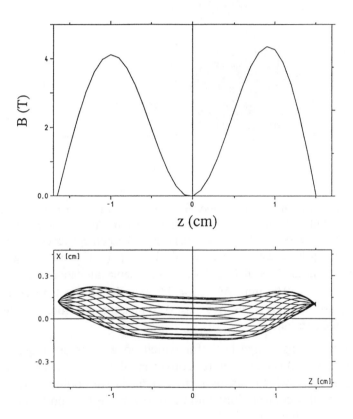

Fig.3: μ⁻ transport between 2 foils in a final cooling stage. a) Axial magnetic field obtained by using small superconducting coils with alternating current excitation. b) Muon trajectories for a point object on the first foil at a radius of 1 mm imaged on a second foil placed at a 3.15 cm distance. The initial muon energy and divergence are $E = 3$ keV and $\theta = 45°$ for all beams. The second foil is placed at a 2.4 kV higher potential. Shown is the x coordinate in function of the axial position z.

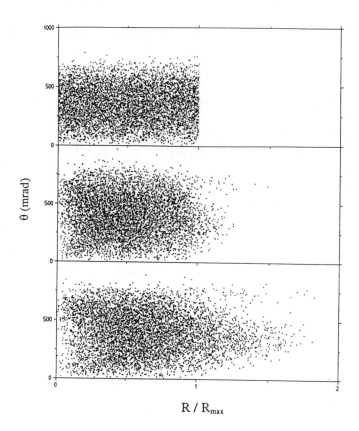

Fig.4: Distribution of divergence angles θ at the second foil for muons coming out from the first foil with the energy distribution given in Fig. 2 (lower curve) and a radial distribution taken for illustrative purposes to be homogeneous in the distance R to the axis up to a distance R_{max}. a) Ideal transport with no demagnification; b)c) Transport in the lens of Fig. 3 with slight demagnification (~0.9); b) $R_{max} = 0.5$ mm; c) $R_{max} = 0.25$ mm. At $R_{max} = 0.5$ mm, the effect of the aberrations on the radial distribution is small. At $R_{max} = 0.25$ mm, the aberrations enlarge appreciably the radial distribution.

An emittance exchange stage [13] (for which no wedges are required but simply some slanted cavities) is then used for converting the enlarged energy spread into an increased beam size which can be further reduced (in cooling stages where from now on RF cavities controls longitudinal phase space transport) until the smallest achievable beam radius is reached.

How small a beam radius can be achieved have been investigated on the basis of a compact high field transport stage where the accelerated muons are focused by magnetic lenses [Fig. 3]. Adopting a 150 kV/cm electric field on the foils and an acceleration of 150 keV the aberration effects can be inspected by looking at the effect of beam transport on the radius and divergence of the muons. It comes out that aberration radii smaller than 0.25 mm are achievable for most of the cooled muons. The effect of the aberrations are illustrated in Fig. 4. A final r.m.s.

Acceleration to Collisions for the μ⁺-μ⁻ Collider

David V. Neuffer,[†] Fermilab, P. O. Box 500, Batavia IL 60510

Abstract. We discuss the problem of transforming muon beam bunches from a low-energy cooled state ($E_\mu \sim 1$ GeV) to short, high-energy bunches matched to high-energy collision conditions ($E_\mu \sim 2$ TeV). In this process the beam energy must increase by ~ three orders of magnitude, while the bunch length must be reduced by ~ two orders of magnitude (to ~ 3mm), while beam emittance dilution and beam losses, particularly through decay, must be minimized. From general considerations, we discuss possible acceleration scenarios including rapid-cycling synchrotron and recirculating linac options. The presently favored choice is a multi-stage recirculating linac system, which is discussed, and initial simulations of possible scenarios are presented. Future directions for development are discussed.

INTRODUCTION

The possibility of muon (μ^+-μ^-) colliders has been introduced by Skrinsky et al.[1] and Neuffer[2]. More recently, intensified investigations with the goal of a practical design for a high-energy high-luminosity $\mu^+\mu^-$ collider have increased the level of conceptual development,[3, 4, 5, 6] and this effort includes the present workshop.[7] A candidate scenario for a collider, with an energy of $E_{cm} = 2E_\mu = 4$ TeV, and a luminosity of L ~ 10^{35}cm^{-2}s^{-1}, has been developed.[8] Table 1 shows parameters for the candidate design. The design consists of a muon source, a muon collection, cooling and compression system, a recirculating linac (or rapid-cycling) system for acceleration, and a full-energy collider with detectors for multiturn high-luminosity collisions.

In this paper we concentrate on the portion of this scenario in which the muons are accelerated from the output of the cooling system to full energy and transferred to storage in the 2 TeV collider ring. Thus, we assume that the muons have been cooled and collected into moderately compact $\mu^+$and μ^- bunches at $E_\mu \sim 1$ GeV. Studies of the cooling system indicate that an energy spread of ~1% at a bunch length of ~30cm at ~1 GeV are reasonable design goals, and we use these as reference initial parameters. The accelerator must accelerate these bunches to 2 TeV and transfer them into the collider, with a final energy spread of ~0.1% and a bunch length reduced to ~0.3cm.

A critical requirement is that the muons must be accelerated before they decay. This sets severe constraints on the acceleration system and these are first discussed

---

[†] on assignment from CEBAF, 12000 Jefferson Avenue, Newport News VA 23606

in detail. We then describe potential acceleration systems, including full-energy linac, rapid-cycling synchrotron and recirculating-linac options. Our currently favored acceleration choice is a sequence or cascade of recirculating linacs, each of which increases beam energy by ~ an order of magnitude and accommodates bunch length reductions by almost as much. We discuss constraints and present candidate scenarios. Particle tracking in a candidate choice provides a "proof of principle" of this general approach. Optimization considerations and possible variations are discussed. We also discuss directions for further development of these acceleration and transport systems.

MUON LIFETIME CONSIDERATIONS

The central difficulty in a $\mu^+\mu^-$ collider is that muons decay with a mean lifetime of $\tau_\mu = 2.2$ μs (in the μ rest frame), and the muons must be collected, cooled, accelerated, and collided within that lifetime. In the lab frame the lifetime is increased by the relativistic factor $\gamma = E_\mu/m_\mu$, where E_μ is the μ energy and m_μ is the mass ($m_\mu = 0.10566$ MeV). The muon decay rate along the beam path length s can be written as:

$$\frac{dN}{ds} = -\frac{1}{L_\mu\gamma}N, \text{ where } L_\mu = c\,\tau_\mu \cong 660 \text{ m}.$$

and where we have used the relativistic approximation $v/c \cong 1$.

In a non-accelerating transport, this implies the usual exponential beam loss:

$$N = N_o e^{-\frac{s}{L_\mu\gamma}}$$

In an accelerating section, γ is not constant:

$$\gamma = \gamma_0 + \gamma's = \gamma_0 + \frac{eV_{rf}'}{m_\mu c^2}s \quad,$$

where eV_{rf}' is the accelerating gradient. Using this in the decay equation obtains the solution :

$$N(s) = N_o\left(\frac{\gamma_0}{\gamma_0 + \gamma's}\right)^{\frac{1}{L_\mu\gamma'}} \quad \text{or} \quad \frac{N(s)}{N_o} = \left(\frac{E_o}{E_{final}}\right)^{\frac{1}{L_\mu\gamma'}} \tag{1}$$

Low losses imply that the exponential factor must be small, which implies that: $L_\mu \gamma' \gg 1$. This can be rewritten as:

$$L_\mu \frac{eV_{rf}'}{m_\mu c^2} \gg 1$$

which means $eV_{rf}' \gg 0.16$ MeV/m is required. This general rule must be followed throughout the entire muon system. For example, beam-cooling and reacceleration must occur in systems whose averaged accelerating gradients (including loss and transport elements) are much greater than 0.16 MeV/m to avoid large decay losses.

All previous acceleration systems have not been concerned with μ decay. However, we can compare existing accelerator and transport systems with this guideline to obtain some sense of the changes necessary in transforming to a μ accelerating system:

LEP II synchrotron: 2 GeV/ 27 km $\Rightarrow$ 0.074 MV/m

CEBAF recirculating linac: 0.8 GeV/ 1.3 km $\Rightarrow$ 0.6 MV/m

SLAC linac: 50 GeV/ 3 km $\Rightarrow$ 17 MV/m

In these examples, the SLAC linac easily meets the gradient requirement by two orders of magnitude, and any linac-based system should have adequate gradient. The CEBAF recirculating linac barely meets the criterion; however, a recirculating linac with somewhat improved gradient per total transport length would also be adequate. Almost all existing synchrotrons do not have adequate gradient, and a synchrotron-based scenario would have to be greatly changed to be acceptable.

For a multiturn μ accelerator, the gradient criterion can be rewritten as:

$$E' \rightarrow \frac{E_{final}}{N_{turns} 2\pi R} \gg 0.16 \text{ MeV/m},$$

where R can be written in terms of the mean bending field B and the magnetic rigidity Bρ as $R = B\rho/B \approx 0.00334\ E_{final}(\text{MeV})/B$, and N_{turns} is the total number of acceleration turns. Inserting this into the previous equation obtains the criterion for any multi-turn accelerator:

$$\frac{N_{turns}}{B(T)} \ll 300.$$

In this expression B refers to the average bending field in the highest energy turn (including straight sections, rf sections, and other non-bending elements).

ACCELERATION OPTIONS AND SCENARIOS

From these constraints, we can develop possible acceleration scenarios. A single-pass linac can easily meet the gradient constraint. However single-pass rf structures are prohibitively expensive and do not exploit a primary advantage in muons: our ability to bend them into multipass devices, enabling multipass use of the accelerating structures. We thus consider two forms of multipass acceleration: rapid-cycling synchrotrons and recirculating linac. These are shown in schematic form in figure 2.

Rapid-Cycling Synchrotrons

A synchrotron consists of rf accelerating structures within a circular magnetic beam transport, and the magnetic fields are increased from low-field to high-field while the beam is accelerated from low to high energies, passing many times through the same transport system. The magnetic fields must change rapidly to follow the beam transport, and with current technology only conventional magnets ($B < 2$ T) can cycle rapidly. The multiturn acceleration criterion can be met (barely) for $N_{turn} < {\sim}100$ and mean bending field $B \sim 1$T. As an example, we can consider a scenario with a final multiturn rapid-cycling cycle in which the beam is accelerated from 100 to 2000 GeV in a ring with R = 5 km (B = 1.33T). This would require a 19 GV/turn rf system (1 km of 19 MV/m rf) for a 100-turn cycle, and would have an acceleration cycle of ~12ms. 46.2% of initial μ's would survive the cycle. We note that this cycle time is reasonably well matched to an ~30 Hz driver, and that the ring circumference is remarkably similar to that of the CERN LEP tunnel.

From equation (1), we can write an expression for beam survival in a multi-turn system:

$$\frac{N(s)}{N_0} \cong \left(\frac{E_o}{E_{final}} \right)^{\frac{2\pi R N_{turn} m_\mu}{L_\mu E_{final}}}$$

We can improve survival by increasing acceleration rate (decreasing N_{turn}). For example, reducing N_{turn} to 50 turns improves survival to 68%.

Recirculating Linacs

Another multiturn approach is the use of recirculating linacs, similar to CEBAF, which accelerates electrons to 4 GeV in a 5-pass system. In a recirculating linac (RLA), the beam is accelerated and returned for several passes of acceleration in the same linac, but a separate return path is provided for each pass. At the end of the linac, the beam passes through dipoles, which sort the beam by energy, directing it to an energy-matched return arc. (A pulsed kicker magnet system may also be used.) The various energy transports are then recombined at the end of the arc for further acceleration, until full energy is reached, when the beam is transferred to another linac or the collider. Thus the magnets are at fixed-field and the beam passes through each transport only once.

Since the beam passes through a separate transport on each turn, the magnets can be at fixed-field, allowing superconducting magnets, and simplified designs. However the requirement for a separate transport on each turn limits the total number of turns that could be practical, to ~10—20 turns. This is very compatible with the lifetime constraint: $N_{turns}B \ll 300$, which then can be met with relatively modest field magnets, and typically beam-survivals of ~95% are obtained in μ RLA's. High-field magnets are not required. The RLA is rather ideally matched to μ-acceleration constraints.

Because of the independence of each return transport, there is an enormous flexibility in RLA design, with only the rf acceleration frequency and voltage remaining constant from linac pass to linac pass. Since return path lengths are independent, the synchronous phase φ_s can be changed arbitrarily from pass to pass. Also the chronicity, $M_{56} = \partial z/\partial(\delta p/p)$, where z is particle position within the bunch, can be changed from turn to turn, by fitting the transport. At CEBAF,[9] an isochronous transport ($M_{56} = 0$) was used, but for the μ-collider some bunching is required and non-zero M_{56} will be needed in some of the transport. Higher order chronicity control (M_{566}) with sextupoles is also possible, and one can consider adding higher-harmonic rf and additional compressor arcs, if needed.

The same RLA system could be used to accelerate both μ^+ and μ^- bunches. The oppositely charged bunches would propagate around the RLAs in opposite directions. If the bunches are injected into opposite sides of each RLA at the beginning of the separate linacs, then energy match of the beams in each arc is obtained, as well as phase matching across the arcs. Separate (but symmetric) transport lines into the higher-energy RLA's and into the collider would be needed.

RLA ACCELERATION SCENARIOS

From the previous discussion, RLA scenarios are currently the preferred μ-acceleration option. In this section we develop in more detail explicit acceleration scenarios for the 2 TeV collider, and then discuss possible variations.

Following cooling and initial bunch compression to ~.1–.3m bunch lengths at ~GeV energies, the beams are accelerated to full energy (2 TeV). In this process, the μ-bunches must be compressed, to a length of ~0.003m at full energy. A factor of a 1000 energy increase in a single RLA is probably not optimum. A sequence of RLAs (i. e., 1–10, 10–100 and 100–2000 GeV), with rf frequency increasing as bunch length decreases, may be used. A factor of ~10 energy increase per stage is a plausible first approximation, before detailed optimization. It is important to obtain the acceleration and bunch compression with minimal phase space dilution, in order to avoid energy-spread blowup and beam losses. The RLA flexibility permits many possible compression scenarios; however, it is also quite easy to obtain very badly matched schemes within that broad tuneability.

Sample scenario - simulation results

As a simplified first example, which we use as a proof of principle, we consider in detail the scenario displayed in Table 2. This is a modularized 3-stage case, and a schematic view of a 3-stage RLA accelerator is displayed in figure 4. In each stage the energy is increased by a factor of 10 (2 to 20 to 200 to 2000 GeV). The rf frequency is also changed by a factor of 4 from RLA to RLA, from 100 to 400 to 1600 MHz. Each RLA consists of two linacs (each at 1 to 10 to 100 GeV) with recirculating arcs connecting them, and a total of ~10 turns in each stage. In this simplified format it is straightforward to scale the design from stage to stage.

We have developed the 1-D program μRLA to simulate the RLA longitudinal motion. In that program particle energy and position offsets are calculated from turn to turn. On each passage through a linac, particle energies change following:

$$\Delta E \rightarrow \Delta E + eV_{rf}(\cos\phi - \cos\phi_s),$$

while the synchronous energy increases by $eV_{rf}\cos\phi_s$. On each pass through an arc, particle phases change by:

$$\phi \rightarrow \phi + M_{56}\frac{2\pi}{\lambda}\frac{\Delta E}{E} + M_{566}\frac{2\pi}{\lambda}\left(\frac{\Delta E}{E}\right)^2 + \cdots$$

where we have included first and second order chronicities M_{56} and M_{566}. Note that ϕ_s, M_{56}, and M_{566} can be changed from turn to turn.

In this initial scenario, the beam is bunched within the injection transport for each RLA, while within the body of the RLA the synchronous phase is kept constant and M_{56} changes to maintain matched bucket conditions for fixed bunch-length. The matching conditions are set by varying ϕ_s and M_{56} to obtain a stable phase-space bucket matched to the beam-phase space area, and maintaining a constant area bucket. We approximate that bucket shape from synchrotron formulae. The matched energy spread of the rf bucket is:

$$\frac{\Delta E}{E} = \pm \sqrt{\frac{eV_{rf}\lambda}{EM_{56}}} \sqrt{\frac{2(\sin\phi_s - \phi_s \cos\phi_s)}{\pi}}$$

Maintaining a matched energy spread for fixed bunch length requires $\Delta E/E$ to decrease as $1/E$, which therefore implies that M_{56} must increase linearly with E. That condition was used in our initial simulations. (We note that the small number of turns in an RLA makes the synchrotron motion approximation somewhat inaccurate.) This matching minimizes bunch lengths within the RLAs, which reduces amplitude-dependent nonlinearities and also reduces bunch length oscillations, both of which can cause phase-space dilution. A similar matching condition on M_{56} occurs naturally in microtron design.

We have simulated this initial scenario using μRLA, and some results are summarized in Table 2, and displayed in figure 4. Some phase-space dilution and mismatch does occur, particularly in transfers between RLAs. However the rms emittance dilution is $<\sim 5\%$ per RLA or 15% over the entire system. Particle loss through the beam dynamics is less than 1%. Particle loss through μ-decay is somewhat larger, but less than $\sim 5\%$ per RLA or $\sim 12\%$ over the entire system. (We have assumed gradients of up to 19 MV/m in the linacs, and mean bending fields of ~ 5T in the arcs.) Bunch compression to $\sigma < 0.003$ m is obtained through rebunching and matching with the frequency increase from RLA to RLA, and is acceptable. Thus the simulation demonstrates that a cascade of RLAs can provide acceptable acceleration with bunching for a $\mu^+\mu^-$ collider, with minimal dynamic and decay beam loss and emittance dilution.

This scenario sets a "proof of principle" baseline for the exploration of acceleration scenarios. It is certainly unoptimized, and does not exploit the full degrees of freedom possible in the RLA scenarios. As initially formulated, it requires a separate rf system for bunching at the entrance of each RLA (0.2 GV of 100 MHz rf before RLA 1, 1.25 GV of 400 MHz rf before RLA 2, and 6 GV of 1600 MHz rf before RLA 3). In future development, these will be integrated with the acceleration rf, perhaps within a more gradual bunching scenario.

Another scenario, presented by Palmer in July 1995,[10] is displayed in table 3, and gives some impression of the possible variations in design. Beam is accelerated from 1 GeV to 2 TeV using 4 RLA steps, with top energies of 8, 75, 250 and 2000 GeV. The 250 GeV step is a suitable accelerator for a 250×250 GeV collider. Similar performance to the initial baseline is obtained, but with slightly larger losses and dilution due to the additional RLA. Beam loss through decay is ~19%. A complete bunching and acceleration sequence for this scenario is not yet developed, however.

COMMENTS ON SCENARIO OPTIMIZATION

In a multiturn RLA system there is a balance between rf acceleration and beam transport cost/requirements. Increasing the number of turns per RLA directly reduces the linac lengths and therefore linac costs, but it also increases the total amount of beam transport, adding cost and complexity. We have not yet developed cost estimates that are adequate to obtain an accurate optimum. In this section we discuss some of the considerations which must be included in developing an optimum design.

rf Considerations

We need a separate rf linac system for each RLA, with lower frequencies for the initial lower-energy RLAs, where the beam has a relatively long bunch length and higher frequencies for the high energy end, where the bunches are shortened, since higher-frequencies are expected to be less expensive. We have not determined whether separate bunching rf systems are desirable.

Very high-gradient is not essential in the acceleration, but rather minimal cost is. The Table 2 scenario requires ~200 GV of rf acceleration, while the table 3 scenario requires ~100GV; these are both quite large and would require ~5—10km at 20 MV/m.

The rf cavities must sustain field throughout the multipass acceleration time, which is ~1ms in the 2 TeV RLA. That implies SRF cavities should be used in the higher-energy RLAs, although we do not have clear guidelines for optimum parameters. We have used TESLA[11] and CEBAF parameters (~1.5GHz) in this study. TESLA is actually designed for 1ms cycles, repeating at 5Hz; these parameters are very close to our requirements. These use low-temperature(~2K) materials; higher temperature alternatives (4K or ??) should also be studied.

Transport Considerations

The beam transports for the recirculation arcs are relatively straightforward, but are nontrivial, since they require good transverse matching throughout the system to avoid emittance dilution. Each transport must be achromatic (matched to zero dispersion), and also must have a chronicity M_{56} matched to the bunching requirements. A transport modeled on the CEBAF RLA could be used. High field is not required, and even conventional fields (B<2T) are adequate.

Since the beam passes through a different return arc on each turn, the total amount of beam transport is relatively large (~85km of arcs in the Table 2 scenario, and 160km for Table 3). The transport can easily become **very** expensive, so cost-saving designs are needed. Multiple-aperture magnets, in which several passes go through separate (different field) apertures in the same magnetic structure are possible. S. Kahn, G. Morgan and E. Willen[12] have proposed 9 and 18 aperture dipoles with this purpose. Other "low-cost" technologies could be used (permanent magnets, super-ferric, etc.).

Hybrid magnets, in which rapid-cycling and high-field magnetic elements are mixed and pulsed so that several passes can go through the same transport, would be a very attractive technology in this application, and could permit more passes. (A scenario requiring only 20 GeV of rf, using an injector and three RLA stages with 200-turn rapid-cycling in the last stage, has been developed.)

Note that at the beginning and end of the arcs beam-separation and beam-recombination transports for all passes must be inserted, and this adds considerable complication. CEBAF has a 5-pass separation and recombination system with carefully matched transports, and it is easy to imagine a 10-pass extrapolation of that system to our case. However many more passes (20?) may lead to impractically congested designs.

There will be some μ-decay in the transport, which will deposit electrons with an average of 1/3 of the μ energy throughout the system. Since the decay rate decreases as the energy increases, the mean beam energy deposition (per μ) per meter is a constant :

$$\frac{dE}{ds} = \frac{m_\mu c^2}{3L_\mu} \text{ per } \mu.$$

This comes to ~0.25 watts/m with a beam of 10^{12} μ's at 30Hz; and this level seems tolerable even in superconducting structures.

COMMENTS AND CONCLUSIONS

We have presented a candidate scenario for a high-energy μ^+-μ^- accelerator. That scenario includes a first proof-of-principle calculation of the design concept. Much further optimization and design and concept development is needed.

The bunch-compression and acceleration scenario must be optimized and further simulated. Variations such as rapid-cycling should be considered. Complete lattices are needed, with designs for the transport arcs, including beam separation and recombination. An accurate cost algorithm for rf and beam transport components is needed to obtain an optimal scenario. rf acceleration development would also be desirable, both in the low-frequency rf systems needed in the first stages and in the high-frequency SRF needed in the high-energy accelerators.

In this paper,we have concentrated on a 2 TeV accelerator. We can obtain a first muon collider (FμC at ~250.GeV) accelerator by stopping with the penultimate RLA. Total transport and rf requirements (now 10—20 GV) are naturally an order of magnitude less. However the rapid-cycling variations we are also considering apply primarily to the last RLA stage (2TeV). The FμC would require rapid-cycling at an order of magnitude larger frequency.

Acknowledgments

We acknowledge important contributions from our colleagues, especially F. Mills, D. Cline, R. Noble, and R. Palmer.

References

1. E. A. Perevedentsev and A. N. Skrinsky, Proc. 12th Int. Conf. on High Energy Accel., 485 (1983), A. N. Skrinsky and V.V. Parkhomchuk, Sov. J. Nucl. Physics **12**,3 (1981).
2. D. Neuffer, Particle Accelerators, **14**, 75 (1983), D. Neuffer, Proc. 12th Int. Conf. on High Energy Accelerators, 481 (1983), D. Neuffer, in *Advanced Accelerator Concepts*, AIP Conf. Proc. **156**, 201 (1987).
3. R. J. Noble, in *Advanced Accelerator Concepts*, AIP Conf. Proc. **279**, 949 (1993).
4. Proceedings of the Mini-Workshop on μ⁺-μ⁻ Colliders: Particle Physics and Design, Napa CA, Nucl. Inst. and Meth. A350, 24-56(1994).
5. Proceedings of the Muon Collider Workshop, February 22, 1993, Los Alamos National Laboratory Report LA-UR-93-866 H. A. Theissen, ed.(1993)
6. Proceedings of the 2nd Workshop on Physics Potential & Development of μ⁺-μ⁻ Colliders, Sausalito, 1994, to appear as AIP Conf. Proc. (1995).
7. Proceedings of the Workshop on Beam Dynamics and Technology Issues for μ⁺-μ⁻ Colliders, Montauk, 1995, to appear as AIP Conf. Proc. (1996).
8. D. Neuffer and R. Palmer, Proc. European Particle Accelerator Conference, EPAC94, London, p. 52, 1994
9. CEBAF Design Report, CEBAF, Newport News VA (1986) (unpublished).
10. R. Palmer, presented at the 2+2 TeV μ⁺-μ⁻ Collider Collaboration meeting, Fermilab, July 1995 (unpublished).
11. Q-S Shu for the TESLA collaboration, presented at the CEC/ICMC Conf, Columbus, July 1995.
12. S. Kahn, G. Morgan, and E. Willen, "Recirculating arc dipole for the 2+2 TeV μ⁺-μ⁻ Collider", Montauk proceedings, October, 1995.
13. R. Palmer, (unpublished) ,1995.

Table 1: Parameter list for a 4 TeV μ^+-μ^- Collider

Parameter	Symbol	Value
Energy per beam	E_μ	2 TeV
Luminosity	$L = f_0 n_s n_b N_\mu^2 / 4\pi\sigma^2$	10^{35} cm^{-2}s^{-1}
Source Parameters		
Proton energy	E_p	30 GeV
Protons/pulse	N_p	$2\times3\times10^{13}$
Pulse rate	f_0	15 Hz
μ-production acceptance	μ/p	.2
μ-survival allowance	N_μ/N_{source}	.33
Collider Parameters		
Number of μ /bunch	$N_{\mu\pm}$	2×10^{12}
Number of bunches	n_B	1
Storage turns	n_s	1000
Normalized emittance	ε_N	3×10^{-5} m-rad
μ-beam emittance	$\varepsilon_t = \varepsilon_N/\gamma$	1.5×10^{-9} m-rad
Interaction focus	β_0	0.3 cm
Beam size at interaction	$\sigma = (\varepsilon_t\beta_0)^{1/2}$	2.1 μm

Table 2: Parameters for an idealized 3-RLA acceleration scenario
(The Bi are bunchers; RLAi are multipass recirculating linacs)

Cycle	Energy (GeV)	rf frequency	Bunch length σ	$\delta E/E$	passes	Time (μs)
B1	2	100 MHz	25→7	1→4%		
RLA 1	2→20	100 MHz	7 cm	4→0.4 %	9	8
B2	20	400 MHz	7→1.5	0.4→2%		
RLA 2	20→200	400 MHz	1.5 cm	2→0.2%	10	65
B3	200	1.6 GHz	1.5→0.3	0.2→1.0%		
RLA 3	200→2000	1.6 GHz	0.3 cm	0.2→0.1%	10	585

Table 3: Parameters for a 4-RLA acceleration scenario

Cycle	Energy (GeV)	rf frequency	Bunch length σ	$\delta E/E$	passes	Time (μs)
B1	1	100 MHz	25→7.5	1.5→ 5%		
RLA 1	1→8	100 MHz	7.5→5.0	5.0→1 %	8	5.6
B2	8	400 MHz	5→1.6	1→4.5%		
RLA 2	8→75	400 MHz	1.6cm	4.5→0.5%	12	30
B3	75	1.3 GHz	1.6→0.5	0.5→1.5%		
RLA 3	75→250	1.3 GHz	0.5	1.5→0.5%	18	96
B4	250	2.0 GHz	0.5→0.3	0.5→0.8%		
RLA 4	250→2000	2.0 GHz	0.3cm	0.8→0.1%	18	662

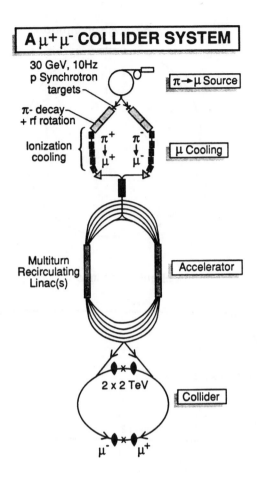

A μ⁺μ⁻ COLLIDER SYSTEM

30 GeV, 10Hz
p Synchrotron
targets

π⁻ decay
+ rf rotation

Ionization
cooling

π⁺   π⁻
↓   ↓
μ⁺   μ⁻

Multiturn
Recirculating
Linac(s)

2 x 2 TeV

μ⁻   μ⁺

π → μ Source

μ Cooling

Accelerator

Collider

Figure 1: Overview of the μ⁺-μ⁻ collider system, showing a muon (μ) source based on a high-intensity rapid-cycling proton synchrotron, with the protons producing pions (π's) in a target, and the μ's are collected from subsequent π decay. The source is followed by a μ-cooling system, and an accelerating system of recirculating linac(s) and/or rapid-cycling synchrotron(s), feeding μ⁺ and μ⁻ bunches into a superconducting storage-ring collider for multiturn high-energy collisions. The entire process cycles at 15 Hz.

Rapid-Cycling Synchrotron

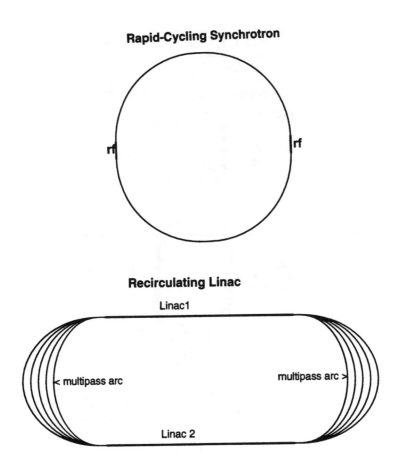

Recirculating Linac

Figure 2. Schematic views of a rapid-cycling synchrotron (RCS) and a recirculating linac (RLA). In the RCS, the beam is accelerated for many turns through the rf, while the magnetic fields in the ring cycle from low-field to high-field, following the beam energy; the beam passes through the same transport on each turn. In the RLA, the beam is accelerated through several passes of the linacs. On each return arc, the beam passes through a different transport path, matched to the increasing beam energy. Magnetic fields are fixed, and the number of return transports (per arc) equals the number of linac passes. (Hybrid schemes, with several return passes, but with some cycling magnets in each pass which track the increasing beam energy, keeping the beam for several passes through the same transport, are also possible.)

μ⁺μ⁻ Accelerator and Collider System

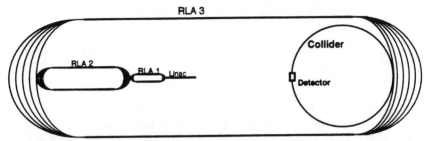

Figure 3. Conceptual view of an RLA-based accelerator, showing a linac feeding beams into a sequence of 3 recirculating linacs (RLA1, RLA2, RLA3) followed by a collider ring. Note that the drawing is not to scale (size change from RLA to RLA would be greater), and the separation between lines in the arcs is exaggerated in this sketch. (There will also be more arc beam lines than displayed, and the separations could be vertical.)

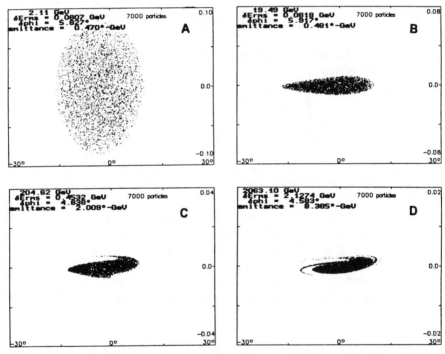

Figure 4. Some simulation results from μRLA. In these simulations a beam is accelerated from 2GeV to 2000 GeV through the three cascaded RLAs of table 2, with bunching at the beginning of each linac. An initially bunched beam for RLA 1 is shown in Fig. 4A, and beam phase-space distributions at the end of RLAs 1, 2, and 3 are shown in 4B, 4C, 4D. The vertical and horizontal scales are δE/E and δφ, respectively. Note that rf frequency increases from 100 to 400 to 1600 MHz from RLA to RLA. The beam is accelerated with very little loss from beam dynamics acceptance and with a longitudinal emittance dilution of ~12%. Beam loss from decay would be ~12%.

329

Working Group Summary:
Machine Design for the μ^+-μ^- Collider

K. Hirata,[*] D. Neuffer,[*] B. Autın, P. Chen, W-H Cheng, D. Cline, P. Dahl, J. Gallardo, M. Green, C. Johnstone, S. Kahn, G. Morgan, K.-Y. Ng, Z. Parsah, J. Peters, D. Summers, A. Tollestrup, D. Trbojevic

Abstract. We summarize the discussions of the working group on Machine Design. The scope of the working group included the entire accelerator system from beam cooling to the collider and the collider ring itself. Particular topics that were discussed in some detail included acceleration options, magnet designs (including the accelerator arcs and collider ring), instabilities, particularly in the collider ring, and lattice issues, particularly in the interaction regions (IRs). Considerable challenges remain in defining a complete machine design for the collider.

INTRODUCTION

The possibility of muon (μ^+-μ^-) colliders has been introduced by Skrinsky et al.[1] and Neuffer[2]. More recently, intensified investigations with the goal of a practical design for a high-energy high-luminosity $\mu^+\mu^-$ collider have increased the level of conceptual development,[3, 4, 5, 6] and this effort includes the present workshop.[7] A candidate scenario for a collider, with an energy of $E_{cm} = 2E_\mu = 4$ TeV, and a luminosity of L ~ 10^{35}cm^{-2}s^{-1}, has been developed.[8] Table 1 shows possible parameters for a candidate design. The design consists of a muon source, a muon collection, cooling and compression system, a recirculating linac (or rapid-cycling or ...) system for acceleration, and a full-energy collider with detectors for multiturn high-luminosity collisions.

In the machine design section, we concentrated on the acceleration system and the collider in this scenario. In the acceleration system, the muons are accelerated from the output of the cooling system to full energy and transferred to storage in the 2 TeV collider ring. In current scenarios the accelerator and collider occupy the bulk of the real estate and the hardware and therefore comprise most of the cost of the collider system. Optimum design is essential in developing a practical collider scenario.

---

[*]Working group leaders.

ACCELERATION OPTIONS

Potential acceleration systems include full-energy linac, rapid-cycling synchrotron and recirculating-linac options, plus possible hybrid approaches. A critical requirement is that the muons must be accelerated before they decay. This sets severe constraints on the acceleration system and these are discussed by D. Neuffer in these proceedings.[9] The results are that the acceleration rate (eV_{rf}') must be much greater than $m_{\mu}c^2/L_{\mu}$, where m_{μ} is the muon mass and $L_{\mu} = c\tau_{\mu}$ is the muon decay length, which implies **eV_{rf}' >> 0.16 MeV/m.** In a multiturn accelerator this can be rewritten as **N_{turns} << 300 B(T)** where N_{turns} is the total number of turns and B(T) is the average bending field in the highest energy turn (including non-bending regions).

Our currently preferred acceleration choice is a sequence or cascade of recirculating linacs, each of which increases beam energy by ~ an order of magnitude and accommodates bunch length reductions by almost as much. Constraints and candidate scenarios were discussed at the workshop, and particle tracking provided a "proof of principle" of this general approach.[9]

In a recirculating linac (RLA), the beam is accelerated and returned for several passes of acceleration in the same linac, but a separate return path is provided for each pass, which are at fixed magnetic fields. The requirement for a separate transport on each turn limits the total number of turns that could be practical, to ~10—20 turns, which is very compatible with the μ lifetime constraint. Even with these limited number of turns, the total amount of beam transport is relatively large (~85—160km in typical scenarios). The transport can easily become **very** expensive, so cost-saving magnet designs are needed, and are discussed below.

While the RLA approach is possible with current technology, its potentially large costs invited discussion of alternatives, some of which require untried technology. These approaches are "rapid-cycling" scenarios, in which the magnetic fields in an accelerator ring are changed to follow the increasing energy of the muons, so that the muons can use the same transport over many turns. In conventional rapid-cycling synchrotrons iron-dominated magnets and ac-resonant power supplies are used, which limit the peak magnetic fields to ~1.5 T and repetition rates to ~ 60 Hz, which may be useful for the 2 TeV μ-accelerator, but would still require a large-circumference ring. D. Summers[10] discussed alternatives which may permit higher fields. These included hybrid magnets in which rapid-cycling (<2T) dipoles are mixed with high-field fixed dipoles, and the intriguing possibility of rotating dipoles. Following the workshop these approaches were extended to include fast pulsed (non-iron) magnetic fields, which could accelerate at much faster rates, and therefore be suitable for lower-energy acceleration.[11]

RF acceleration systems will be needed in all scenarios. The RLA requirements are closely matched to those of the TESLA superconducting RF program, with very similar total currents, and pulse-length and frequency requirements, but with higher peak currents.[12] Optimal adaptation to the μ^+-μ^- system has not yet been established, and rf for variations has not yet been determined.

MAGNET DESIGNS

The working group also discussed several magnet designs which could be used for the accelerator and for the collider. S. Kahn presented possible magnet designs for RLAs in which several passes of transport are placed within the same cryostat. 9- and 18- pass multiaperture magnets have been designed, and approximate cost estimates have been developed.[13] The multiaperture magnets should be significantly cheaper than completely separated magnets.

The collider ring magnets were also discussed. A serious problem in the collider ring is the decay products from μ decay in the ring, which can place a large heat load on cryogenic components. M. Green presemted a warm-iron, high-field dipole magnet design, with no conductor on the midplane, which could be an adequate collider dipole.[14] It remained to be studied whether quadrupoles and other magnets could be added to form a complete ring design.

INSTABILITIES

Progress in design of μ production and cooling has reached a point where it is conceivable that one might actually have too many muons and collective instability limitations may be reached. High-luminosity μ^+-μ^- colliders will indeed require intensities at or above stability thresholds. Because of muon decay, long-term stability is not required and only instabilities with growth rates faster than that decay time can be important. Ng[15] and Cheng[16] presented results of studies of instabilities which could occur in the collider ring. These included beam breakup due to strong head-tail interactions, resistive-wall and broad-band impedance generated instabilities, longitudinal microwave instabilities, and wake field limitations. The stability strategy is to reduce impedances and intensities so that growth rates are below the μ lifetime, and to use BNS damping if needed.

Wake field effects can also be undesirably large in the accelerators (RLA or rapid-cycling). It is important that the wake-fields be much smaller than the accelerating voltages to avoid large effects. Further study and simulation is needed.

Another important effect in the collider may be beam-beam interaction related effects. P. Chen[17] presented the first analysis of beam-beam effects in the µ-collider, and these received further discussion in the working group. The relatively high beam-beam tune shifts in a high-luminosity collider could lead to distortion, which could increase over the beam lifetime. While direct beamstrahlung production is small, e^+-e^- production in the beam-beam fields through decay $e^\pm$ or virtual photons may be significant. Further study is needed and will be reported in this proceedings and future studies.

LATTICE ISSUES - IR CONSTRAINTS

The Collider ring must be nearly isochronous in order to maintain short bunch lengths, and must also focus the beam to small spot sizes at the interaction region (IR), with a design goal of the $\beta^* = 0.003$ m, where β^* is the IR focusing parameter. Ng and Trbojevic[18] led discussions of lattices for the Collider, and suggested the use of a "flexible momentum compaction"-type lattice [19] to obtain small chronicity. The small IR β^* value implies large-β in the IR quads with correspondingly large chromatic and geometric aberrations. Large corrections using special correction insertions are required. K. Ng reported that correction could be improved by increasing the dispersion in the insertion, and reducing maximum β. B. Autin suggested adding small-aperture permanent magnets close to the IR and using these to minimize the beam size in the IR. Further study and lattice optimization is needed, and should include tracking with realistic beam sizes and energy spreads. Integration of IR with detector was also discussed, including the goal of minimizing the effects of decay-induced backgrounds in the detector.

SUMMARY AND DISCUSSION

We have presented some of the discussions of the machine design working group, which covered some but not all of the key issues involved in developing an accelerator and collider system. The acceleration scenario must be developed and optimized, and rapid-cycling variations should be developed. Complete lattices are needed, with designs for the transport arcs, including beam separation and recombination, as well as injection from the end of the µ-cooling section. RF acceleration development is needed, both in the low-frequency RF systems needed in the first stages and in the high-frequency SRF needed in the high-energy accelerators. Collider lattice development is needed, with integration of the constraints of small-β^* and decay loss effects. Wakefields, instability limitations, and beam-beam effects should be studied. Accurate cost algorithms will also be needed. While encouraging initiatives have been made in these areas, many exciting challenges remain.

References

1. E. A. Perevedentsev and A. N. Skrinsky, Proc. 12th Int. Conf. on High Energy Accel., 485 (1983), A. N. Skrinsky and V.V. Parkhomchuk, Sov. J. Nucl. Physics **12**,3 (1981).
2. D. Neuffer, Particle Accelerators, **14**, 75 (1983), D. Neuffer, Proc. 12th Int. Conf. on High Energy Accelerators, 481 (1983), D. Neuffer, in *Advanced Accelerator Concepts*, AIP Conf. Proc. **156**, 201 (1987).
3. R. J. Noble, in *Advanced Accelerator Concepts*, AIP Conf. Proc. **279**, 949 (1993).
4. Proceedings of the Mini-Workshop on μ^+-μ^- Colliders: Particle Physics and Design, Napa CA, Nucl. Inst. and Meth. A350, 24-56(1994).
5. Proceedings of the Muon Collider Workshop, February 22, 1993, Los Alamos National Laboratory Report LA-UR-93-866 H. A. Theissen, ed.(1993)
6. Proceedings of the 2nd Workshop on Physics Potential & Development of μ^+-μ^- Colliders, Sausalito, 1994, to appear as AIP Conf. Proc. (1995).
7. Proceedings of the Workshop on Beam Dynamics and Technology Issues for μ^+-μ^- Colliders, Montauk, 1995, to appear as AIP Conf. Proc. (1996).
8. D. Neuffer and R. Palmer, Proc. European Particle Accelerator Conference, EPAC94, London, p. 52, 1994
9. D. Neuffer, these Proceedings (1996).
10. D. Summers, these Proceedings (1996).
11. R. Palmer and E. Willen, unpublished communication (1996).
12. Q-S Shu for the TESLA collaboration, presented at the CEC/ICMC Conf, Columbus, July 1995.
13. S. Kahn, G. Morgan, and E. Willen, 'Recirculating arc dipole for the 2+2 TeV μ^+-μ^- Collider", these Montauk proceedings, October, 1995.
14. M. Green, these proceedings, (1996).
15. K.-Y. Ng, these Proceedings (1996).
16. W.-H. Cheng et al., these Proceedings (1996).
17. P. Chen, these Proceedings (1996).
18. D. Trbojevic et al., these Proceedings (1996).
19. S. Y. Lee, K. Y. Ng, and D. Trbojevic, Physical Review E 48, p. 3040 (1993).

Table 1: Parameter list for a 4 TeV μ^+-μ^- Collider

Parameter	Symbol	Value
Energy per beam	E_μ	2 TeV
Luminosity	$L = f_0 n_s n_b N_\mu^2/4\pi\sigma^2$	10^{35} cm^{-2}s^{-1}
Source Parameters		
Proton energy	E_p	30 GeV
Protons/pulse	N_p '	$2\times3\times10^{13}$
Pulse rate	f_0	15 Hz
μ-production acceptance	μ/p	.2
μ-survival allowance	N_μ/N_{source}	.33
Collider Parameters		
Number of μ /bunch	$N_{\mu\pm}$	2×10^{12}
Number of bunches	n_B	1
Storage turns	n_s	1000
Normalized emittance	ε_N	3×10^{-5} m-rad
μ-beam emittance	$\varepsilon_t = \varepsilon_N/\gamma$	1.5×10^{-9} m-rad
Interaction focus	β_0	0.3 cm
Beam size at interaction	$\sigma = (\varepsilon_t\beta_0)^{1/2}$	2.1 μm

II. APPENDICES

List of Participants

Bruno Autin	CERN
Valerie I. Balbekov	Branch Inst. Nuclear Physics-Protvino
Vernon D. Barger	Univ. of Wisconsin
Yong-Chul Chae	Argonne National Laboratory
Pisin Chen	Stanford Linear Accelerator Center
Wen-Hao Cheng	Lawrence Berkeley National Laboratory
David B. Cline	Univ. of California, Los Angeles
Per F. Dahl	U.S. Department of Energy
Sally L. Dawson	Brookhaven National Laboratory
Rashid M. Djilkibaev	Institute for Nuclear Research
Deborah M. Errede	Univ. of Illinois
Richard C. Fernow	Brookhaven National Laboratory
William R. Frisken	York University
Juan C. Gallardo	Brookhaven National Laboratory
Howard A. Gordon	Brookhaven National Laboratory
William S. Graves	Brookhaven National Laboratory
Michael A. Green	Lawrence Berkeley National Laboratory
Katherine C. Harkay	Argonne National Laboratory
Kohji Hirata	KEK- Nat. Lab. for High Energy Phys.
Carol Johnstone	Fermilab
Stephen A. Kahn	Brookhaven National Laboratory
Harold G. Kirk	Brookhaven National Laboratory
Thaddeus F. Kycia	Brookhaven National Laboratory
David Lissauer	Brookhaven National Laboratory
Alfredo U. Luccio	Brookhaven National Laboratory
Frederick E. Mills	Fermilab
Alfred Moretti	Fermilab
Gerry H. Morgan	Brookhaven National Laboratory
Michael J. Murtagh	Brookhaven National Laboratory
David V. Neuffer	CEBAF/Fermilab
King-Yuen Ng	Fermilab
Robert J. Noble	Fermilab
Robert B. Palmer	Brookhaven National Laboratory
Zohreh Parsa	Brookhaven National Laboratory
Gerald J. Peters	U.S. Department of Energy
Venetios A. Polychronakos	Brookhaven National Laboratory
David C. Rahm	Brookhaven National Laboratory
Laura Reina	Brookhaven National Laboratory
Thomas Roser	Brookhaven National Laboratory
Nicholas P. Samios	Brookhaven National Laboratory
Yannis K. Semertzidis	Brookhaven National Laboratory
Andrew M. Sessler	Lawrence Berkeley National Laboratory

Gregory I. Silvestrov	Budker Inst. of Nuclear Physics
Alexandr N. Skrinsky	Budker Inst. of Nuclear Physics
Iuliu Stumer	Brookhaven National Laboratory
Don Summers	Univ. of Mississippi
David F. Sutter	U.S. Department of Energy
Hiroshi Takahashi	Brookhaven National Laboratory
Helio Takai	Brookhaven National Laboratory
David Taqqu	Paul Scherrer Institut
Alvin V. Tollestrup	Fermilab
Yagmur Torun	Brookhaven National Laboratory/SUNY
Dejan Trbojevic	Brookhaven National Laboratory
Tatiana A. Vsevolozhskaya	Budker Inst. of Nuclear Physics
Erich H. Willen	Brookhaven National Laboratory
David R. Winn	Fairfield University
David U.L. Yu	DULY Research, Inc.
Shou-Yuan Zhang	Brookhaven National Laboratory
Yongxiang Zhao	Brookhaven National Laboratory
Max Zolotorev	Stanford Linear Accelerator Center

Author Index

A

Autin, B., 190, 330

B

Balbekov, V. I., 140
Ball, R., 146
Baltz, A. J., 3
Barger, V., 269

C

Chae, Y.-C., 31
Chen, P., 3, 330
Chen, X., 87
Cheng, W.-H., 3, 206, 330
Cho, Y., 3, 31
Cline, D., 279, 330
Courant, E., 3
Crosbie, E., 31

D

Dahl, P., 330
Dawson, S., 285
Djilkibaev, R. M., 53

F

Fernow, R. C., 3, 146
Friedsam, H., 31

G

Gallardo, J. C., 3, 146, 330
Garren, A., 3
Green, M. A., 3, 100, 257, 330

H

Harkay, K., 31
Hirata, K., 330
Horan, D., 31

J

Johnstone, C., 178, 330

K

Kahn, S., 3, 330
Kirk, H. G., 3, 146
Kustom, R., 31
Kycia, T., 146

L

Lee, Y. Y., 3, 146
Lessner, E., 31
Littenberg, L., 146
Lobashev, V. M., 53

M

Marx, M., 146
McDowell, W., 31
McGhee, D., 31
Mills, F., 3
Moe, H., 31
Mokhov, N. V., 3, 61, 234
Morgan, G., 3, 330

N

Neuffer, D. V., 3, 146, 315, 330
Ng, K.-Y., 178, 198, 224, 330
Nielsen, R., 31
Noble, R. J., 3, 61
Norek, G., 31
Norem, J., 3

341

P

Palmer, R. B., 3, 146
Parsa, Z., 330
Peters, J., 330
Peterson, K., 31
Polychronakos, V., 146
Popovic, M., 3

Q

Qian, Y., 31

R

Reina, L., 293
Roser, T., 47

S

Schachinger, L., 3
Sessler, A. M., 3, 206
Silvestrov, G. I., 3, 168
Skrinsky, A. N., 3, 133
Striganov, S. I., 234
Stumer, I., 3, 146
Summers, D., 3, 330
Syphers, M., 3

T

Takahashi, H., 87
Taqqu, D., 301
Thompson, K., 31
Thun, R., 146
Tollestrup, A., 3, 330
Torun, Y., 3, 146
Trbojevic, D., 3, 178, 330
Turner, W. C., 3, 108

V

Van Ginneken, A., 3, 61, 267
Vsevolozhskaya, T. A., 3, 159

W

Weggel, R., 3
White, M., 31
Willen, E., 3
Winn, D. R., 3, 146
Wurtele, J. S., 3, 206

Z

Zeller, M., 146
Zhang, S. Y., 218